Generation of Energy from Municipal Solid Waste

Abdelkader Anouzla · Salah Souabi
Editors

Generation of Energy from Municipal Solid Waste

Circular Economy and Sustainability

Springer

Editors
Abdelkader Anouzla
Faculty of Science and Technology
University of Hassan II Casablanca
Rabat, Morocco

Salah Souabi
Faculty of Science
and Techniques-Mohammadia
University of Hassan II Casablanca
Mohammedia, Morocco

ISBN 978-3-031-74333-7 ISBN 978-3-031-74334-4 (eBook)
https://doi.org/10.1007/978-3-031-74334-4

This Springer imprint is published by the registered company Springer Nature Switzerland AG
The registered company address is: Gewerbestrasse 11, 6330 Cham, Switzerland

Preface

Waste such as biomass, household waste, and sludge as an energy carrier and in combination with thermochemical techniques for converting it into combustible gases, could be an exciting energy carrier if it is produced sustainably in light of the ever-increasing demand for energy and the environmental issues that come along with it. Making the most of renewable energy sources like biomass and trash is becoming increasingly important due to the likelihood of declining oil supply and rising demand from emerging industrial powers. Establishing CO2 emission permits completes the judicial and economic framework nations put in place in response to the Kyoto Protocol.

Renewable energies are an essential alternative to the depletion of fossil fuels and a vital means of combating the harmful effects of climate change, which is partly linked to environmental pollution by solid waste, in particular by reducing greenhouse gas emissions, eliminating pollution linked to these types of waste, and protecting the environment. Solid waste can be made up of biomass, urban waste made up of household refuse, sludge from sewage treatment plants, green waste, waste from commercial activities, etc. It can also be an industrial waste produced by the agri-food and agro-industry industries comprising non-recovered co-products. Faced with the pollution linked to non-renewable energies, countries worldwide seek other, cleaner energy resources, such as renewable energies.

The most promising of these clean, renewable energy sources is biomass, which includes sewage sludge and household trash. The book provides examples of various methods for recovering solid waste in light of this.

- Energy: Biogas is a renewable energy source that may be used to power electricity generators, heat houses, or provide hot water and fuel alternative vehicles.
- Environmental: Recycling garbage as an energy source helps protect the environment by reducing greenhouse gas emissions and preventing pollution.

- Economic: the energy recovery of solid waste offers advantages for companies responsible for waste management and agricultural activities, which can economically and energetically recover their waste from plant or animal sources. Introducing this technology will contribute to sustainable development and economic autonomy for renewable energy producers.

Technical: waste-to-energy technology can be easily installed on an industrial scale.

This book aims to evaluate the energy potential of garbage so that it may be seen positively as an energy source rather than merely a source of pollution. It offers a database to assist in selecting a method for trash disposal while generating renewable energy.

The text comprises 16 chapters on waste management with clean energy generation (heat, CH4, H2, diesel, petrol, methanol, ethanol, etc.) that experts in the field have suggested.

Energy from trash may be recovered, which results in a decrease in greenhouse gas emissions and the creation of new recovery technologies. Lowering environmental pollution is an intelligent approach to ensure national energy security and combat the trend toward global warming.

Rabat, Morocco Abdelkader Anouzla

Contents

Valorisation of Animal Waste for Energy and Material Recovery

Catarina Viegas

Abstract Animal waste has the potential to yield a variety of renewable energy sources and material applications. Biogas can be produced from waste and used as a renewable energy source, while the solid fraction can be used as a fertilizer or a source of organic matter for soil improvement. The combustion of animal waste can produce heat and electricity, making it a promising source of renewable energy. Furthermore, animal fat can be converted into biodiesel, which can be used as a transportation fuel. Additionally, animal waste can undergo thermochemical conversion processes such as pyrolysis and gasification, to produce syngas or biochar, which can be utilized as feedstock for the production of biofuels or as a soil amendment for agriculture.

Animal by-products and waste can also be a source of valuable biomolecules, which can be utilized in a range of applications. For instance, animal blood, bones, and hides contain collagen, which can be extracted and used in the production of gelatine, a widely used protein in food, pharmaceuticals, and cosmetics. In addition, animal fat, particularly from fish waste, contains high levels of polyunsaturated fatty acids such as omega-3, which are essential nutrients for human health. These fatty acids can be fractionated and used as an ingredient in functional foods and nutraceuticals. Moreover, animal waste can also be a source of enzymes, hormones, and growth factors, which can be utilized in various industrial and biomedical applications. The development of these biomolecules not only adds value to animal by-products but also offers new opportunities for the production of sustainable products with potential health benefits.

The development of these renewable energy sources can mitigate the reliance on non-renewable energy sources, can contribute to the development of a circular economy and offer a sustainable solution for waste management in the animal farming industry.

C. Viegas (✉)
LEAF Linking Landscape, Environment, Agriculture and Food, Associate Laboratory TERRA, School of Agriculture, University of Lisbon, Lisbon, Portugal
e-mail: catarina.viegas.de.sousa@gmail.com

A. Anouzla and S. Souabi (eds.), *Generation of Energy from Municipal Solid Waste*,
https://doi.org/10.1007/978-3-031-74334-4_1

1 Introduction

Animal waste, like other waste, remains a global concern for which there is still no effective, sustainable, and inexpensive solution. The growing global population and its escalating demand for animal-derived food products have been exerting pressure on the agricultural sector's production capacity. Therefore, the generation of waste produced by animals and their management has also been increasing significantly.

Livestock used in food production, including animals such as cattle, poultry, and fish farming, produce significant quantities of aqueous waste containing manure, slurry, and wastewater from production facilities, and also remains from the animals, such as skins and viscera. These agro-industrial effluents pose a significant pollution risk to both marine and freshwater ecosystems due to their high organic content. Before being discharged, these wastes need to undergo physicochemical treatments, which often involve the use of stabilization ponds and anaerobic digestion reactors.

The treatment processes undergone by animal waste can generate energy and commercially valuable products. Several chemical, biological, and thermochemical processes yield energy in the form of heat and gaseous, liquid, or solid biofuels. Additionally, these processes can produce biofertilizers and biomolecules with applications in various industries, such as cosmetics, pharmaceuticals, and food, among others.

2 Characterization of Animal Waste

Animal waste comprehends a diverse array of residues originating from livestock production. These residues comprise all animal byproducts not directly utilized as food or feed, such as sick animals, viscera, skins, bones, and feathers, among others. The wastes associated with livestock production and related industries include all aqueous effluents produced by the animals, including manure, slurry, chicken litter, effluents from cleaning operations, processing deceased animals, and wastes from dairy industries. It also includes the waste obtained in aquaculture production (Viegas, 2022).

Figure 1 outlines the origins and the different residues that compose the animal wastes.

Animal residues tend to possess a substantial nitrogen burden, particularly when they contain excreta (Amini et al., 2020). Conversely, all byproducts resulting from animal processing exhibit elevated protein and fat concentrations, which could potentially be reclaimed for utilization in food, animal feed, or the production of biofuels. Recent data indicates that seafood items constitute approximately 20% of the human diet, yet a significant 50–70% of the raw materials derived from seafood are converted into waste annually. This phenomenon arises due to the relatively lower proportion of usable matter in marine organisms, leading to the generation of substantial seafood waste in the form of shells, carapaces, spines, and exoskeletons (Kumar et al., 2018).

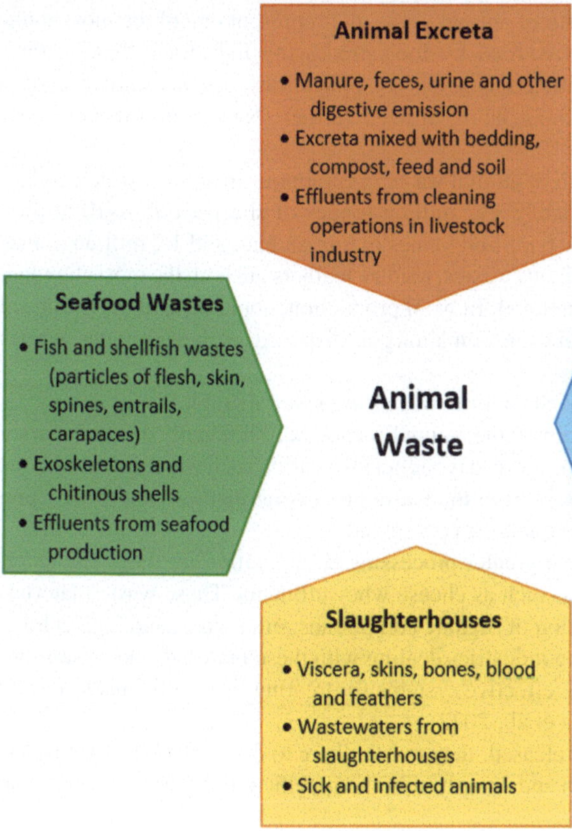

Fig. 1 Animal waste: Type of waste and origins

Another category of waste generated on a global scale is that arising from dairy processing, primarily comprising cheese whey, lactose, and dairy waste permeate. On average, dairy industries consume 4–8 L of water for every litre of milk processed, with an estimated 2% of all processed milk being lost through drainage (equivalent to 17.1 million tons in 2018). In Europe, a standard dairy facility produces roughly 180,000 cubic meters of waste effluents annually (Ekka et al., 2022; Krishna et al., 2022). This type of waste constitutes approximately 26% of the total food waste and is enriched with mineral salts, nitrogen, phosphorus, protein, and lipids (Ganesh et al., 2022). These residual materials possess a high chemical oxygen demand (COD) and a substantial biochemical oxygen demand (BOD_5), rendering them susceptible to vigorous microbial proliferation. Consequently, effective treatment is imperative to avoid the dissemination of pathogenic microorganisms in the environment, necessitating prompt intervention shortly after waste production (Krishna et al., 2022; Simón et al., 2021).

Eggshells are another form of animal waste and represent one of the most abundant agricultural residues discarded by food processing industries. Despite their remarkable properties and numerous potential applications, eggshells are discarded in significant quantities without being utilized further (Ngayakamo & Onwualu, 2022).

An often-overlooked form of animal waste is keratinous materials, which collectively account for approximately 1–2 million tonnes of sheep wool, 8–10 million tonnes of poultry feathers, 4–6 million tonnes of human hair, and 1.2 million tonnes of horns. Among these keratinous wastes, poultry feathers are both the most abundant and the most extensively generated. In wool production, about 60% of the produced wool goes into apparel, while the remaining is disposed of as keratinous wastes (Ossai et al., 2022).

Different compositions of the animal wastes are given in Table 1.

As indicated in the preceding table, animal wastes exhibit notably distinct characteristics, varying not only across waste types but also within each waste category. This variability may arise not solely from the waste processing methods and production conditions, but also from the quantities processed.

Animal wastes that present greater processing complexities are those characterized by elevated COD levels, such as cheese whey effluents. These waste materials contain a notable concentration of organic compounds, oil, grease, and high acidity. Consequently, manufacturing industries dealing with these materials face a substantial challenge in maintaining effective systems for treating these effluents to ensure their safe discharge (Krishna et al., 2022).

For these effluents to be released, they must adhere to discharge limits set by the respective countries of origin and comply with the specifications outlined by effluent treatment facilities.

3 Environmental Impact of Animal Waste

The largest amount of animal waste produced refers to the aqueous effluents that include the liquid excreta of animals and all the washing water from agro-industrial facilities. In 1998 the global meat production was 218 million tonnes (FAO, 1999) and it was estimated that the annual global animal waste (manure) was to be over 1,188 billion tonnes, and 2.0 billion tonnes was produced in the USA. alone (Haşimoğlu & Aksoy, 1998). In 2013, the global volume of food wastage was estimated to be 1.6 billion tonnes of primary product equivalents, and the wastage for the edible part was nearly 1.3 billion tonnes. The carbon footprint of food wastage was estimated to be 3.3 billion tonnes of CO_2 equivalent to greenhouse gases released into the atmosphere per year (FAO, 2013). In 2021 the total meat production was 356 million tonnes, mainly from chicken, pig, and cattle (FAO, 2022), therefore it can be estimated that the production of animal waste is approximately 1,940 billion tonnes. Within the European Union (EU-27) and the UK, between 2016 and 2019, animal farming generated over 1.4 billion tonnes of manure on an annual basis (Köninger et al., 2021).

Table 1 Composition of different animal wastes

	N (mg/L)	P (mg/L)	COD (g O_2/L)	BOD_5 (g O_2/L)	Biodegradability	Total solids (g/L)	Suspended solids (g/L)	Reference
Animal Excreta								
Bovine Crude Manure	3463.07	810.50	43.86	23.50	0.54	118.37	–	Viegas et al. (2021b)
Cow manure anaerobic fermentation effluence	424.90	39.80	6.89	–	–	–	–	Gong et al., (2020)
Poultry manure wastewater	2820.00	450.00	12.10	5.90	0.49	8.28	5.02	Yetilmezsoy and Sakar (2008)
	2455.00	–	5.80	–	–	–	–	Li et al., (2016)
Swine wastewater	1250.00	225.00	7.99	–	–	13.10	–	Mendonça and Santos (2022)
Swine wastewater (mainly urine)	860.00	26.20	2.30	1.25	0.54	6.30	–	Viegas and Gonçalves (2021)
Swine wastewater (clarified by a dissolved air floatation)	195.00	19.00	0.68	–	–	0.19	–	Rossi et al., (2022)
Seafood Wastes								
Brown crabs' transportation effluent	737.80	22.10	5.20	3.05	0.59	40.90	–	Viegas et al. (2021a)

(continued)

Table 1 (continued)

	N (mg/L)	P (mg/L)	COD (g O_2/L)	BOD_5 (g O_2/L)	Biodegradability	Total solids (g/L)	Suspended solids (g/L)	Reference
Freshwater catfish cultivation	50.10 (NH_3–N)	2.80 (PO_4-P)	0.14	–	–	–	0.49	Kurniawan et al., (2022)
Grouper fish cultivation	60.00	6.80	0.11	–	–	–	–	Kuo et al., (2016)
Rearing tank of shrimp	6.81	0.42	–	0.09	–	–	–	Gao et al., (2016)
Intensive shrimp production	5.81	4.50	0.14	–	–	–	–	Lukwambe et al., (2018)
Slaughterhouses								
Bovine slaughterhouse	293.50		5.11	–	–	1.92	–	Menezes Lima et al., (2020)
Poultry slaughterhouse	244.00	–	3.87	–	–	3.04	–	Menezes Lima et al., (2020)
	271.00	59.00	5.60	2.40	0.43	5.70	–	Viegas et al. (2021c)
Swine slaughterhouse	235.00	–	3.74	–	–	1.55	–	Menezes Lima et al., (2020)
Livestock and Animal Processing								
Dairy wastewater	154.00	56.60	7.65	4.59	0.60	–	0.48	Pereira et al., (2020)

(continued)

Table 1 (continued)

	N (mg/L)	P (mg/L)	COD (g O_2/L)	BOD_5 (g O_2/L)	Biodegradability	Total solids (g/L)	Suspended solids (g/L)	Reference
Dairy processing wastewater	7.90	5.30	9.77	4.90	–	–	2.37	Finnegan et al., (2018)
Butter waste effluent	–	–	8.93	2.42	0.27	–	5.07	Shi et al., (2021)
Cheese whey effluent	333.00	6.90	17.20	11.25	0.65	7.65	3.24	Shi et al., (2021)
Hard cheese whey effluent	–	–	73.45	29.48	0.40	–	7.15	Shi et al., (2021)
Ice cream waste effluent	60.00	14.00	5.20	2.45	0.47	3.90	–	Shi et al., (2021)
Beef processing wastewater	0.42	–	4.42	–	–	–	–	Durkin et al., (2022)
Pork processing wastewater	0.27	–	4.31	–	–	–	–	Durkin et al., (2022)
Fresh fish processing wastewater	0.26	–	2.42	–	–	–	–	Durkin et al., (2022)
Canned fish processing wastewater	0.25	–	4.20	–	–	–	–	Durkin et al., (2022)
Wool dyeing effluent	–	–	2.39	0.81	0 34	–	0.063	Ossai et al. (2022)

(continued)

Table 1 (continued)

	N (mg/L)	P (mg/L)	COD (g O_2/L)	BOD_5 (g O_2/L)	Biodegradability	Total solids (g/L)	Suspended solids (g/L)	Reference
Tannery effluent from bovine and ovine skins	1287.00	22.20	23.96	6.33	0.26	35.50	–	Balakrishnan et al., (2020)
Tannery wastewater	375.00	–	2.17	1.11	0.51	–	0.74	Cokgor et al., (2008)

Another way to assess the global production of animal waste would be through the phosphorus cycle. In this context, the worldwide production of animal residues would range between 16 and 20 million metric tonnes of phosphorus per year, considering an average concentration of 0.8–1% of phosphorus for both confined and unconfined animal waste (Liu & Chen, 2014).

The environmental consequences associated with these aqueous effluents primarily stem from their potential to cause eutrophication (Oryschak & Beltranena, 2020). Moreover, utilizing agro-industrial effluents directly as fertilizer can lead to negative impacts, such as emitting unpleasant odours, contaminating soils with pathogenic microorganisms, and polluting groundwater with compounds from these effluents (Markou et al., 2018).

The majority of animal solid residues are typically disposed of in landfill sites and incinerated, while liquid residues are primarily directed to be processed at wastewater treatment facilities and in stabilization ponds. Nonetheless, a portion of these waste materials is also repurposed for the production of animal feed and compost intended for agricultural use. Inadequate management of these waste materials results in the emission of substantial quantities of methane (CH_4), a potent contributor to climate change, while neglecting the possible utilization of these valuable resources (Karić et al., 2022).

In traditional productions with small herds, the mixture of urine, feces, and wastewater is often discharged into the soil as fertilizer. However, with the rise of intensive farming practices, this option is no longer viable (European Commission, 2015).

In developed countries, there is concern regarding waste disposal and its treatment; however, considerable gaps still exist, and a significant portion of the waste does not undergo proper treatment. However, in many less developed countries, these agro-industrial effluents are often discharged directly into water bodies, rivers, groundwater, and soils without any treatment, leading to a tremendous impact on the environment and public health. Shockingly, over 80% of wastewater worldwide is released into water bodies without undergoing any treatment (United Nations - Water, 2021). Therefore, besides being a potential source of pollution, its complete utility is not being fully harnessed, particularly concerning its energy content or inherent composition.

European Union guidelines and strategy for waste management, presented in Directive 2008/98/EC, state that the waste hierarchy that should be applied in waste prevention, management legislation, and policy has the following priority order: prevention, preparing for re-use, recycling, energy recovery and disposal (European Commission, 2015). This concept establishes the priority of valorisation processes over disposal methods such as landfilling.

4 Treatment of Animal Waste

Current technologies used to treat animal wastes and wastewater are complex and expensive processes. However, these treatments are absolutely necessary because these effluents carry a high organic load and serve as a significant source of nutrients and pollutants in both freshwater and marine ecosystems (Mo & Zhang, 2012).

Poultry, piggery, and dairy farms generate vast amounts of manure and wastewater on a daily basis from cleaning operations. These effluents cannot be discharged into conventional wastewater treatment plants, and complete incorporation into the soil as fertilizer is also not feasible (Markou et al., 2018). Treating these effluents typically involves employing multiple methods to effectively reduce their chemical oxygen demand and microbiological contamination, imposing a substantial economic burden on animal producers (Viegas et al., 2021b). Landfilling and incineration are commonly employed methods for waste management, even though their lack of sustainability gives rise to significant concerns regarding resource conservation and environmental protection (Wang et al., 2022).

Currently, several types of physicochemical treatments can be applied to the effluents of animal waste before releasing it into the environment. The two most commonly used systems for animal waste effluents are stabilization ponds and anaerobic digestion. The treatments performed in stabilization ponds are time-consuming and require significant energy input for aeration to facilitate bacteria oxidation processes. During this treatment, various compounds, including CO_2, CH_4, and H_2S, are lost to the atmosphere through degradation. On the other hand, anaerobic digestion, while effective in decomposing these effluents, is costly, complex, and time-consuming, and it can only handle specific ranges of chemical and organic loads. Moreover, the resulting digestate needs to be properly disposed of (Siddique & Wahid, 2018). For the treatment of solid animal waste, such as keratinous wastes, and eggshells, among others, the most used resource is incineration and landfilling.

Other conventional methods can be used for treating animal waste; however, these methods were initially developed for treating urban wastewater and, therefore, need to be modified and improved to suit animal waste effluent treatment. Commonly, these treatments involve chemical and electrochemical processes. These methods include coagulation/flocculation, precipitation, biodegradation, filtration, and adsorption using chromatography processes (AC). These processes are commonly employed in wastewater treatment due to their capability to remove various contaminants, moderate investment requirements, and effectiveness with most effluents (Wang et al., 2022).

In recent decades, the utilization of electrochemical processes for wastewater treatment has seen a significant rise. The main advantages of employing these processes include their durability, automation, energy efficiency, selectivity, reduced chemical usage, and compatibility with environmental concerns. The procedure is straightforward, involving the application of an electric current or electrode potential, leading to the elimination of a wide range of compounds. The effectiveness of pollutant removal largely depends on the type of anode electrode material and various electrochemical parameters, such as current density, time, and agitation (Salmerón, Oller, & Malato, et al., 2021; Suman et al., 2021). Various electrochemical methods exist, such as electrochemical oxidation, electrochemical reduction, electrocoagulation, indirect electrooxidation with strong oxidants, electrodeionization, capacitive deionization, photo or ultrasound-assisted electrochemical methods, and electrokinetics (Sillanpää & Shestakova, 2017).

Despite the efficiency of these methods, the treated effluents still contain nutrients that are not recycled or utilized to create any usable end-products.

5 Valorisation of Animal Waste

The existing circumstances are sufficiently concerning to impose the immediate advancement of technologies capable of managing waste through environmentally conscious and economically viable means. Moreover, promoting lifestyles that curtail waste generation at its origin is crucial. Among the potential solutions, waste valorisation stands out as a highly promising approach that should be synergistically integrated with the principles of the Circular Economy. This integration holds the potential to aid nations and regions worldwide in mitigating the detrimental effects of waste, ultimately leading to the attainment of both environmental and economic sustainability (Wang et al., 2021).

Waste valorisation, as a representative waste-to-wealth strategy, includes various industrial processes aimed at reusing or recycling waste materials, transforming them into valuable products or energy sources. The core objective of waste valorisation is to establish closed material flow loops using innovative and highly efficient technologies. Depending on the source, waste valorisation can incorporate anything from simple treatments for potential reuse to comprehensive processes within biorefinery facilities, significantly improving the worth of waste outputs. Through valorisation, components within waste streams are converted into valuable chemicals, materials, and fuels, thereby minimizing their harmful impact on the environment, human health, and the economy (Banerjee & Arora, 2021; Le Pera et al., 2022a).

Tailored valorisation strategies for local conditions, considering the distinct categories of generated waste and the available resources for remediation, will enhance the efficient transformation of waste streams into value-added products that align with the demands of the local market (Wang et al., 2022).

The primary goal of animal waste valorisation is to recover valuable products, such as nutritional substances and bioactive compounds, for use in industries like food, cosmetics, packaging, and pharmaceuticals. Traditional methods involving organic solvent extraction of bioactives are time-consuming and environmentally unfriendly. However, it is possible to optimize their utilization using various methods, for example: enzymatic hydrolysis before solvent extraction aids bioactive compound release. On the other hand, supercritical fluids like CO_2 present a greener option but might be economically unfeasible on an industrial scale. Cost-effective techniques such as ultrasound- and microwave-assisted extractions and pulsed electric fields are emerging alternatives (Le Pera et al. 2022a).

Presently, incineration and landfilling are the predominant methods employed for the disposal of solid animal waste, with the exception of manure which is mostly applied to agricultural land. However, concerns regarding economic viability and environmental sustainability have arisen. Consequently, there has been a focus on exploring biological strategies to recover both nutrients and energy from food waste, a practice referred to as food waste valorisation. Nevertheless, these biological approaches possess limitations and are unable to offer a comprehensive solution to managing food waste effectively. In this context, various sustainable technologies are being examined, employing intelligent methodologies such as zero-solid waste discharge, anaerobic digestion, and bio-fermentation, among others, to facilitate the recovery of resources and energy. Given the substantial organic content, availability, and rich nutritional composition of animal waste, it presents itself as a cost-effective and appropriate raw material for the production of value-added substances like biohydrogen, biomethane, biodiesel, biogas, and fertilizers. Composting, bioethanol synthesis, feed fermentation, anaerobic digestion, and other bioprocesses are being researched for the purpose of recovering resources and energy from animal waste. However, these processes designed to transform animal waste into valuable products do come with certain drawbacks. Issues such as a scarcity of skilled labour, high operational expenses, and the need for the post-processing of solid residues are some of the challenges associated with these approaches (Said et al., 2023; Wang et al., 2022).

A substantial proportion of animal waste consists of biodegradable materials that can serve as a bioresource for the reclamation of value-added products like biochemicals, bioplastics, biofuels, enzymes, and more. This is achieved through bioprocessing techniques such as anaerobic digestion, fermentation utilizing microorganisms, and bioprocess-based electrochemical systems (Banerjee & Arora, 2021). Among these approaches, bioprocessing of urban waste emerges as the most economically viable waste management solution in the current landscape, owing to the potential for generating non-toxic byproducts. Bacteria employed in waste bioprocessing have demonstrated a considerable capacity for breaking down the organic components within waste fractions (Tamasiga et al., 2022).

Biofuels derived from animal waste show promise as sustainable fossil fuel alternatives, being anaerobic digestion a popular method for generating gaseous fuels, like methane and hydrogen, due to its low operating costs (Banerjee & Arora, 2021).

Animal waste valorisation technologies can be classified into three main groups: biological processes (such as bioconversion and biosorption), thermochemical processes (including carbonization, pyrolysis, liquefaction, and gasification) and physicochemical processes (transesterification).

Table 2 gives some examples of waste valorisation technologies.

Table 2 Examples of animal waste valorisation technologies

Valorisation Technology	Animal Waste/ Feedstock	Products	Reference
Anaerobic digestion and Co-digestion	Cow manure	Biogas and biocompost	Gong et al. (2020)
	Deep bedding and Wastewater (pig farms)	Biogas and biofertilizer	Mendonça and Santos (2022)
	Pig slurry	Biogas and biofertilizer	Wang et al. (2022)
Bio-adsorption	Manure and Wastewater (livestock production)	Animal feed and biofertilizer (from microalgae)	Viegas and Gonçalves (2021), Viegas et al. ()
	Aquaculture Wastewater		Viegas et al. (2021a)
	Manure and Wastewater	Lipids and carotenoids (from microalgae)	Wang et al. (2021)
Fermentation	Cheese whey	Hydrogen, Ethanol, Butanediol	Yaashikaa et al. (2022)
	Agro-industrial wastes	Bio-etanol	Yaashikaa et al. (2022)
	Livestock manure		Kanani et al. (2020)
Enzymatic hydrolysis	Poultry feather waste	Methane	Kanani et al. (2020)
Solid-state fermentation (SSF)	Poultry processing	Enzymes, animal feedstuff, biofertilizers	Yazid et al. (2017)
	Marine products processing	Enzymes, bioactive compounds	
	Animal products, Tannery, Slaughterhouse	Enzymes, animal feedstuff, glues, surfactants, lubricants, filler	
	Cattle, broiler	Animal glue, animal feed supplement, methane production, biochar, activated carbon, biofertilizers, biopesticide	
Torrefaction	Horse manure	Biochar	Isemin et al. (2019)
Carbonisation	Livestock manures	Biochar	Kanani et al. (2020)
Pyrolysis	Poultry waste	Bio-oil, Biochar	Kanani et al. (2020)
	Livestock manures		Kanani et al. (2020)
	Agro-industrial wastes		Yaashikaa et al. (2022)
Hydrothermal liquefaction	Livestock manures	Hydrochar	Kanani et al. (2020)

(continued)

Table 2 (continued)

Valorisation Technology	Animal Waste/ Feedstock	Products	Reference
Gasification	Agro-industrial wastes	Syngas and Biochar	Yaashikaa et al. (2022), Kanani et al. (2020)
Combustion	Agro-industrial wastes	Electricity, Heat	Yaashikaa et al., (2022)
Calcination	Eggshells	Calcium oxide (CaO)	Kanani et al. (2020)
Transesterification	Seafood waste	Biodiesel	Venugopal (2022)
	Waste Animal Tallow (livestock)		Hariprasath et al. (2019)
	Pig and chicken fat		Andreo-Martínez et al. (2022)
Direct utilization	Feathers	Sound-absorbing nonwoven materials	Wang et al. (2022)
	Greasy wool (collected on the sheep farm)	Sweater and lanolin (byproduct)	

5.1 Bioconversion and Biosorption of Animal Waste

Proposed techniques for the valorisation of animal waste include biological treatment strategies and bioprocessing technologies such as anaerobic digestion and microbial fermentation. This involves utilizing enzymatic treatment of waste to yield liquid or methane gas. Anaerobic digestion entails transforming waste effluents into energy and nutrient-rich natural fertilizers. Meanwhile, microbial fermentation involves extracting chemical energy from organic matrices without oxygen utilization, resulting in the production of carbon and ethanol. Another recent bioprocess technology, microbial electrochemistry, has been specifically suggested for urban liquid waste streams. This technology harnesses the capacity of bacteria to generate electric current (Banerjee & Arora, 2021).

Within anaerobic digestion (AD), microorganisms facilitate the conversion of organic matter into biogas without the presence of oxygen. This biogas consists of a mixture primarily composed of methane (CH_4) (~65%) and carbon dioxide (CO_2) (~35%). The residual product of this procedure is termed digestate, containing both solid and liquid organic remnants. The strengths of AD as a waste valorisation technology lie in its capacity to handle moist biomass without prior treatment, and its straightforward and dependable approach to biogas production. Biogas can serve as a source of heat and/or electricity, with emission levels varying based on the specific system. Its applicability spans across small and large scales. Notably, in the United States, the prevalence of anaerobic digesters on commercial livestock farms has grown, rising from 241 digesters in 2016 to 280 in 2018. Among these facilities, dairy farms accounted for 78%, while poultry farms represented only 3%. An illustrative example of progress in this field can be seen in Europe, where the first AD plant fuelled exclusively by poultry litter was successfully established for commercial purposes in Northern Ireland. Operational since 2018, the facility has the capacity to convert up to 40,000 tonnes of poultry manure annually into biogas. This biogas is then utilized in two 1.5 mw gas engines that produce on-site electricity for integration into the national grid. Comparatively, chicken manure boasts a higher energy content than cow and pig manure. However, the elevated nitrogen levels in chicken manure lead to the conversion of undigested proteins into total ammonia nitrogen during the AD process (Kanani et al., 2020; Le Pera et al., 2022b).

In the process of anaerobic co-digestion, the generation of biogas from various organic materials hinges on multiple factors, including the ratio of substrates convertible into biogas, their chemical makeup, and their biodegradability. Co-digestion presents the opportunity to enhance biogas production by diluting ammonia levels and minimizing hydraulic retention times. Utilizing complementary substrates such as dairy manure, wheat straw, and fruit and vegetable waste, which possess higher volatile solids content compared to chicken manure, facilitates the accommodation of elevated nitrogen levels. This approach permits a greater proportion of chicken manure in the feed mix, leading to a reasonable methane yield. Notably, dairy manure stands out as a particularly effective co-substrate for chicken manure, as their combined digestion results in the lowest recorded concentrations of total ammonia nitrogen. However, a significant practical hurdle in the implementation of co-digestion revolves around logistical considerations. The success of co-digestion hinges on the availability of appropriate co-substrates in close proximity to the digestion facility (Kanani et al., 2020; Mendonça & Santos, 2022).

Two distinct approaches are employed in ethanol production: separate hydrolysis and fermentation (SHF) and simultaneous saccharification and fermentation (SF). The SHF technique, although the most established, comes with relatively higher costs due to the requirement for enzymes to hydrolyse biomass and pre-treatment. On the other hand, the SF process, as presently utilized, accomplishes both product fermentation and the incorporation of glucose-rich substrates produced during saccharification. Saccharification denotes the breakdown of starch from biomass or alternative feedstocks into glucose through enzymatic hydrolysis. SF proves to be more energy-efficient when compared to SHF, as it consistently converts hydrolysed sugars into ethanol and mitigates the presence of glucose, which tends to inhibit cellulose activity. Livestock manure contains substantial quantities of lignocellulosic materials, polysaccharides, proteins, and other organic compounds (comprising 30–50% of dry weight), of which 50–84% can be transformed into readily usable simple sugars (Kanani et al., 2020).

Solid-state fermentation (SSF) has garnered significant attention in recent years due to its potential for bioprocess development. The success of bioprocesses within SSF is contingent upon various factors, encompassing the compatibility of microorganisms, substrates, and process parameters. Microorganisms used in SSF include fungi, yeasts, and bacteria, with fungi and yeasts being the most prevalent choices due to the resemblance of SSF conditions to their natural habitat characterized by low water activity in the fermentation media. Although SSF is traditionally associated with fungi and yeast growth, there's a growing body of evidence showcasing the successful application of bacteria in producing bio-products via SSF. Numerous studies highlight the preference for filamentous fungi in SSF, primarily due to their capacity to generate high-value, thermostable enzymes. Improving SSF efficiency involves optimizing process variables such as initial moisture, particle size, pH, temperature, media composition, sterilization, water activity, inoculum density, agitation, aeration, product extraction, and downstream processing. SSF has introduced a novel approach to convert organic solid waste into biologically active metabolites, both on laboratory and industrial scales. This method finds application in producing a diverse array of bio-products, ranging from enzymes, organic acids, and biofertilizers to biopesticides, biosurfactants, bioethanol, aroma compounds, animal feed, pigments, vitamins, and antibiotics (Yafetto, 2022; Yazid et al., 2017).

Industries like leather and slaughterhouses generate substantial quantities of protein-rich organic waste, including animal fleshing, hair wastes, skin trimming, keratin wastes, chrome shaving, and buffing residues, which often remain underutilized. Authors have explored the utilization of animal fleshing in SSF for protease production. Combining hair wastes with activated sludge or anaerobically digested sludge has shown promising results in yielding protease (Yazid et al., 2017).

Crustacean by-products, a byproduct of industrial seafood processing, have been harnessed for the production of chitinase and chitosanase, which find diverse applications in the biomedical, food, and agrochemical sectors. SSF utilizing fish processing waste is advantageous due to its low cost and availability, and it provides a conducive environment for microorganism growth. Fish processing waste, rich in lipids and proteins, is well-suited for producing esterases, versatile products with applications in organic chemical processing, detergent formulations, and the surfactant and oleochemical industry (Yazid et al., 2017).

5.1.1 Using Microalgae

Instead of relying almost exclusively on physicochemical methods, the organic load and elevated levels of nitrogen and phosphorus in these effluents could be effectively managed through the use of microalgae (Viegas et al., 2021a). This approach offers the opportunity to mitigate water bodies' eutrophication potential while efficiently recovering and recycling nitrogen and phosphorus for the production of algal biomass. This method enables the incorporation of the generated biomass into the composition of agricultural fertilizers and animal feed, thereby decreasing the carbon footprint of the agri-food industries. Furthermore, it contributes to a circular economy by repurposing the byproducts generated in these industries, aiming for zero waste and converting them into value-added products (Viegas, 2022).

The autotrophic assimilation of nutrients from wastewater through biological processes is not only four times more cost-effective but also demands less energy, offers greater efficiency, and carries fewer ecological risks compared to conventional wastewater treatment methods (Acién et al., 2016). By combining the bioremediation of residual effluents with biomass and energy production, it's possible to address the current challenges of unsustainability (Mendonça et al., 2021).

In addition to the valuable biomass generated through phytoremediation, microalgae introduce oxygen into wastewater, promoting aerobic degradation by other microorganisms. This in turn enhances the levels of COD and BOD_5 in the effluents, playing a vital role in tertiary treatment. These organisms also break down recalcitrant compounds such as hydrocarbons, antibiotics, heavy metals, as well as other common components found in wastewater (Patel et al., 2017; Suganya et al., 2016).

The effectiveness of microalgae-based remediation varies depending on the specific agro-industrial waste being treated and the microalgae species selected. Trials conducted on diverse agro-industrial effluents have demonstrated the capability of microalgae to treat these wastewater. For instance, piggery effluents were effectively treated using *Chlamydomonas oblonga*, *Tetradesmus obliquus*, and *Chlorella vulgaris* (Abou-Shanab et al., 2013; Viegas & Gonçalves, 2021). Similarly, effluents from cattle farming exhibited promising results when treated with *Chlorella* sp. and *Scenedesmus* sp. (Gramegna et al., 2020; Ledda et al., 2016, 2017; Viegas et al., 2021b). Microalgae such as *Chlorella* sp. and *Scenedesmus* sp. have also shown potential in treating effluents from poultry production (Ferreira et al., 2017; Markou et al., 2018; Viegas et al., 2021c), while *Tetradesmus obliquus* and *Chlorella sorokiniana* proved effective for aquaculture production effluents (Ansari et al., 2017; Apandi et al., 2019; Viegas et al., 2021a). Additionally, *Tetradesmus obliquus* demonstrated remediation potential for effluents generated from oil extraction (Hodaifa et al., 2013).

The cultivation of microalgae can be linked to industrial facilities, like cement production plants, creating an opportunity to capture the CO_2 emissions generated in those areas. The integration of carbon dioxide fixation from flue gases with nutrient removal from wastewater during microalgae production represents a highly promising alternative, both economically and environmentally (Mendonça et al., 2021).

Using algal–bacterial systems appears to be the best solution to obtain the finest levels of remediation (Zhong et al., 2020).

5.2 *Thermochemical Processes for Animal Waste Valorisation*

Thermochemical treatment entails subjecting animal wastes to heat, resulting in the generation of heat energy or the conversion of the waste into products such as biochar, gas, or oxygenated fuel oil (bio-oil).

Thermochemical processes include combustion or incineration of waste, gasification, pyrolysis, hydrothermal liquefaction (HTL), and carbonization.

Incineration, a long-standing technique, involves thermally combusting animal waste to generate heat and energy. This process reduces waste volume by approximately 80–85% while also enabling energy recovery from the waste materials. Occurs in the presence of oxygen and involves temperatures of up to 1450°C. The incineration process produces gases such as CO_2, CH_4, NH_3, HCN, CO, and H_2, leading to significant environmental hazards. It generates energy in the form of heat, but all the material that could be used to generate new products is lost (Nayak & Bhushan, 2019).

Gasification stands as a thermochemical procedure that transforms carbon-rich substances into syngas, a mixture encompassing carbon monoxide (CO, 30–60%), hydrogen (H_2, 25–30%), and typically varying amounts of carbon dioxide (CO_2, 5–15%) and methane (CH_4, 0–5%) and some solid residue in the form of ash. This transformation occurs with the aid of a gasification agent (which could be steam, oxygen, and/or air) and a catalyst. It takes place at temperatures between 500 and 1100ºC with the limitation of the oxidizing agent. The resultant gases generated find utility as energy sources for heating, transportation, and power generation. This approach proves superior to conventional combustion due to its capacity to minimize emissions of nitrogen oxides (NOx) and sulfur oxides (SOx), attributed to the reduced oxygen levels and operating temperatures within the process. Gasification is not a novel concept in the realm of biomass utilization. Its roots trace back to the 1800s when it was initially employed for lighting and cooking purposes. Moreover, it can be effectively combined with carbon-rich biomass like wood waste to yield higher heating values of syngas. For example, for poultry feathers, fixed-bed gasification emerges as the most suitable technique, considering that liquefaction of feathers poses challenges in terms of energy intensity and cost inefficiency. Among the notable setbacks of gasification, however, is the concurrent formation of tar within the syngas during the scrubbing process aimed at impurity removal and gas enhancement. Moreover, an incomplete combustion may yield char, a solid carbonaceous material, and ash. The presence of biochar could lead to defluidization, and undesired plant shutdowns, ultimately diminishing overall yield efficiency rates (Kanani et al., 2020; Sakheta et al., 2023).

Pyrolysis involves the thermal decomposition of organic matter, between 400 and 750ºC occurring entirely without the presence of an oxidizing agent that would otherwise restrict intensive gasification. This process transforms biowaste into three distinct material states: bio-oil, sometimes referred to as tar (comprising condensable gases), char (solid residue), and syngas (consisting of non-condensable gases). The balance between gaseous and solid products depends mainly on the process temperature. The resulting bio-oil comprehends a diverse array of organic compounds, characterized by H/C molar ratios exceeding 1.5. This bio-oil can serve as a precursor for fuels and bulk chemicals, with various methodologies such as zeolite cracking, hydrogenation, and aqueous phase processing employed for conversion. The gaseous byproducts, syngas, present opportunities for ethanol production, and under pressurized or steam-reformed conditions, syngas can yield hydrogen (Kanani et al., 2020; Sakheta et al., 2023)

Hydrothermal liquefaction stands as a distinctive conversion method that employs the inherent water content within manure as both a reactant and reaction medium, all under subcritical conditions (ranging from 275 to 350ºC and 5 to 18 MPa). This process results in the production of an oil-like substance. The utilization of fresh manure in this process serves to curtail the duration of manure storage, consequently minimizing airborne emissions, with a particular focus on reducing ammonia emissions. Presently, hydrothermal liquefaction seems to exhibit greater suitability for dairy or swine manure, primarily due to its elevated water content when compared to poultry manure (Kanani et al., 2020).

Carbonization takes place in a temperature range above torrefaction. The product obtained, biochar, can have several uses, including as a biofuel, as a soil fertilizer and as a bioadsorbent (Sakheta et al., 2023).

5.3 *Transesterification Process for Animal Waste Valorisation*

The most commonly employed technique for the synthesis of biodiesel on a commercial scale is the transesterification reaction. This process involves converting triglycerides (TG) and/or free fatty acids (FFAs) into esters using organic solvents, typically short-chain alcohols. A wide range of lipid sources are utilized for this purpose, including both edible and non-edible oils, discarded and recycled greases, animal fats, and edible oil waste, each presenting unique profiles of fatty acid composition. In the production of biodiesel from vegetable oils and animal fats, methanol or ethanol is commonly used as the alcohol reactant. Methanol is the preferred choice due to its widespread availability and cost-effectiveness, although other short-chain alcohols such as isopropanol can also be employed. The transesterification reaction's outcome is influenced by various factors, including reaction temperature, reaction time, alcohol-to-oil ratio, type and quantity of catalyst, and feedstock composition (Andreo-Martínez et al., 2022).

This process can be carried out with either acid or alkali catalysts under mild conditions, or alternatively, in the absence of a catalyst under supercritical conditions. The initial two methods are constrained by the feedstock's composition, particularly in terms of free fatty acids (FFA) and water content. The presence of these compounds in the raw material can lead to saponification during biodiesel production, ultimately reducing the yield of the reaction and accelerating catalyst deactivation (Hariprasath et al., 2019).

5.4 *Some Cases of Animal Waste Recovery*

5.4.1 Eggshell Wastes

Eggshell waste has been used as an additive in a variety of applications such as in calcium phosphate bio-ceramics, calcium supplements, as adsorbent material, as a solid catalyst for biodiesel production, a substitute for bone grafts, a solution for environmental remediation, and a method for heavy metal removal. Nonetheless, the primary applications of eggshells in both food and industry predominantly revolve around the calcite component, which is utilized for creating fertilizers, medicines, and construction materials (Ngayakamo & Onwualu, 2022).

Research indicates that the production of cement from non-renewable sources necessitates a high calcination temperature of approximately 1450 C, resulting in the emission of around 0.85 tons of carbon dioxide for each ton of cement manufactured. These environmental implications stemming from conventional construction material production underscore the unsustainability of the construction industry. However, the utilization of bio-waste, like eggshells, has the potential to mitigate CO_2 emissions. This is due to the fact that calcination at temperatures below 750 C for biomass does not generate CO_2, rendering the process economically viable, sustainable, and environmentally friendly. The utilization of eggshell waste for developing sustainable construction materials namely, green concrete, clay bricks, ceramic tiles and stabilized masonry blocks has been explored (Ngayakamo & Onwualu, 2022).

Concrete, clay bricks, ceramic tiles and compressed stabilized blocks are materials widely used in construction that can incorporate eggshells. This integration has not only ensured their sustainable nature but also led to cost reduction and bolstered environmental preservation efforts. Substituting supplementary cementitious materials in amounts ranging from 5 to 30% has resulted in improved physical and mechanical properties in cement-based materials (like concrete and sandcrete) and clay-based materials (including clay bricks and tiles). However, excessive incorporation of supplementary cementitious materials has been observed to diminish the mechanical strength of these selected materials (Ngayakamo & Onwualu, 2022).

Conversely, calcium oxide (CaO) can be generated through the sequence of calcination, hydration, and dehydration, employing eggshells as the source material. This nano-catalyst finds application in the biodiesel production process, particularly in transesterification involving dry biomass or discarded cooking oil (Kanani et al., 2020).

5.4.2 Keratinous Wastes

The valorisation of keratinous wastes involves strategies and operational processes that include extraction, solubilization, reuse, recycling, and processing of these wastes, along with their residues or by-products. These processes aim to transform keratinous waste into valuable raw materials suitable for energy generation and to convert these wastes into value-added products that can be successfully marketed commercially, as hydrogels, nanogels, meshes, sponges, foams, scaffolds and glues (Ossai et al., 2022).

6 Biorefineries

Biorefinery is a growing concept within the sphere of waste management, with the objective of converting diverse waste-derived materials into various forms of energy, biofuels, chemicals, and materials. This is accomplished through a series of economically viable and environmentally gentle technical processes (Said et al., 2023).

Algal biomass biorefineries hold immense potential for generating a diverse variety of high-value bio-chemicals and biofuels, including products like biodiesel, bioethanol, biogas, protein additives, pigments, biofertilizers, biochars, and bioplastics. However, several critical challenges must be overcome. These include addressing production costs, managing complexities related to multi-product biorefining, ensuring the stability and robustness of genetically modified strains, and tackling technical difficulties during the scale-up process (Nishshanka et al., 2022). Nonetheless, it's essential to assess these integrated processes to uncover strategies that can mitigate production costs and the associated emissions.

An assessment of a microalgae-based biorefinery's life cycle highlighted that the foremost contributor to greenhouse gas emissions is the biomass production phase, accounting for 42.3% of total emissions. By substituting fossil fuels utilized for transportation and energy generation with biofuels, it's possible to significantly curtail the carbon footprint of the microalgae production stage, thus enhancing the sustainability of the entire process (Mobarezkhoo et al., 2022).

In the context of the circular waste-based bioeconomy, biorefineries assume a critical role by employing valorisation technologies to bioconvert both conventional and hazardous materials into bio-based products and bioenergy. Establishing sustainable valorisation pathways that prioritize high technical efficiency, and technology synergies, and include environmental as well as socio-economic aspects (human health, well-being, and social inclusion) is imperative for achieving closed loops within the integrated biorefinery concept. In essence, adopting a holistic approach that integrates environmental, economic, social, and technical considerations is recommended on a global and interdisciplinary scale to ensure the sustainable development of biorefineries (Wang et al., 2022).

7 Current Limitations of Waste Valorisation

In general, the constraints on the large-scale implementation of animal waste valorisation stem from several factors. A key limitation is associated with the upstream supply chain, involving waste collection and transportation from production sites to biorefineries. However, there has been limited research on transportation costs, emissions, waste seasonality, and quantities available for valorisation, necessitating comprehensive assessments (Wang et al., 2022).

Components that impair process efficiency, such as phenolics or other compounds that induce toxicity and inhibit microbial growth and downstream fermentation, require identification and disposal before further recovery. Challenges posed by inherent components, secondary metabolites, or co-contaminants in feedstocks, leading to process inefficiencies in various biorefinery models, must be examined, considered, and resolved (Banerjee & Arora, 2021).

Prioritizing animal waste valorisation measures generating byproducts that can substitute resource-intensive market products takes precedence over those derived from the largest waste quantities. It is evident that valorisation technologies are rapidly advancing in terms of efficiency and sustainability. The success and market reach of bio-based products from waste streams through valorisation techniques heavily rely on public policy implementation. Public perception of bio-based product safety and affordability in comparison to conventional counterparts is a significant concern. Governmental financial support and well-defined legislation can effectively address waste valorisation limitations (Le Pera et al., 2022a; Wang et al., 2022).

Hence, there's a pressing need for updated policies and waste management strategies to align with the fast-evolving landscape and achieve sustainability goals. Challenges that might dissuade decision-makers from adopting and leveraging these technologies include (1) being entrenched in outdated technologies, (2) unfavourable local economic conditions, (3) regulatory landscapes, and (4) hesitancy in investing in sustainable valorisation methods.

8 Conclusions and Future Perspectives

The treatment of animal effluents and wastes is currently one of the greatest challenges facing science globally because of the huge amounts of resources and energy required by these processes and their vast impacts on the environment.

The advancement of eco-friendly, innovative, secure, and enduring methodologies enhances the feasibility of embracing sustainable technologies for the transformation of food waste into valuable commodities and energy sources.

Various approaches can valorise animal waste, producing diverse end-products like biofuels, biochemicals, and biomaterials while reducing the original waste's hazards. Nonetheless, developing resilient microbial strains that can efficiently produce target products under treatment conditions remains a research challenge. Effective pretreatment methods that release valuable end-products at low costs and with minimal environmental impact are also essential. Future research should explore innovative fermentation technologies, including consortium-based bioprocessing.

Nonetheless, the practical utilization of waste valorisation has been largely unexploited. The actual implementation of waste valorisation hinges on economic feasibility—whether the costs associated with equipment, operations, and maintenance align with commercial-scale production. In light of this, findings from the literature review suggest that governmental policies offering financial backing (such as tax incentives, investment loans, and land concessions) should be contemplated to incentivize the engagement of businesses and enterprises in waste valorisation endeavours.

Bibliography

Abou-Shanab, R. A. I., et al. (2013). Microalgal species growing on piggery wastewater as a valuable candidate for nutrient removal and biodiesel production. *Journal of Environmental Management, 115*, 257–264.

Acién, F. G., et al. (2016). Wastewater treatment using microalgae: How realistic a contribution might it be to significant urban wastewater treatment? *Applied Microbiology and Biotechnology, 100*(21), 9013–9022. https://doi.org/10.1007/s00253-016-7835-7

Amini, E., et al. (2020). Municipal wastewater treatment by semi-continuous and membrane algal-bacterial photo-bioreactors. *Journal of Water Process Engineering, 36*(April), 101274. https://doi.org/10.1016/j.jwpe.2020.101274

Andreo-Martínez, P., et al. (2022). Waste animal fats as feedstock for biodiesel production using non-catalytic supercritical alcohol transesterification: A perspective by the PRISMA methodology. *Energy for Sustainable Development, 69*, 150–163.

Ansari, F. A., Singh, P., Guldhe, A., & Bux, F. (2017). Microalgal cultivation using aquaculture wastewater: Integrated biomass generation and nutrient remediation. *Algal Research, 21*, 169–177. https://doi.org/10.1016/j.algal.2016.11.015

Apandi, N., et al. (2019). Scenedesmus biomass productivity and nutrient removal from wet market wastewater, a bio-kinetic study. *Waste and Biomass Valorization, 10*(10), 2783–2800. https://doi.org/10.1007/s12649-018-0313-y

Balakrishnan, A., Kanchinadham, S., & Kalyanaraman, C. (2020). Studies on the effect of pre-treatment of vegetable tanning process wastewater prior to biological treatment. *Journal of Environmental Chemical Engineering, 8*(4), 104020. https://doi.org/10.1016/j.jece.2020.104020

Banerjee, S., & Arora, A. (2021). Sustainable bioprocess technologies for urban waste valorization. *Case Studies in Chemical and Environmental Engineering, 4*, 100166. https://doi.org/10.1016/j.cscee.2021.100166

Cokgor, E. U., Karahan, O., & Orhon, D. (2008). The effect of mixing pharmaceutical and tannery wastewaters on the biodegradation characteristics of the effluents. *Journal of Hazardous Materials, 156*(1–3), 292–299.

Commission, E. (2015). Directive of the European Parliament and of the council—amending directive 2008/98/EC on waste. *Official Journal of the European Union, 0275*, 1–5.

Durkin, A., Guo, M., Wuertz, S., & Stuckey, D. C. (2022). Resource recovery from food-processing wastewaters in a circular economy: A methodology for the future. *Current Opinion in Biotechnology, 76*(May), 102735. https://doi.org/10.1016/j.copbio.2022.102735

Ekka, B., et al. (2022). Quantification of different fatty acids in raw dairy wastewater. *Cleaner Engineering and Technology, 7*(April 2021), 100430.

FAO. (1999, February). FAO *Food Outlook No.1*. https://www.fao.org/3/x1116e/x1116e07.htm

FAO. (2013). Food wastage footprint *Food Wastage Footprint: Impacts on Natural Resources—Summary Report*. https://www.fao.org/publications

FAO. (2022). 60 FAOSTAT Analytical Brief Series No. 60 *Agricultural Production Statistics 2000–2021*. https://doi.org/10.4060/cc3751en

Ferreira, A., et al. (2017). Scenedesmus obliquus mediated brewery wastewater remediation and CO_2 biofixation for green energy purposes. *Journal of Cleaner Production, 165*, 1316–1327. https://doi.org/10.1016/j.jclepro.2017.07.232

Finnegan, W., et al. (2018). DairyWater: Striving for sustainability within the dairy processing industry in the Republic of Ireland. *Journal of Dairy Research, 85*(3), 366–374.

Ganesh, K. S., Sridhar, A., & Vishali, S. (2022). Utilization of fruit and vegetable waste to produce value-added products: conventional utilization and emerging opportunities—A review. *Chemosphere, 287*(P3), 132221. https://doi.org/10.1016/j.chemosphere.2021.132221

Gao, F., et al. (2016). Continuous microalgae cultivation in aquaculture wastewater by a membrane photobioreactor for biomass production and nutrients removal. *Ecological Engineering, 92*, 55–61. https://doi.org/10.1016/j.ecoleng.2016.03.046

Gong, W., et al. (2020). Cow manure anaerobic fermentation effluent treatment by oxygen-based membrane aerated biofilm reactor. *Chemical Engineering Journal, 395*(April), 125116. https://doi.org/10.1016/j.cej.2020.125116

Gramegna, G., et al. (2020). Exploring the potential of microalgae in the recycling of dairy wastes. *Bioresource Technology Reports, 12*, 100604. https://doi.org/10.1016/j.biteb.2020.100604

Hariprasath, P., et al. (2019). Some studies on waste animal tallow biodiesel produced by modified transesterification method using heterogeneous catalyst. *Materials Today: Proceedings, 16*, 1271–1278. https://doi.org/10.1016/j.matpr.2019.05.224

Haşimoğlu, S., & Aksoy, A. (1998). Animal industry its waste (manure), human health and environmental concerns. *Atatürk Üniversitesi Ziraat Fakültesi Dergisi, 29*(1), 1–10.

Hodaifa, G., Sánchez, S., Martínez, E., & Órpez, R. (2013). Biomass production of scenedesmus obliquus from mixtures of urban and olive-oil mill wastewaters used as culture medium. *Applied Energy, 104*, 345–352.

Isemin, R., et al. (2019). Application of torrefaction for recycling bio-waste formed during anaerobic digestion. *Fuel, 243*(September 2018), 230–239. https://doi.org/10.1016/j.fuel.2019.01.119

Kanani, F., Heidari, M. D., Gilroyed, B. H., & Pelletier, N. (2020). Waste valorization technology options for the egg and broiler industries: A review and recommendations. *Journal of Cleaner Production, 262*, 121129. https://doi.org/10.1016/j.jclepro.2020.121129

Karić, N., et al. (2022). Bio-Waste valorisation: Agricultural wastes as biosorbents for removal of (in)organic pollutants in wastewater treatment. *Chemical Engineering Journal Advances, 9*, 100239.

Köninger, J., et al. (2021). Manure management and soil biodiversity: Towards more sustainable food systems in the EU. *Agricultural Systems, 194*(December), 103251. https://doi.org/10.1016/j.agsy.2021.103251

Krishna, R., et al. (2022). Different treatment methodologies and reactors employed for dairy effluent treatment—A review. *Journal of Water Process Engineering, 46*(December 2021), 102622. https://doi.org/10.1016/j.jwpe.2022.102622

Kumar, A., et al. (2018). A process for complete biodegradation of shrimp waste by a novel marine isolate Paenibacillus sp. AD with simultaneous production of chitinase and chitin oligosaccharides. *International Journal of Biological Macromolecules, 109*, 263–272. https://doi.org/10.1016/j.ijbiomac.2017.12.024

Kuo, C. M., et al. (2016). Simultaneous microalgal biomass production and CO2 fixation by cultivating Chlorella sp. GD with aquaculture wastewater and boiler flue gas. *Bioresource Technology, 221*(75), 241–250. https://doi.org/10.1016/j.biortech.2016.09.014

Kurniawan, S. B., et al. (2022). Treatment of real aquaculture effluent using bacteria-based bioflocculant produced by *Serratia marcescens*. *Journal of Water Process Engineering, 47*(March), 102708. https://doi.org/10.1016/j.jwpe.2022.102708

Labbé, J. I., et al. (2017). Microalgae growth in polluted effluents from the dairy industry for biomass production and phytoremediation. *Journal of Environmental Chemical Engineering, 5*(1), 635–643.

Ledda, C., et al. (2016). Integration of microalgae production with anaerobic digestion of dairy cattle manure: An overall mass and energy balance of the process. *Journal of Cleaner Production, 112*, 103 112.

Li, C., Zhang, Z., Cao, J., & Li, Yi. (2016). Study on poultry manure wastewater treatment by two stage aerobic coupled process and its microbial community analysis. *Biochemical Engineering Journal, 110*, 107–114.

Lima, M., Aline, J., Filho, F. J. C. M., Constantino, M., & Formagini, E. L. (2020). Techno-Economic and performance evaluation of energy production by anaerobic digestion in Brazil: Bovine, swine and poultry slaughterhouse effluents. *Journal of Cleaner Production, 277*, 123332.

Liu, Y., & Chen, J. (2014). Phosphorus cycle. In B. Fath (Ed.), *Encyclopedia of ecology* (2nd ed., pp. 181–191). Elsevier.

Lukwambe, B., et al. (2018). Bioturbation by the razor clam (*Sinonovacula constricta*) on the microbial community and enzymatic activities in the sediment of an ecological aquaculture Wastewater treatment system. *Science of the Total Environment, 643*(818), 1098–1107. https://doi.org/10.1016/j.scitotenv.2018.06.251

Markou, G., Wang, L., Ye, J., & Unc, A. (2018). Using agro-industrial wastes for the cultivation of microalgae and duckweeds: contamination risks and biomass safety concerns. *Biotechnology Advances, 36*(4), 1238–1254. https://doi.org/10.1016/j.biotechadv.2018.04.003

Mendonça, H. V., et al. (2021). Microalgae in a global world: New solutions for old problems? *Renewable Energy, 165*, 842–862. https://doi.org/10.1016/j.renene.2020.11.014

Mendonça, H. V., & Mônica, S. S. (2022). Co-Digestion of deep bedding and wastewater from pig farming: A new strategy for bioenergy increase and biofertilizer recovery. *Journal of Environmental Management, 304*(November 2021).

Mo, W., & Zhang, Q. (2012). Can municipal wastewater treatment systems be carbon neutral? *Journal of Environmental Management, 112*, 360–367.

Mobarezkhoo, H., Saidi-Mehrabad, M., & Sahebi, H. (2022). Life cycle evaluation of microalgae based biorefinery supply network: A sustainable multi-criteria approach. *Algal Research, 66*(August), 102816. https://doi.org/10.1016/j.algal.2022.102816

Nayak, A., & Brij B. (2019). An overview of the recent trends on the waste valorization techniques for food wastes. *Journal of Environmental Management, 233*(December 2018), 352–370. https://doi.org/10.1016/j.jenvman.2018.12.041

Ngayakamo, B., & Onwualu, A. P. (2022). Recent advances in green processing technologies for valorisation of eggshell waste for sustainable construction materials. *Heliyon, 8*(6), e09649. https://doi.org/10.1016/j.heliyon.2022.e09649

Nishshanka, G. K., Hasara, S., et al. (2022). Marine Microalgae as sustainable feedstock for multi-product biorefineries. *Biochemical Engineering Journal, 187*(August), 108593. https://doi.org/10.1016/j.bej.2022.108593

Oryschak, M. A., & Beltranena, E. (2020). Reconsidering the contribution of canadian poultry production to anthropogenic greenhouse gas emissions: Returning to an integrated crop-poultry production system paradigm. *Poultry Science, 99*(8), 3777–3783. https://doi.org/10.1016/j.psj.2020.05.004

Ossai, C. Innocent, F. S. H., & Hassan, A. (2022). Valorisation of keratinous wastes: A sustainable approach towards a circular economy. *Waste Management,* 151(February): 81–104. https://doi.org/10.1016/j.wasman.2022.07.021

Patel, A., Gami, B., Patel, P., & Patel, B. (2017). Microalgae: Antiquity to era of integrated technology. *Renewable and Sustainable Energy Reviews, 71*, 535–547.

Le Pera, A., Miriam, S., Egidio, B., & Francesco, D'. A. (2022a). Environmental sustainability of an integrate anaerobic digestion-composting treatment of food waste: Analysis of an Italian plant in the circular bioeconomy strategy. *Waste Management, 139*(July 2021), 341–351. https://doi.org/10.1016/j.wasman.2021.12.042

Le Pera, A., Miriam, S., Egidio, B., & Francesco, D'. A. (2022b). Environmental sustainability of an integrate anaerobic digestion-composting treatment of food waste: Analysis of an Italian Plant in the Circular Bioeconomy Strategy. *Waste Management, 139*(January), 341–351. https://doi.org/10.1016/j.wasman.2021.12.042

Pereira, M. d. S. et al. (2020). Dissolved air flotation optimization for treatment of dairy effluents with organic coagulants. *Journal of Water Process Engineering,* 36(April). 101270. https://doi.org/10.1016/j.jwpe.2020.101270.

Rossi, S., et al. (2022). Piggery wastewater treatment with algae-bacteria consortia: Pilot-scale validation and techno-economic evaluation at farm level. *Bioresource Technology, 351*(February), 127051. https://doi.org/10.1016/j.biortech.2022.127051

Said, Z., et al. (2023). Intelligent approaches for sustainable management and valorisation of food waste. *Bioresource Technology, 377*(February), 128952. https://doi.org/10.1016/j.biortech.2023.128952

Sakheta, A., Nayak, R., O'Hara, I., & Ramirez, J. (2023). A review on modelling of thermochemical processing of biomass for biofuels and prospects of artificial intelligence-enhanced approaches. *Bioresource Technology, 386*(July), 129490. https://doi.org/10.1016/j.biortech.2023.129490

Salmerón, I., Oller, I., & Malato, S. (2021). Solar photo-assisted electrochemical processes applied to actual industrial and urban wastewaters: A practical approach based on recent literature. *Chemosphere*, 130560. https://doi.org/10.1016/j.chemosphere.2021.130560

Shi, W., et al. (2021). Dairy processing sludge and co-products: A review of present and future re-use pathways in agriculture. *Journal of Cleaner Production, 314*(June), 128035. https://doi.org/10.1016/j.jclepro.2021.128035

Siddique, N. I., & Zularisam, A. W. (2018). Achievements and perspectives of anaerobic co-digestion: A review. *Journal of Cleaner Production, 191.*

Sillanpää, M., & Marina, S. (2017). Electrochemical water treatment methods. In M. Sillanpää & M., Shestakova. *Electrochemical water treatment methods: Fundamentals, methods and full scale applications* (pp. 47–130). Butterworth-Heinemann.https://doi.org/10.1016/B978-0-12-811462-9.00002-5

Simón, D., et al. (2021). Valorization of sludge from the effluent treatment of the dairy industry as clay substitutes in building bricks. *Construction and Building Materials, 307*(July).

Suganya, T., Varman, M., Masjuki, H. H., & Renganathan, S. (2016). Macroalgae and microalgae as a potential source for commercial applications along with biofuels production: A biorefinery approach. *Renewable and Sustainable Energy Reviews, 55*, 909–941.

Suman, H., Sangal, V. K., & Vashishtha, M. (2021). Treatment of tannery industry effluent by electrochemical methods: A review. *Materials Today: Proceedings.* https://doi.org/10.1016/j.matpr.2021.03.300

Tamasiga, P., Taghi, M., Helen O., & Abarasi, H. (2022). Food waste and circular economy: Challenges and opportunities. *Sustainability (Switzerland), 14*(16).

United Nations—Water. (2021). UN-Water *Summary Progress Update 2021 : SDG 6—Water and Sanitation for All.* Geneve. https://www.unwater.org/new-data-on-global-progress-towards-ensuring-water-and-sanitation-for-all-by-2030/

Venugopal, V. (2022). Green processing of seafood waste biomass towards blue economy. *Current Research in Environmental Sustainability, 4*(December 2021), 100164. https://doi.org/10.1016/j.crsust.2022.100164

Viegas, C. (2022). *Bioremediation of agro-industrial effluents mediated by microalgae.* NOVA University Lisbon.

Viegas, C., & Gonçalves, M. (2021). Integrated treatment of pig production wastewaters using pre-treatment with biomass ash and bioremediation by microalgae. *Acta Scientific Agriculture, 5*(7), 44–57.

Viegas, C., Luísa G., & Margarida G. (2021a). Aquaculture wastewater treatment through microalgal. Biomass potential applications on animal feed, agriculture, and energy. *Journal of Environmental Management, 286*(November 2020), 112187. https://doi.org/10.1016/j.jenvman.2021.112187

Viegas, C., Luísa G., & Margarida G. (2021b). Bioremediation of cattle manure using microalgae after pre-treatment with biomass ash. *Bioresource Technology Reports, 14*(March), 100681. https://doi.org/10.1016/j.biteb.2021.100681

Viegas, C., Luísa G., & Margarida G. (2021c). Evaluation of microalgae as bioremediation agent for poultry effluent and biostimulant for germination. *Environmental Technology and Innovation, 24*, 102048. https://doi.org/10.1016/j.eti.2021.102048

Wang, X., et al. (2021). Synergistic bioconversion of lipids and carotenoids from food waste by *Dunaliella salina* with fulvic acid via a two-stage cultivation strategy. *Energy Conversion and Management, 234*(February), 113908. https://doi.org/10.1016/j.enconman.2021.113908

Wang et al. (2022). Emerging waste valorisation techniques to moderate the hazardous impacts, and their path towards sustainability. *Journal of Hazardous Materials, 423*(PA), 127023. https://doi.org/10.1016/j.jhazmat.2021.127023

Yaashikaa, P. R., Senthil Kumar, P., & Varjani, S. (2022). Valorization of agro-industrial wastes for biorefinery process and circular bioeconomy: A Critical review. *Bioresource Technology, 343*(August 2021), 126126. https://doi.org/10.1016/j.biortech.2021.126126

Yafetto, L. (2022). Application of solid-state fermentation by microbial biotechnology for bioprocessing of agro-industrial wastes from 1970 to 2020: A review and bibliometric analysis. *Heliyon, 8*(3), e09173. https://doi.org/10.1016/j.heliyon.2022.e09173

Yazid, N. A., Barrena, R., Komilis, D., & Sánchez, A. (2017). Solid-State fermentation as a novel paradigm for organic waste valorization: A review. *Sustainability (Switzerland), 9*(2), 1–28.

Yetilmezsoy, K., & Sakar, S. (2008). Improvement of COD and color removal from UASB treated poultry manure wastewater using fenton's oxidation. *Journal of Hazardous Materials, 151*(2–3), 547–558.

Zhong, Y. et al. (2020). Does exogenous carbon source always promote algal biomass and nutrients removal in algal-bacterial wastewater remediation? *Journal of Cleaner Production*, 104743.

Waste Valorization Techniques

Zakia Gueboudji

Abstract Waste valorization is the process of converting waste materials into valuable products or energy, with the goal of reducing waste and improving resource efficiency. This technique can help to address the growing global waste problem while also promoting a more sustainable and circular economy. There are several techniques of waste valorization that have been developed, each with their own benefits and limitations. One of the most commonly used techniques is recycling, which involves collecting and processing waste materials such as plastics, paper, and metal to produce new products. Recycling can significantly reduce the amount of waste sent to landfills and can help to conserve natural resources by using recycled materials instead of virgin resources. Another technique is composting, which involves the decomposition of organic waste materials such as food and yard waste to create a nutrient-rich soil amendment. Composting can reduce the amount of organic waste sent to landfills, promote soil health, and reduce the need for synthetic fertilizers. Bioconversion is another technique that can be used to valorize waste. This process involves the use of microorganisms to convert organic waste into biofuels, bioplastics, and other value-added products. Bioconversion can help to reduce greenhouse gas emissions, promote energy independence, and reduce dependence on fossil fuels. Pyrolysis and gasification are thermal treatment techniques that can be used to convert waste into energy. Pyrolysis involves heating waste in the absence of oxygen to produce a liquid oil, while gasification involves heating waste in the presence of oxygen to produce a gas that can be used for energy. These techniques can help to reduce waste and provide a source of renewable energy. Finally, landfill mining is a technique that involves excavating and processing waste materials that have been buried in landfills. This process can recover valuable resources such as metals and plastics, reduce the amount of waste in landfills, and mitigate the environmental impacts of landfilling. In conclusion, waste valorization techniques have the potential to reduce waste generation, recover valuable resources, and minimize environmental

Z. Gueboudji (✉)
Biotechnology, Water, Environment and Health Laboratory, Abbes Laghrour University of Khenchela, Khenchela, Algeria
e-mail: zakia.gueboudji@univ-batna2.dz

Department of Microbiology and Biochemistry, Faculty of Natural and Life Sciences, University of Batna 2 Mostefa Ben Boulaid, Batna, Algeria

A. Anouzla and S. Souabi (eds.), *Generation of Energy from Municipal Solid Waste*,
https://doi.org/10.1007/978-3-031-74334-4_2

impacts. The choice of the appropriate technique depends on the type of waste material, the desired product, and the economic feasibility of the process. It is important to promote and encourage the adoption of waste valorization practices to ensure sustainable waste management and resource recovery.

1 Introduction

The surge in global solid waste generation can be attributed to rapid population expansion, escalated urbanization, agricultural requirements, and industrial progress. Forecasts indicate that the world's population will likely reach 8.5 billion by 2030, leading to an estimated 2.59 billion tons of solid waste production. This concerning trend poses further strain on the already compromised environment and climate conditions. Consequently, it is imperative to urgently explore and implement effective approaches for solid waste recycling (Peng et al., 2023). Waste constitutes a noteworthy global ecological concern, as the production of waste presents a substantial impediment to the enduring progress of societies. The administration of waste constitutes a pivotal facet of sustainability, and the diminution, repurposing, and recycling of waste have garnered paramount attention across numerous nations (Said et al., 2023). Nevertheless, novel waste valorization methodologies have materialized as an innovative modus operandi within waste administration, transcending established conventions. Waste valorization encompasses the conversion of waste into invaluable commodities or substances, consequently engendering a circular economy while mitigating the volume of waste consigned to landfill facilities (Abdelbasir, 2021; Ashokkumar et al., 2022).

In the contemporary era, the effective control and proper disposal of waste have arisen as pressing obstacles, owing to the swift expansion of the worldwide populace, urbanization, and industrial advancement (Liu & Liu, 2023). Conventional methods of waste management, including landfilling and incineration, not only exhibit constrained capabilities but also present notable risks to the environment and public well-being (Fang et al., 2023). Consequently, the notion of waste valorization has garnered noteworthy recognition as a forward-looking and environmentally viable strategy for tackling the predicament of waste (Pujara et al., 2019).

In recent times, waste valorization methodologies have garnered extensive attention as a mechanism for waste reduction and the advancement of sustainable practices (More et al., 2022). These methodologies encompass the transformation of waste into valuable products or materials, spanning energy, chemicals, fuels, and construction components. The central objective of waste valorization is to foster a circular economy, wherein waste is repurposed or recycled, leading to diminished waste accumulation in landfills, reduced dependency on virgin resources, and the promotion of sustainability (Nandhini et al., 2022; Shahbaz et al., 2023). Various approaches exist for the valorization of waste, encompassing thermal, biological, and chemical procedures (Gueboudji et al., 2024). Thermal processes entail the conversion of waste into energy via methods such as incineration, gasification, or pyrolysis (Shahbaz et al., 2023). Biological methods entail the utilization of microorganisms to disintegrate waste into beneficial outputs, such as compost or biogas (Gao et al., 2022). Chemical processes involve the conversion of waste into valuable chemicals or fuels through intricate chemical reactions (Zheng et al., 2023). The potential advantages associated with waste valorization methodologies are multifaceted. Firstly, these methodologies present a sustainable alternative to established waste management practices by curtailing the necessity for landfill disposal and incineration, both of which are linked to emissions of greenhouse gases, contamination through leachates, and depletion of resources. Secondly, waste valorization holds the potential to reclaim valuable resources, encompassing metals, organic compounds, and energy, thereby fostering the evolution of a circular economic model and diminishing reliance on finite natural resources. Moreover, waste valorization techniques offer the capacity to ameliorate environmental contamination by treating and transforming hazardous waste substances into non-toxic or less detrimental forms (Nayak & Bhushan, 2019; Yaashikaa et al., 2022). Nevertheless, the extensive-scale execution of waste valorization approaches presents several challenges. These encompass intricacies of technology, economic feasibility, regulatory structures, social acceptance, and the imperative for cohesive waste management systems. Surmounting these challenges mandates interdisciplinary partnerships among scientists, engineers, policy architects, and stakeholders, coupled with robust endeavors in research and development (Mandal et al., 2023; Roy et al., 2023).

This chapter book aims to present a comprehensive survey of diverse waste valorization techniques, delineating their respective benefits and drawbacks, while also analyzing their suitability across diverse waste management scenarios. The paper will additionally address the complexities linked to waste valorization challenges and delve into the imperative realm of policies and regulations crucial for fostering the uptake of waste valorization methods.

2 Waste Generation and Composition

Waste can be categorically differentiated by its source, intrinsic attributes, and capacity for ecological repercussions. The ensuing sections elucidate the frequently acknowledged classifications of waste.

- **Municipal Solid Waste (MSW)**

This classification pertains to the waste emanating from domestic residences, commercial enterprises, and institutional establishments. It encompasses a spectrum of materials such as food waste, paper, plastics, glass, metals, textiles, and assorted sundries (Aziz et al., 2023).

- **Industrial Waste**

This category corresponds to waste originating from industrial processes and manufacturing endeavors. Its composition can exhibit considerable diversity and encompasses solids, liquids, and gases. Illustrative examples comprise chemical residues, sludges, discarded materials, and packaging detritus (Das et al., 2022).

- **Hazardous Waste**

Hazardous waste engenders notable risks to both human well-being and the environment owing to its toxic, inflammable, corrosive, or reactive attributes. This classification encompasses waste stemming from industrial operations, healthcare facilities, laboratories, and households. Instances encompass chemical byproducts, batteries, electronic refuse, and pharmaceutical discards (Stoett et al., 2021).

- **Construction and Demolition (C&D) Waste**

C&D waste materializes during construction, refurbishment, and demolition undertakings. It includes elements such as concrete, wood, metals, bricks, asphalt, and insulation. The unwieldy volume and diverse composition of C&D waste pose challenges in its effective management (Córdoba et al., 2019).

- **Agricultural and Organic Waste**

This grouping comprises waste emerging from agricultural pursuits, encompassing crop remnants, animal excrement, and packaging materials. Organic waste also subsumes culinary refuse originating from households, dining establishments, and food processing industries (Garcia et al., 2017).

- **Electronic Waste (e-waste)**

E-waste encompasses discarded electronic and electrical apparatuses, encompassing computers, televisions, smartphones, refrigerators, and batteries. These items harbor perilous substances and valuable resources that necessitate proper handling and treatment (Shahabuddin et al., 2023).

3 Waste Valorization: Concepts and Principles

Waste valorization, also acknowledged as waste utilization or resource recovery, pertains to the procedure of deriving value from waste materials by converting them into beneficial products, energy, or resources. This entails the transformation of waste streams into valuable commodities through diverse physical, chemical, and biological processes. The ambit of waste valorization is expansive, encompassing an array of waste materials such as municipal solid waste, industrial residuals, agricultural remains, wastewater sludge, and more (Arancon et al., 2013).

The concept of waste valorization aligns harmoniously with the tenets of the circular economy, where waste is reconceptualized as a latent resource rather than a liability. This approach aims to deviate from the conventional linear waste management paradigm, characterized by the sequence "take-make-dispose," and instead embraces a more ecologically sound and resource-efficient framework. Through the extraction of value from waste, this paradigm shift diminishes dependence on virgin resources, curtails ecological footprints, and advances the adoption of a self-sustaining, closed-loop system (Priyadarshini & Abhilash, 2020).

3.1 *Waste Management Hierarchy*

Waste valorization is commonly situated within the waste management hierarchy, a structured framework that aids in prioritizing waste management strategies according to their ecological implications. This hierarchy is typically organized into the following tiers, sequenced based on preference:

- **Waste Prevention**

The optimal approach involves averting the generation of waste from the outset. This entails curbing waste at its source through actions like innovative product design, material substitution, and the advocacy of sustainable consumption behaviors (Priefer et al., 2016).

- **Minimization and Reduction**

If complete waste prevention proves unfeasible, the subsequent step is to mitigate and decrease waste production. This is achieved through methods such as recycling, reutilization, and the enactment of waste reduction schemes (Garcia-Garcia et al., 2017).

- **Recycling and Resource Recovery**

Recycling encompasses the processing of waste materials to extract valuable elements or transform them into novel products. This practice conserves resources, diminishes energy utilization, and curtails the demand for raw materials (Gaur et al., 2022; Gueboudji, 2022).

- **Energy Recovery**

When waste materials cannot be effectively recycled or reused, energy recovery emerges as a viable alternative. This involves the conversion of waste into energy via processes like incineration, anaerobic digestion, or gasification, resulting in the production of heat, electricity, or biofuels (Lee et al., 2023).

- **Treatment and Disposal**

The ultimate stratum of the hierarchy involves the secure and ecologically responsible treatment and disposal of waste that cannot be valorized. This encompasses techniques such as controlled landfilling, where waste is deposited in designated sites under regulated conditions to mitigate environmental impacts.

The waste management hierarchy serves as a directive for decision-making, championing waste valorization as a favored option whenever practicable. By doing so, it endorses the adoption of sustainable waste management methodologies that align with principles of environmental stewardship (Artiola, 2019).

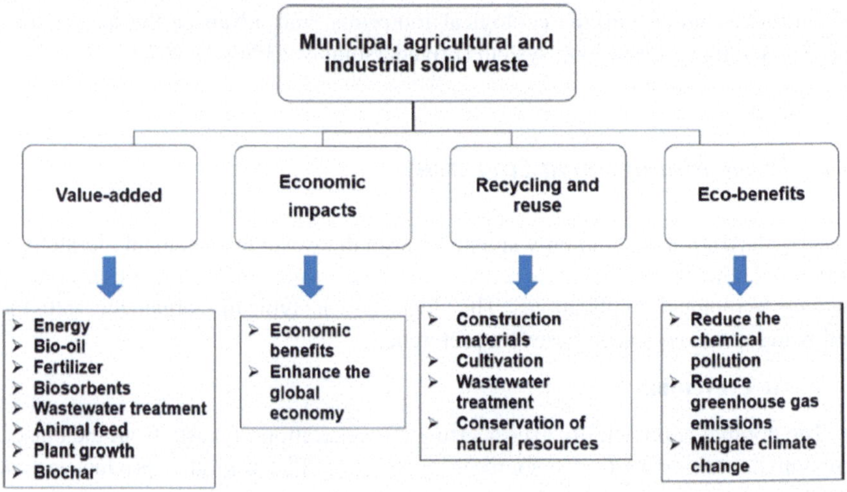

Fig.1 Valorization of Solid Wastes (Peng et al., 2023)

Figure 1 depicts the potential reuse and recycling of various types of solid waste, such as municipal, agricultural, and industrial waste, leading to numerous advantageous outcomes. The process of waste valorization results in the generation of value-added byproducts, environmental benefits, recycling and reusing practices, and favorable economic outcomes, creating a sequential cascade of benefits.

3.2 *Challenges and Opportunities in Waste Valorization*

While waste valorization holds substantial potential, it confronts several obstacles that necessitate resolution for its widespread adoption. Firstly, the technological intricacy of waste valorization methods frequently demands sophisticated technologies and infrastructure, thereby posing technical and operational hurdles. The development of efficient and scalable processes capable of accommodating diverse waste streams emerges as a pivotal focus area. Secondly, the economic feasibility of waste valorization initiatives constitutes a pivotal determinant. The presence of cost-effective solutions and substantial market demand for valorized goods or energy stands as imperative prerequisites for the financial viability of such undertakings. Thirdly, regulatory and policy frameworks play a fundamental role in fostering and incentivizing waste valorization endeavors. Clearly delineated guidelines, standards, and supportive structures are indispensable to expedite the execution of waste valorization projects and to ensure conformity with environmental and safety stipulations. Fourthly, the perception and acceptance of waste valorization technologies by the public can profoundly influence their successful integration. Active engagement and education of the public regarding the benefits and safety considerations linked to these methodologies are essential for garnering public confidence and endorsement. Lastly, the harmonious incorporation of waste valorization within comprehensive waste management systems encompassing waste reduction, recycling, energy recovery, and appropriate treatment and disposal is paramount. The formulation of integrated strategies that optimize the entirety of the waste management process constitutes a challenge necessitating close collaboration among stakeholders (Leong & Chang, 2022; Swain, 2023).

4 Waste Valorization Techniques

The management of municipal solid waste encompasses several approaches: disposal in landfills, incineration, as well as waste-to-energy methods such as biological processes, chemical processes, biochemical conversion, and thermochemical processes (Nandhini et al., 2022), as illustrated in **Fig. 2**.

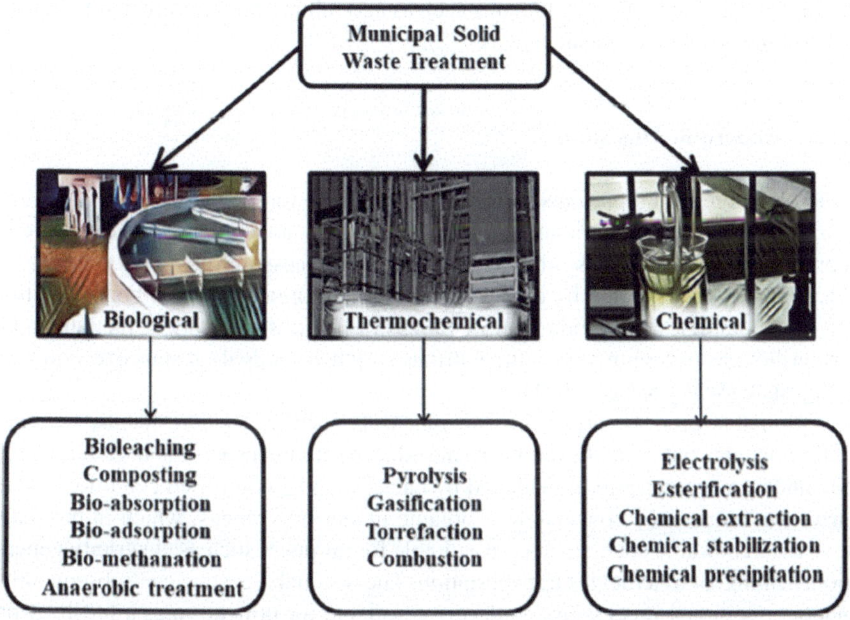

Fig.2 Categorization of Thermal and Non-Thermal Techniques for Energy Generation from Municipal Solid Waste (Nandhini et al., 2022)

4.1 Biological Waste Valorization Techniques

4.1.1 Composting

Composting serves as a biological technique for the valorization of waste, involving the decomposition of organic waste components into a nutrient-enriched soil amendment recognized as compost. This process is inherently natural, orchestrated by microorganisms comprising bacteria, fungi, and earthworms. Composting exhibits versatility in its implementation, spanning from modest backyard composting endeavors to expansive commercial operations (Amuah et al., 2022; Nandhini et al., 2022).

Throughout the composting progression, organic waste materials such as food remnants, yard clippings, and agricultural residues undergo decomposition in the presence of oxygen. The microorganisms catalyze the breakdown of organic substances, leading to their conversion into a stable, humus-like substance that enhances both soil fertility and structure. The outcome of composting is not solely a reduction in the volume of organic waste but also the generation of a valuable end-product applicable in diverse domains such as agriculture, horticulture, landscaping, and soil remediation (Amuah et al., 2022).

4.1.2 Anaerobic Digestion

Anaerobic digestion represents a biological technique for the valorization of waste, characterized by the degradation of organic waste constituents in an oxygen-deprived environment. This intricate biochemical process unfolds within an anaerobic digester, orchestrated by a consortium of microorganisms. The consequential outcome comprises the production of biogas, a composite blend of methane and carbon dioxide, in conjunction with a nutrient-enriched residual substance recognized as digestate (Wang & Lee, 2021).

Within this process, organic waste materials such as food residuals, agricultural remnants, and animal excreta are introduced into the anaerobic digester. Here, they undergo decomposition mediated by bacteria under controlled conditions. The bacteria effectuate the conversion of organic matter into biogas, which in turn can serve as a renewable energy resource, suitable for purposes such as electricity generation, heating, or as a fuel for transportation. The residual digestate can subsequently undergo additional processing, rendering it suitable for utilization as a fertilizer or as a means to amend soil quality (Wang & Lee, 2021).

4.1.3 Vermicomposting

Vermicomposting constitutes a variant of composting that harnesses the activity of earthworms to expedite the process of decomposition. This technique entails the deliberate provision of organic waste materials to earthworms, specifically species like Eisenia fetida or Lumbricus rubellus, within an appropriate environment. The earthworms engage in the consumption of organic matter and subsequently release nutrient-enriched castings, thereby furnishing a valuable amendment for soil enrichment (Amuah et al., 2022).

Vermicomposting presents distinct advantages vis-à-vis conventional composting methodologies. The involvement of earthworms enhances the disintegration of organic waste through their feeding and tunneling activities, thus expediting the decomposition process. The resultant vermicompost is distinguished by its abundant content of beneficial microorganisms, enzymes, and plant nutrients, thereby facilitating the cultivation of robust soil health and fostering optimal plant growth. Remarkably, vermicomposting can be executed on a smaller scale, rendering it particularly suitable for household and community-level waste management initiatives (Swati & Hait, 2017).

4.1.4 Bioconversion and Bioremediation

Bioconversion and bioremediation stand as biological methodologies within the realm of waste valorization, leveraging microorganisms to effectuate the conversion of waste materials or the remediation of polluted environments.

Bioconversion entails the utilization of microorganisms to transmute specific waste substances into valuable commodities. To illustrate, particular strains of bacteria and fungi can be harnessed to convert agricultural residues or food waste into biofuels, enzymes, organic acids, or bioplastics. This process harnesses the metabolic proficiencies of microorganisms to metamorphose waste materials into compounds of elevated value (Salehi & Wang, 2022).

In contrast, bioremediation involves the employment of microorganisms to disintegrate or detoxify contaminants within polluted settings. Certain bacterial and fungal strains exhibit the capacity to decompose or metabolize pollutants, such as petroleum-derived hydrocarbons, heavy metals, pesticides, and solvents, into less injurious states. Bioremediation advances a sustainable and ecologically conscientious strategy to ameliorate predicaments of soil, water, and air contamination (Salehi & Wang, 2022).

Both bioconversion and bioremediation exemplify promising techniques within the ambit of waste valorization, yielding dual outcomes of abating environmental pollution and facilitating the retrieval of resources.

4.2 Thermal Waste Valorization Techniques

4.2.1 Incineration

Incineration represents a thermal technique for the valorization of waste, characterized by the controlled combustion of waste materials at elevated temperatures. This process is typically conducted within specialized facilities termed waste-to-energy plants. Throughout incineration, both organic and combustible waste constituents undergo combustion, resulting in the liberation of thermal energy. This released heat energy can be harnessed to generate steam, which subsequently propels turbines for the generation of electricity.

The practice of incineration yields a reduction in waste volume and can constitute an efficacious means of managing specific waste categories, such as municipal solid waste and hazardous waste. Nevertheless, it is imperative to institute stringent emission control mechanisms to curtail the discharge of pollutants—encompassing particulate matter, heavy metals, and dioxins—into the environment (Shahbaz et al., 2023).

4.2.2 Pyrolysis

Pyrolysis denotes a thermal method of waste valorization, involving the disintegration of organic materials devoid of oxygen. This procedure entails subjecting waste to elevated temperatures, typically ranging between 400°C and 800°C, within a controlled environment. The absence of oxygen impedes full combustion, resulting in the generation of distinct outputs, including biochar, liquid bio-oil, and combustible gases such as syngas. The products engendered by pyrolysis find diverse applications. Biochar, a solid residue rich in carbon, can be deployed as a soil amendment to augment soil fertility and sequester carbon. Liquid bio-oil can undergo further refinement for conversion into biofuels or as a precursor for chemical synthesis. The syngas produced can function as a fuel source for heat and electricity generation, or alternatively serve as a foundational material for the production of both chemicals and fuels (Shahbaz et al., 2023).

4.2.3 Gasification

Gasification constitutes a thermal strategy for the valorization of carbonaceous materials, effecting their conversion into a synthetic gas termed syngas. This process involves the partial oxidation of waste materials at elevated temperatures, typically exceeding 800°C, within a controlled environment featuring limited oxygen provision. The resultant syngas comprises primarily of carbon monoxide, hydrogen, and other combustible gases (Lee et al., 2023).

Syngas emerges as a versatile energy resource applicable to diverse purposes. It can be engaged in the combustion process within gas engines or turbines to generate both electricity and heat. Alternatively, the syngas can undergo further refinement to yield biofuels, chemicals, or other value-added commodities (Lee et al., 2023).

The technique of gasification presents a host of benefits, encompassing elevated energy efficiency, diminished greenhouse gas emissions when compared to conventional combustion, and the potential to employ a wide spectrum of waste feedstocks, spanning from biomass and coal to waste plastics (Shahbaz et al., 2023).

4.2.4 Plasma Arc Technology

Plasma arc technology represents a sophisticated thermal approach for the valorization of waste materials, harnessing plasma, an intensely ionized gas, to effect the transformation of waste into valuable outputs. This technique necessitates subjecting waste to exceedingly high temperatures, reaching several thousand degrees Celsius, inducing the disintegration of waste materials into their elemental constituents (Hazra et al., 2019).

Within plasma arc technology, an electric arc is engendered between two electrodes, engendering a plasma zone characterized by its extreme heat. In this context, waste materials are introduced into the plasma zone, where they undergo vaporization and ionization. The elevated temperatures and the presence of reactive species within the plasma medium synergistically facilitate the conversion of waste materials into products such as syngas, slag, and various other secondary outputs (Hazra et al., 2019).

Of significance, plasma arc technology showcases the capability to accommodate an extensive array of waste streams, spanning from hazardous to non-hazardous waste, as well as contaminated soil and electronic waste. Among its notable advantages are its pronounced conversion efficiency, the obliteration of hazardous components, and the retrieval of valuable metals from the waste stream (Hazra et al., 2019).

4.3 Chemical and Physical Waste Valorization Techniques

4.3.1 Hydrothermal Processing

Hydrothermal processing represents a chemical approach to the valorization of waste materials, entailing the treatment of said materials within an environment characterized by elevated temperature and pressure, typically involving hot, pressurized water or an aqueous solution. This method capitalizes on the distinctive attributes of water under conditions of heightened temperature and pressure to expedite the decomposition, solubilization, and transformation of diverse waste constituents (Song et al., 2021).

Throughout the course of hydrothermal processing, organic waste materials are subject to various chemical reactions, including hydrolysis, liquefaction, and depolymerization. These reactions collectively yield outcomes such as bio-oil, gas, and solid residues. Notably, this process can be tailored to selectively address specific waste streams, encompassing biomass, organic sludge, and food waste. The resultant products have the potential to be deployed for energy generation, chemical production, or as raw material for further industrial procedures (Nayak & Bhushan, 2019).

Hydrothermal processing garners several merits, including the capacity to accommodate an extensive range of feedstocks, effective utilization of water as a reaction medium, and the ability to convert intricate waste materials into outputs of substantive value (Song et al., 2021).

4.3.2 Solvent Extraction

Solvent extraction serves as a physical methodology within the sphere of waste valorization, concentrating on the extraction and reclamation of valuable constituents from waste materials through the utilization of appropriate organic solvents. This technique hinges on the variance in solubility exhibited by waste components within particular solvents, allowing for the selective extraction of desired compounds or metals (Gueboudji et al., 2022; Muhsan & Zahoor, 2023).

In the domain of waste valorization, solvent extraction finds prevalent application in the retrieval of valuable metals from sources such as electronic waste, mining byproducts, and industrial residues. The procedure entails subjecting the waste material to a solvent, promoting the dissolution of targeted components. This is succeeded by steps involving separation and refinement, ultimately yielding the coveted end products (Muhsan & Zahoor, 2023; Nayak & Bhushan, 2019).

Solvent extraction endows a streamlined and ecologically conscious avenue for resource recuperation. It serves to curtail the necessity for raw material extraction, thereby ameliorating environmental repercussions associated with waste disposal practices.

4.3.3 Pyrolysis and Gasification for Synthesis Gas Production

Pyrolysis and gasification are prominent techniques in the realm of waste valorization, encompassing both chemical and physical processes aimed at converting organic waste materials into a mixture of gases commonly referred to as synthesis gas or syngas. These methods are executed at elevated temperatures in environments characterized by either the absence (pyrolysis) or restricted supply (gasification) of oxygen (Lee et al., 2023; Shahbaz et al., 2023).

In the pyrolysis process, waste materials undergo thermal decomposition, engendering a diverse array of outputs encompassing biochar, bio-oil, and syngas. Gasification, in contrast, effects the conversion of waste materials into syngas via partial oxidation. This syngas can fulfill the role of a fuel source, catering to heat and electricity generation, or it can undergo subsequent refinement to yield transportation fuels and assorted chemicals.

Pyrolysis and gasification exhibit a versatile capacity to address a multitude of waste streams, including biomass, plastics, and organic residues. Their utility extends to encompass energy recovery and the generation of valuable chemical products through the treatment of waste materials (Nguyen-Thi el al., 2023).

4.3.4 Recycling and Upcycling Technologies

Recycling and upcycling stand as prominent physical methods within the ambit of waste valorization, directed toward the retrieval and reintegration of waste materials into the production cycle, thereby mitigating resource depletion and the accrual of waste (Ellis et al., 2021).

Recycling orchestrates the collection, sorting, and processing of waste materials, with the intention of generating fresh products or raw materials. Established recycling techniques encompass activities like shredding, melting, or chemical treatments to reshape waste materials into usable configurations. This practice finds widespread application for materials such as paper, glass, metals, and plastics. (Ellis et al., 2021; Jung et al., 2023).

In contrast, upcycling entails the conversion of waste materials into products of augmented value or materials endowed with heightened quality or functionality. This approach pivots around imaginative reuse and inventive approaches, thereby imbuing waste materials with renewed purpose. Instances of upcycling entail the metamorphosis of discarded textiles into fashion accessories or the repurposing of shipping containers into modular housing units (Jung et al., 2023).

The technologies of recycling and upcycling play pivotal roles in the reduction of waste generation, the preservation of resources, and the cultivation of a circular economy. By facilitating the closure of material loops and by diminishing reliance on virgin resources, these methodologies align with the objectives of waste minimization and sustainable resource management.

Figure 3 represents the potential for creating value from solid waste through various avenues such as application diversification, methods for assessing economic feasibility, and the direction of sorting techniques for recycling. The enhancement of waste sorting technology is also a focal point. Furthermore, there exists a necessity to broaden the scope of valuable applications derived from solid waste. It is recommended to introduce novel evaluation approaches and economic indicators to enhance the likelihood of achieving economic sustainability (Peng et al., 2023).

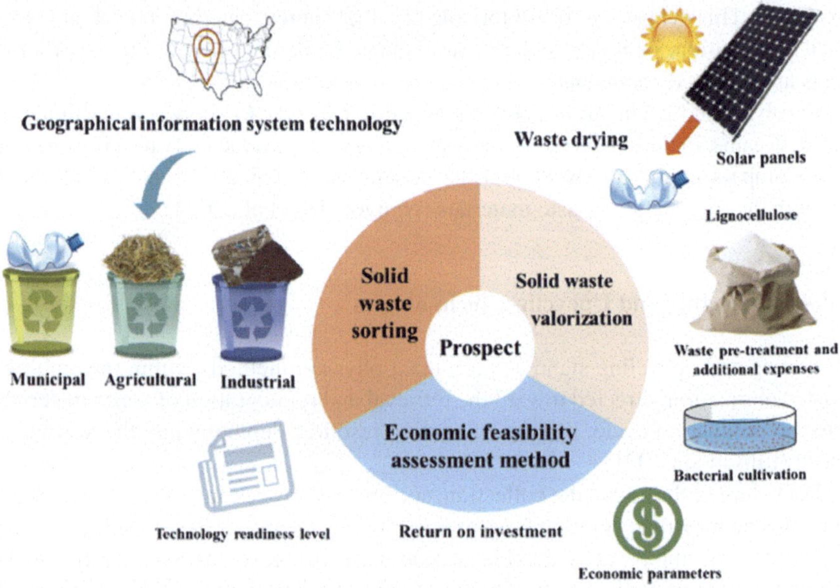

Fig.3 Enhanced Solid Waste Recycling and Sorting Techniques (Peng et al., 2023)

4.4 *Waste-To-Energy Conversion*

4.4.1 Energy Recovery from Waste

Energy recovery from waste, also recognized as waste-to-energy conversion, encompasses a procedure wherein non-recyclable waste materials are transformed into usable energy forms. This strategy contributes to the reduction of waste volume, the mitigation of landfill utilization, and the harnessing of the latent energy content within waste.

The energy recovery process is typically enacted through the combustion of waste materials, conducted within specialized facilities known as waste-to-energy plants. The heat derived from the combustion procedure is subsequently channeled to produce steam, which propels turbines, thereby generating electricity. In certain instances, waste can be converted into heat or integrated into combined heat and power (CHP) systems, engendering both electricity and heat outputs applicable to district heating schemes or industrial processes (Castaldi et al., 2017).

4.4.2 Power Generation Technologies

Diverse power generation technologies find application within waste-to-energy conversion, facilitating the generation of electricity from the heat engendered during waste combustion. Among these technologies are:

- Steam Turbines: Steam turbines constitute a prevalent choice in waste-to-energy plants. The high-temperature steam stemming from waste combustion is directed onto turbine blades, inducing rotation and the subsequent generation of electricity (Ibikunle et al., 2021).
- Gas Turbines: Gas turbines harness the heated gases resulting from waste combustion to propel a turbine, thereby producing electricity. This technology is frequently integrated with steam turbines in a combined cycle arrangement, augmenting overall efficiency (Indrawan et al., 2020).
- Internal Combustion Engines: Internal combustion engines, encompassing reciprocating engines or gas engines, possess the capability to directly convert the hot gases originating from waste combustion into mechanical energy. This energy is subsequently channeled to generate electricity (Khan & Kabir, 2020).
- Organic Rankine Cycle (ORC): ORC technology capitalizes on an organic working fluid characterized by a lower boiling point than water. The heated gases emerging from waste combustion elevate the temperature of the working fluid, thereby prompting its vaporization and the ensuing rotation of a turbine, culminating in the generation of electricity (Rathod et al., 2019).

4.4.3 Energy Conversion Efficiency and Environmental Impacts

The energy conversion efficiency of waste-to-energy processes is variable, contingent on the adopted technology and the composition of the waste. As a general assessment, contemporary waste-to-energy plants tend to achieve electrical efficiency levels spanning approximately 20%–30%. Notably, combined heat and power configurations often yield superior overall energy efficiencies.

While waste-to-energy conversion confers the advantage of energy recovery, it concurrently introduces environmental ramifications. The combustion of waste materials yields emissions encompassing greenhouse gases, particulate matter, and other pollutants. Nevertheless, advanced waste-to-energy technologies are outfitted with sophisticated emission control mechanisms, including flue gas cleaning systems and selective catalytic reduction, designed to curtail the release of pollutants and assure conformity with rigorous environmental standards.

Furthermore, waste-to-energy conversion can play a role in diminishing reliance on fossil fuels and contribute to the trajectory toward a more sustainable and low-carbon energy paradigm (Foster et al., 2021).

4.5 *Resource Recovery*

4.5.1 Extraction of Valuable Materials from Waste

The extraction of valuable materials from waste constitutes a pivotal facet of resource recovery and the circular economy paradigm. This process entails the discernment, isolation, and reclamation of valuable components from waste streams, thereby ensuring their reintroduction into the production cycle. A gamut of techniques is harnessed to extract valuable materials from diverse types of waste.

- Mechanical Sorting: Mechanical sorting techniques, encompassing approaches like magnetic separation, eddy current separation, and density-based segregation, are implemented to segregate ferrous and non-ferrous metals, plastics, and other recyclable materials from waste streams (Vermeşan et al., 2019).
- Chemical Leaching: Chemical leaching entails the use of solvents or acids to selectively dissolve and recover valuable metals from electronic waste, mining by-products, or industrial residuals (Mishra et al., 2021).
- Biological Treatment: Biological treatment methodologies, including bioleaching and biosorption, capitalize on microorganisms to recuperate metals from low-grade ores, mining refuse, and electronic waste (Saravanan et al., 2022).
- Advanced Separation Technologies: Advanced separation technologies, which incorporate sophisticated sensors, optical sorting mechanisms, and automated systems, are increasingly employed to enhance the efficiency and precision of material separation and recovery processes (Vermeşan et al., 2019).

4.5.2 Recovery of Metals and Minerals

The recuperation of metals and minerals from waste materials emerges as a pivotal focal point within resource recovery initiatives. Valuable metals like copper, aluminum, and precious metals hold the potential for economic recovery from diverse waste origins, encompassing electronic waste, industrial waste, and mining by-products. Assorted technologies are instrumental in the retrieval of metals and minerals (Krishnan et al., 2021; Mavukkandy et al., 2019).

- Smelting and Refining: Smelting encompasses the application of heat to induce the melting of metals, succeeded by refining procedures that refine and regain the sought-after metals. This approach finds prevalent use in extracting metals from electronic waste and other waste streams containing metals.
- Hydrometallurgical Processes: Hydrometallurgical processes capitalize on aqueous solutions and chemical reactions to dissolve and recover metals from waste materials. Such processes are frequently harnessed for the retrieval of precious metals like gold and silver from electronic waste and expended catalysts.
- Ion Exchange and Adsorption: Ion exchange and adsorption techniques employ specific resins or adsorbents to preferentially capture and recover targeted metals

from waste streams. These methods exhibit efficacy in recuperating valuable metals from dilute solutions or intricate waste streams.

4.5.3 Plastic Recycling and Upcycling

Plastic recycling and upcycling assume pivotal roles within the context of the circular economy, championing the cause of mitigating the environmental repercussions associated with plastic waste while concurrently advocating for the sustainable utilization of plastics (Ellis et al., 2021).

Plastic recycling encompasses the stages of collection, sorting, and processing of plastic waste, culminating in the creation of reusable materials or novel products. Common recycling methodologies encompass mechanical recycling, wherein plastics undergo shredding and melting for subsequent reprocessing, as well as chemical recycling, a process wherein plastics are subjected to chemical transformations yielding monomers or other valuable chemicals amenable for further utilization (Gueboudji, 2024; Jung et al., 2023; Horodytska et al., 2019).

Plastic upcycling centers on the metamorphosis of plastic waste into commodities of augmented value characterized by enhanced quality or heightened functionality. Illustrative instances encompass the conversion of plastic bottles into textile fibers for clothing purposes, or the repurposing of plastic packaging into sturdy construction materials. These endeavors collectively facilitate the alignment of plastic waste management with the principles of sustainability and resource conservation (Blazquez-Martín et al., 2023; Balu et al., 2022).

5 Environmental and Sustainability Assessment

5.1 Life Cycle Assessment (LCA) of Waste Valorization Techniques

The Life Cycle Assessment (LCA) represents a widely adopted methodology for the evaluation of environmental implications linked to waste valorization approaches. LCA engrosses a comprehensive overview of the complete life cycle of a process or product, commencing from the extraction of raw materials and extending through to the ultimate disposal phase, encompassing all facets spanning production, conveyance, and energy utilization.

LCA furnishes a thorough examination of the environmental ramifications associated with waste valorization techniques, including facets such as greenhouse gas emissions, energy utilization, water consumption, and waste creation. Its utility resides in pinpointing pivotal junctures within the life cycle wherein enhancements can be instituted to attenuate environmental burdens and amplify sustainability objectives (Dastjerdi et al., 2021).

Through the execution of LCAs upon various waste valorization techniques, researchers and policymakers gain the capacity to gauge their environmental performance and consequently make informed decisions pertaining to their adoption and implementation. This analytical approach furnishes a scientific foundation for assessing the ecological implications of diverse waste management strategies (Saravanan et al., 2023).

5.2 Environmental Impact Evaluation

Beyond Life Cycle Assessment (LCA), a spectrum of tools and methodologies is engaged to scrutinize the environmental consequences linked to waste valorization techniques. These encompass:

- Environmental Impact Assessment (EIA): EIA constitutes a systematic procedure that delineates and evaluates the conceivable environmental ramifications accompanying a proposed undertaking or endeavor. It embraces an array of factors, including air quality, water contamination, effects on ecosystems, and noise pollution.
- Ecological Footprinting: Ecological footprinting quantifies the environmental influence of a waste valorization technique by numerically gauging the quantum of land, water, and resources requisite for its sustenance. Furthermore, it gauges waste generated and emissions discharged.
- Risk Assessment: Risk assessment appraises the potential hazards tied to waste valorization methods. This encompasses assessing risks to worker health, adjacent communities, and potential impacts on ecosystems and natural resources.
- Toxicity Assessment: Toxicity assessment scrutinizes the potential toxicological implications associated with waste valorization techniques. It identifies and quantifies the liberation of harmful substances, assessing their impact on human health and the environment.

These methodologies dedicated to evaluating environmental impacts facilitate the evaluation of prospective hazards and advantages linked to waste valorization techniques. In so doing, they contribute to the advancement of waste management practices underpinned by ecological sustainability and social responsibility (Iacovidou et al., 2017).

6 Conclusions

Amidst an era marked by escalating environmental concerns and the imperative of sustainable resource management, the adoption of waste valorization methodologies emerges as a beacon of promising innovation. These methodologies, characterized by their intrinsic ability not only to curtail waste generation but also to recuperate valuable resources and alleviate ecological consequences, herald a profound transformation in the paradigms of waste management. At the heart of this transformative potential lies the inherent adaptability of waste valorization techniques, which readily accommodate the distinct composition of waste materials, the envisioned outcomes, and the economic feasibility intrinsic to the endeavor.

The essence of successful waste valorization resides in the astute selection of the most apt technique. This decision resonates with considerations of paramount importance, encompassing the nuanced attributes of the waste material itself, the end product embodying the extracted value, and the economic viability that underscores the feasibility of the enterprise. This strategic alignment of technique, material, and intended outcome establishes the groundwork for an efficient and proficient waste valorization process, optimizing the utilization of resources while minimizing environmental implications.

Of pivotal significance is the enduring advocacy and propagation of waste valorization practices, which assume a central role in this narrative. By mobilizing a collective dedication to the assimilation and integration of these methodologies within the framework of waste management, stakeholders can guide our trajectory towards a sustainable and harmonious coexistence with the environment. As the global community grapples with escalating waste production and finite resource reservoirs, the amalgamation of waste valorization not only bolsters resource retrieval but also functions as a potent instrument in nurturing an ethos of environmental stewardship.

In summation, the advent of waste valorization methodologies signifies a critical juncture in our approach to waste management and the preservation of resources. Endowed with the capability to harness the latent potential of waste and rejuvenate it into valuable resources, these methodologies illuminate a path towards a future characterized by the convergence of sustainability, resource optimization, and environmental equilibrium. The onus now rests on individuals, industries, and policymakers to seize this opportunity, advocate for transformation, and propel the transition towards a realm wherein waste assumes renewed purpose, resources are reclaimed, and the impact on the environment is mitigated (Krishnan et al., 2021; Park et al., 2023; Soliman &Moustafa, 2020).

Acknowledgements Many thanks to the leader editors Doctor Anouzla Abdelkader and Professor Salah Souabi, Hassan II University Casablanca, Faculty of Sciences and Techniques of Mohammedia, Morocco, for the suggestion to contribute to this book.

References

Abdelbasir, S. M. (2021). Recycling, management, and valorization of industrial solid wastes. *Waste Recycling Technologies for Nanomaterials Manufacturing,* 25–63.https://doi.org/10.1007/978-3-030-68031-2_2

Amuah, E. E. Y., Fei-Baffoe, B., Sackey, L. N. A., Douti, N. B., & Kazapoe, R. W. (2022). A review of the principles of composting: Understanding the processes, methods, merits, and demerits. *Organic Agriculture, 12*(4), 547–562. https://doi.org/10.1007/s13165-022-00408-z

Arancon, R. A. D., Lin, C. S. K., Chan, K. M., Kwan, T. H., & Luque, R. (2013). Advances on waste valorization: New horizons for a more sustainable society. *Energy Science & Engineering, 1*(2), 53–71. https://doi.org/10.1002/ese3.9

Artiola, J. F. (2019). Industrial waste and municipal solid waste treatment and disposal. In *Environmental and pollution science* (pp. 377–391). Academic Press. https://doi.org/10.1016/B978-0-12-814719-1.00021-5

Ashokkumar, V., Flora, G., Venkatkarthick, R., SenthilKannan, K., Kuppam, C., Stephy, G. M., & Ngamcharussrivichai, C. (2022). Advanced technologies on the sustainable approaches for conversion of organic waste to valuable bioproducts: Emerging circular bioeconomy perspective. *Fuel, 324,* 124313. https://doi.org/10.1016/j.fuel.2022.124313

Aziz, H. A., Ramli, S. F., & Hung, Y. T. (2023). Physicochemical technique in municipal solid waste (MSW) landfill leachate remediation: A review. *Water, 15,* 1249. https://doi.org/10.3390/w15061249

Balu, R., Dutta, N. K., & Roy Choudhury, N. (2022). Plastic waste upcycling: A sustainable solution for waste management, product development, and circular economy. *Polymers, 14*(22), 4788. https://doi.org/10.3390/polym14224788

Blazquez-Martín, A., Verde-Sesto, E., Arbe, A., & Pomposo, J. A. (2023). Metamorphosis of a Commodity Plastic like PVC to Efficient Catalytic Single-Chain Nanoparticles. *Angewandte Chemie International Edition, 62*(46), e202313502. https://doi.org/10.1002/anie.202313502

Castaldi, M., van Deventer, J., Lavoie, J. M., Legrand, J., Nzihou, A., Pontikes, Y., & Verstraete, W. (2017). Progress and prospects in the field of biomass and waste to energy and added-value materials. *Waste and Biomass Valorization, 8,* 1875–1884. https://doi.org/10.1007/s12649-017-0049-0

Córdoba, R. E., Marques, J. D. C., Santiago, C. D., Pugliesi, E., & Schalch, V. (2019). Alternative construction and demolition (C&D) waste characterization method proposal. *Engenharia Sanitaria e Ambiental, 24,* 199–212. https://doi.org/10.1590/S1413-41522019179720

Das, O., Babu, K., Shanmugam, V., Sykam, K., Tebyetekerwa, M., Neisiany, R. E., & Ramakrishna, S. (2022). Natural and industrial wastes for sustainable and renewable polymer composites. *Renewable and Sustainable Energy Reviews, 158,* 112054. https://doi.org/10.1016/j.rser.2021.112054

Dastjerdi, B., Strezov, V., Rajaeifar, M. A., Kumar, R., & Behnia, M. (2021). A systematic review on life cycle assessment of different waste to energy valorization technologies. *Journal of Cleaner Production, 290,* 125747. https://doi.org/10.1016/j.jclepro.2020.125747

Ellis, L. D., Rorrer, N. A., Sullivan, K. P., Otto, M., McGeehan, J. E., Román-Leshkov, Y., ... & Beckham, G. T. (2021). Chemical and biological catalysis for plastics recycling and upcycling. *Nature Catalysis, 4*(7), 539–556. https://doi.org/10.1038/s41929-021-00648-4

Fang, W., Ding, Y., Geng, J., Liu, Z., Huang, Y., Yang, J., & Qu, Z. (2023). High potential of coupling the source-separation and incineration promotion to reduce costs based on city-level cost-benefit analysis of municipal solid waste management strategies in China. *Resources, Conservation and Recycling, 197,* 107099. https://doi.org/10.1016/j.resconrec.2023.107099

Foster, W., Azimov, U., Gauthier-Maradei, P., Molano, L. C., Combrinck, M., Munoz, J., & Patino, L. (2021). Waste-to-energy conversion technologies in the UK: Processes and barriers–A review. *Renewable and Sustainable Energy Reviews, 135,* 110226. https://doi.org/10.1016/j.rser.2020.110226

Gao, Z., Alshehri, K., Li, Y., Qian, H., Sapsford, D., Cleall, P., & Harbottle, M. (2022). Advances in biological techniques for sustainable lignocellulosic waste utilization in biogas production. *Renewable and Sustainable Energy Reviews, 170*, 112995. https://doi.org/10.1016/j.rser.2022.112995

Garcia, D., & You, F. (2017). Systems engineering opportunities for agricultural and organic waste management in the food–water–energy nexus. *Current Opinion in Chemical Engineering, 18*, 23–31. https://doi.org/10.1016/j.coche.2017.08.004

Garcia-Garcia, G., Woolley, E., Rahimifard, S., Colwill, J., White, R., & Needham, L. (2017). A methodology for sustainable management of food waste. *Waste and Biomass Valorization, 8*(6), 2209–2227. https://doi.org/10.1007/s12649-016-9720-0

Gaur, A., Gurjar S. K., & Chaudhary, S. (2022). Circular system of resource recovery and reverse logistics approach: Key to zero waste and zero landfill. In *Advanced organic waste management* (pp. 365–381). Elsevier. https://doi.org/10.1016/B978-0-323-85792-5.00008-3

Gueboudji, Z. (2022). *Effect of storage time on the quality of olive oil mill wastewaters obtained from the cold extraction of olive oil in the region of Khenchela and their biological properties* (Doctoral dissertation). Abbes Laghrour University-Khenchela, p. 250.

Gueboudji, Z. (2024). From trash to treasure: Transforming landfill tyres, plastics, and glass into sustainable construction resources. *Technical Landfills and Waste Management: Municipal Solid Waste Management* (Vol. 2, pp. 295–313). https://doi.org/10.1007/978-3-031-55665-4_12

Gueboudji, Z., Kadi, K., & Nagaz, K. (2022). Anti-Inflammatory activity of polyphenols from olive oil mill wastewaters. *Jordanian Journal of Engineering & Chemical Industries, 5*(1).

Gueboudji, Z., Mahmoudi, M., Kadi, K., & Nagaz, K. (2024). Characteristics and Impacts of Municipal Solid Waste (MSW): A Review. *Technical Landfills and Waste Management: Landfill Impacts, Characterization and Valorisation* (Vol. 1, pp. 115–134). https://doi.org/10.1007/978-3-031-52633-6_4

Hazra, A., Das, S., Ganguly, A., Das, P., Chatterjee, P. K., Murmu N. C., & Banerjee, P. (2019). Plasma arc technology: A potential solution toward waste to energy conversion and of GHGs mitigation. In *Waste Valorisation and Recycling: 7th IconSWM—ISWMAW 2017* (Vol. 2, pp. 203–217). Springer Singapore. https://doi.org/10.1007/978-981-13-2784-1_19

Horodytska, O., Cabanes, A., & Fullana, A. (2019). Plastic waste management: Current status and weaknesses. In *Plastics in the Aquatic Environment-Part I: Current Status and Challenges* (pp. 289–306). Cham: Springer International Publishing. https://doi.org/10.1007/698_2019_408

Krishnan, S., Zulkapli, N. S., Kamyab, H., Taib, S. M., Din, M. F. B. M., Abd Majid, Z., ... & Othman, N. (2021). Current technologies for recovery of metals from industrial wastes: An overview. *Environmental Technology & Innovation, 22*, 101525. https://doi.org/10.1016/j.eti.2021.101525

Iacovidou, E., Millward-Hopkins, J., Busch, J., Purnell, P., Velis, C. A., Hahladakis, J. N., & Brown, A. (2017). A pathway to circular economy: Developing a conceptual framework for complex value assessment of resources recovered from waste. *Journal of Cleaner Production, 168*, 1279–1288. https://doi.org/10.1016/j.jclepro.2017.09.002

Ibikunle, R. A., Titiladunayo, I. F., & Akinnuli, B. O. (2021). Development of a software system for selecting steam power plant to convert municipal solid waste to energy. *Sustainability, 13*(21), 11665.

Indrawan, N., Kumar, A., Moliere, M., Sallam, K. A., & Huhnke, R. L. (2020). Distributed power generation via gasification of biomass and municipal solid waste: A review. *Journal of the Energy Institute, 93*(6), 2293–2313. https://doi.org/10.1016/j.joei.2020.07.001

Jung, H., Shin, G., Kwak, H., Hao, L. T., Jegal, J., Kim, H. J., Oh, D. X. (2023). Review of polymer technologies for improving the recycling and upcycling efficiency of plastic waste. *Chemosphere*, 138089. https://doi.org/10.1016/j.chemosphere.2023.138089

Khan, I., & Kabir, Z. (2020). Waste-to-energy generation technologies and the developing economies: A multi-criteria analysis for sustainability assessment. *Renewable Energy, 150*, 320–333. https://doi.org/10.1016/j.renene.2019.12.132

Krishnan, S., Zulkapli, N. S., Kamyab, H., Taib, S. M., Din, M. F. B. M., Abd Majid, Z., & Othman, N. (2021). Current technologies for recovery of metals from industrial wastes: An overview. *Environmental Technology & Innovation, 22*, 101525.

Lee, U., Cai, H., Ou, L., Benavides, P. T., Wang, Y., & Wang, M. (2023). Life cycle analysis of gasification and Fischer-Tropsch conversion of municipal solid waste for transportation fuel production. *Journal of Cleaner Production, 382*, 135114.

Liu, C., & Liu, C. (2023). Exploring plastic-management policy in China: Status Challenges and Policy Insights. *Sustainability, 15*(11), 9087. https://doi.org/10.3390/su15119087

Mandal, D. D., Singh, G., Majumdar, S., & Chanda, P. (2023). Challenges in developing strategies for the valorization of lignin—A major pollutant of the paper mill industry. *Environmental Science and Pollution Research, 30*(5), 11119–11140. https://doi.org/10.1007/s11356-022-24022-4

Mavukkandy, M. O., Chabib, C. M., Mustafa, I., Al Ghaferi, A., & AlMarzooqi, F. (2019). Brine management in desalination industry: From waste to resources generation. *Desalination, 472*, 114187. https://doi.org/10.1016/j.desal.2019.114187

Mishra, S., Panda, S., Akcil, A., Dembele, S., & Agcasulu, I. (2021). A Review on chemical versus microbial leaching of electronic wastes with emphasis on base metals dissolution. *Minerals, 11*(11), 1255. https://doi.org/10.3390/min11111255

More, P. R., Jambrak, A. R., & Arya, S. S. (2022). Green, environment-friendly and sustainable techniques for extraction of food bioactive compounds and waste valorization. *Trends in Food Science & Technology*. https://doi.org/10.1016/j.tifs.2022.08.016

Muhsan, M. A., & Zahoor, A. F. (2023). Reclamation of mineral acids from various waste streams using solvent extraction technique: A review. *Geosystem Engineering, 26*(5), 1–21. https://doi.org/10.1080/12269328.2023.2202206

Nandhini, R., Berslin, D., Sivaprakash, B., Rajamohan, N., & Vo, D. V. N. (2022). Thermochemical conversion of municipal solid waste into energy and hydrogen: A review. *Environmental Chemistry Letters, 20*(3), 1645–1669. https://doi.org/10.1007/s10311-022-01410-3

Nayak, A., & Bhushan, B. (2019). An overview of the recent trends on the waste valorization techniques for food wastes. *Journal of Environmental Management, 233*, 352–370. https://doi.org/10.1016/j.jenvman.2018.12.041

Nguyen-Thi, T. X., Nguyen, P. Q. P., Tran, V. D., Ağbulut, Ü., Nguyen, L. H., Balasubramanian, D., & Pham, N. D. K. (2023). Recent advances in hydrogen production from biomass waste with a focus on pyrolysis and gasification. *International Journal of Hydrogen Energy*. https://doi.org/10.1016/j.ijhydene.2023.05.049

Park, E., Kim, M., Pin, M. W., Park, H., & Kim, Y. H. (2023). Precious metal recovery from waste electrical and electronic equipment through oxidative refining. *Recycling, 8*(5), 80. https://doi.org/10.3390/recycling8050080

Peng, X., Jiang, Y., Chen, Z., Osman, A. I., Farghali, M., Rooney, D. W., & Yap, P. S. (2023). Recycling municipal, agricultural and industrial waste into energy, fertilizers, food and construction materials, and economic feasibility: A review. *Environmental Chemistry Letters, 21*(2), 765–801. https://doi.org/10.1007/s10311-022-01551-5

Priefer, C., Jörissen, J., & Bräutigam, K. R. (2016). Food waste prevention in Europe–A cause-driven approach to identify the most relevant leverage points for action. *Resources, Conservation and Recycling, 109*, 155–165. https://doi.org/10.1016/j.resconrec.2016.03.004

Priyadarshini, P., & Abhilash, P. C. (2020). Circular economy practices within energy and waste management sectors of India: A meta-analysis. *Bioresource Technology, 304*, 123018. https://doi.org/10.1016/j.biortech.2020.123018

Pujara, Y., Pathak, P., Sharma, A., & Govani, J. (2019). Review on Indian municipal solid waste management practices for reduction of environmental impacts to achieve sustainable development goals. *Journal of Environmental Management, 248*, 109238. https://doi.org/10.1016/j.jenvman.2019.07.009

Rathod, D., Xu, B., Filipi, Z., & Hoffman, M. (2019). An experimentally validated, energy focused, optimal control strategy for an organic rankine cycle waste heat recovery system. *Applied Energy, 256*, 113991. https://doi.org/10.1016/j.apenergy.2019.113991

Roy, P., Mohanty, A. K., Dick, P., & Misra, M. (2023). A review on the challenges and choices for food waste valorization: Environmental and economic impacts. *ACS Environmental Au, 3*(2), 58–75. https://doi.org/10.1021/acsenvironau.2c00050

Said, Z., Sharma, P., Nhuong, Q. T. B., Bora, B. J., Lichtfouse, E., Khalid, H. M., & Hoang, A. T. (2023). Intelligent approaches for sustainable management and valorisation of food waste. *Bioresource Technology, 309*, 128952. https://doi.org/10.1016/j.biortech.2023.128952

Salehi, B., & Wang, L. (2022). Critical Review on nanomaterials for enhancing bioconversion and bioremediation of agricultural wastes and wastewater. *Energies, 15*(15), 5387. https://doi.org/10.3390/en15155387

Saravanan, A., Kumar, P. S., Nhung, T. C., Ramesh, B., Srinivasan, S., & Rangasamy, G. (2022). A review on biological methodologies in municipal solid waste management and landfilling: Resource and energy recovery. *Chemosphere, 309*, 136630. https://doi.org/10.1016/j.chemosphere.2022.136630

Saravanan, A., Karishma, S., Kumar, P. S., & Rangasamy, G. (2023). A review on regeneration of biowaste into bio-products and bioenergy: Life cycle assessment and circular economy. *Fuel, 338*, 127221. https://doi.org/10.1016/j.fuel.2022.127221

Shahabuddin, M., Uddin, M. N., Chowdhury, J. I., Ahmed, S. F., Uddin, M. N., Mofijur, M., & Uddin, M. A. (2023). A review of the recent development, challenges, and opportunities of electronic waste (e-waste). *International Journal of Environmental Science and Technology, 20*(4), 4513–4520. https://doi.org/10.1007/s13762-022-04274-w

Shahbaz, M., Rashid, N., Saleem, J., Mackey, H., McKay, G., & Al-Ansari, T. (2023). A review of waste management approaches to maximise sustainable value of waste from the oil and gas industry and potential for the State of Qatar. *Fuel, 332*, 126220. https://doi.org/10.1016/j.fuel.2022.126220

Soliman, N. K., & Moustafa, A. F. (2020). Industrial solid waste for heavy metals adsorption features and challenges: A review. *Journal of Materials Research and Technology, 9*(5), 10235–10253. https://doi.org/10.1016/j.jmrt.2020.07.045

Song, B., Lin, R., Lam, C. H., Wu, H., Tsui, T. H., & Yu, Y. (2021). Recent advances and challenges of inter-disciplinary biomass valorization by integrating hydrothermal and biological techniques. *Renewable and Sustainable Energy Reviews, 135*, 110370. https://doi.org/10.1016/j.rser.2020.110370

Stoett, P., & Omrow, D. A. (2021). The transnationalization of hazardous waste. In *Spheres of transnational ecoviolence*. Palgrave Macmillan. https://doi.org/10.1007/978-3-030-58561-7_3

Swati, A., & Hait, S. (2017). Fate and bioavailability of heavy metals during vermicomposting of various organic wastes—A review. *Process Safety and Environmental Protection, 109*, 30–45. https://doi.org/10.1016/j.psep.2017.03.031

Vermeşan, H., Tiuc, A. E., & Purcar, M. (2019). Advanced recovery techniques for waste materials from IT and telecommunication equipment printed circuit boards. *Sustainability, 12*(1), 74.

Wang, W., & Lee, D. J. (2021). Valorization of anaerobic digestion digestate: A prospect review. *Bioresource Technology, 323*, 124626. https://doi.org/10.1016/j.biortech.2020.124626

Yaashikaa, P. R., Kumar, P. S., & Varjani, S. (2022). Valorization of agro-industrial wastes for biorefinery process and circular bioeconomy: A critical review. *Bioresource Technology, 343*, 126126. https://doi.org/10.1016/j.biortech.2021.126126

Zheng, K., Wu, Y., Hu, Z., Wang, S., Jiao, X., Zhu, J., & Xie, Y. (2023). Progress and perspective for conversion of plastic wastes into valuable chemicals. *Chemical Society Reviews*. https://doi.org/10.1039/D2CS00688J

Swain, B. (2023). Challenges and opportunities for sustainable valorization of rare earth metals from anthropogenic waste. *Reviews in Environmental Science and Bio/technology, 22*(1), 133–173. https://doi.org/10.1007/s11157-023-09647-2

Leong, Y. K., & Chang, J. S. (2022). Valorization of fruit wastes for circular bioeconomy: Current advances, challenges, and opportunities. *Bioresource Technology, 359*, 127459. https://doi.org/10.1016/j.biortech.2022.127459

Wastes as Novel Source of Energy: Technical and Economic Aspects

Wafa Hassen, Bilel Hassen, Marwa El Ouaer, and Abdennaceur Hassen

Abstract Faced with the rate at which modern societies have been exploiting natural resources over the last few decades, through economic growth, galloping demographics, increasing resource consumption, the progressive generation of waste, and the unprecedented boom in industrial production, mainly through the consumption of energy and raw materials, and the emergence of climate change phenomena and pollution of natural resources, it is becoming imperative to resort to new resources to sustain the regenerative capacity of the planet Earth and to meet the needs of future generations. Waste management can be a significant part of the solution. Reducing, reusing, recycling, and reclaiming waste contributes to prolonging the residence time of these residual materials in systems and lessening the demand on natural resources, on the one hand, and to making these residual materials economical beneficial and limiting their potential effects, often of a harmful and problematic nature, on the other. In this chapter, we will look at the different aspects of management based on the principles of energy conversion and conservation from a residual source. We will also look at urban waste problems. Management aspects are clarified and dissected according to several scientific, social, economic, and political approaches. This management is therefore in line with the principles of sustainable socio-economic development based on various scientific and socio-economic foundations and approaches. Waste is a residual substance that poses problems for mankind and the environment. It is exploitable biomass that stores chemical energy from agricultural and industrial effluents, plant sludge, and many urban wastes. They are an appreciable source of renewable energy, primarily represented by their incineration in well-established plants, or their biotransformation into biogas in well-established facilities. The principles of these two processes, their installation, and the procedure for their implementation

W. Hassen · M. E. Ouaer · A. Hassen (✉)
Laboratory of Treatment and Wastewater Valorization, Center of Research and Water Technologies (CERTE), Tourist Road of Soliman, P.O. Box 273, 8020 Soliman, Tunisia
e-mail: abdohas@gmail.com

W. Hassen
e-mail: hassen.wafa@gmail.com

B. Hassen
Laboratories of Microbiology and Pathology of Aquatic Organisms, National Institute of Navy Sciences and Technologies (INSTM), Street, March 2nd, 1934, 2025 Salammbô, Tunisia

A. Anouzla and S. Souabi (eds.), *Generation of Energy from Municipal Solid Waste*,
https://doi.org/10.1007/978-3-031-74334-4_3

are described, sometimes in significant detail. In addition, operating characteristics and expected energy yields are described. They are indicated alongside some of the weak points of the various processes listed and quoted. In this way, waste, once seen simply as problematic refuse, now represents a resource that can replace a fraction of the raw material used. This procedure for the energetic reuse of waste is well-advised and contributes to the reinforcement of natural resources, but cannot provide a replacement source for these resources since it has long been known that it remains inexhaustible.

1 Introduction

The word plural waste would have been defined in several almost similar ways: waste is scrap, the remains of food unfit for consumption or use, disposed of waste materials discarded as not having an immediate value or left as residues of a household process, gardening, farming, or an industrial operation, also known as urban, farmer, or industrial residues; they are also non-usable and useless products of the metabolism of living cells, intended either to be released (such as carbon dioxide, excreta, urine, sweat, etc.) or, especially in plants, to be accumulated away from the general circulation (latex, oleoresins, alkaloids, etc.). Industrial waste is well known for its large quantities or its greater or lesser toxicity. As an example, heavy industries, nuclear and radioactive industries generate radioactive materials through nuclear reactors.

Waste may also mean any residue generated from a production, processing, or use process. It may also mean any substance, material, product, or, more generally, any property or furniture abandoned or intended for abandonment by its holder. Any element abandoned is waste. Not that this element is unusable, as it is, or after modification. Only those described as ultimate waste are unusable and must be stored to avoid environmental pollution. In each urban community, each inhabitant can dispose of a few kg of waste per year in their garbage. In addition, there is waste generated and discharged by various industries and enterprises; agricultural and agricultural waste produced and discharged by various agricultural activities; and finally, residues generated by healthcare activities and hospitals. The average quantity of waste generated in a community may vary from one country to another, from one city to another, urbanization of the community, etc. These quantities will be much lower in a small village or rural area, etc. There are numerous ways that waste and residues can have a positive effect, such as recycling, composting, or methanization, but large quantities (measured by an average of one-third) will end up in landfills (Ben Ayed et al., 2005; El Ouaer et al., 2022); Hassen et al., 2002)

In the same sense, waste or wastes are considered residual parts or components resulting from a process of production, processing, or use of substances, materials, products, or more generally, any abandoned or abandoned property.

The term "household waste" should also be distinguished from "household waste." The sharing and differentiation between these two concepts are often delicate; household waste is the waste produced only by households, while household waste is the waste collected as part of a collection organized by local authorities. The two terms do not fully cover each other, so some household waste is disposed of by the inhabitants themselves. In contrast, local authorities collect waste not only from households but also from traders and craftsmen.

It is also important to define the concept of ultimate waste, which is waste resulting from waste treatment or not, that is no longer likely to be treated under current technical and economic conditions, in particular by extracting the recoverable part or by reducing its polluting or dangerous nature. Hence, ultimate waste refers to the waste from which the recoverable part has been extracted as well as the waste that pollutes, such as batteries and accumulators.

Production activities and processes generate three categories of substances or waste products: useful, inert, and harmful. Residual substances or products range from recoverable to useful, as their names indicate, and interest man and his environment. But residual substances or products described as inert or banal are of no interest to man or his environment and have neutral effects. Hazardous waste has harmful and deleterious effects.

2 Different Waste Categories

Depending on its nature, physical–chemical characteristics, origin, or toxic nature, waste can be divided into different categories. But, the main and common classifications are those based on waste origin and physical–chemical properties and are as follows:

- Municipal wastes: Wastes whose collection, treatment, and processing are the community's responsibility. They are defined as wastes of commercial, industrial, or other urban origins similar to household wastes.
- Enterprise wastes: Wastes are inert industrial waste essentially composed of stable minerals (overburden), common industrial waste (CIW) identical to household waste, and special industrial hazardous waste (asbestos, lead, etc.) whose toxicity requires special treatment and attention.
- Agricultural, livestock, and food industry wastes: This category of waste includes cane, straw, leaves, manure, slurry, slaughterhouse waste, delicatessen waste, tanneries (bones, blood, hides, leather, fat, lard, processing waste milling waste, etc.), as well as all packaging waste associated with these related to these activities.
- Special hazardous wastes: These wastes are from economic and healthcare activities that could pose a danger to humans and the environment and should be treated or stored in specialized centers; this category of special hazardous waste includes radioactive wastes that should be entrusted to the National Agency for Radioactive Waste Management.

Waste can be classified as either hazardous or non-harmful. Hazardous waste does not determine its potential recovery: recycling channels exist to specifically treat toxic and polluting waste and recover the components that can be recovered. But some non- dangerous household waste cannot be recycled because it has no value as a raw material. Hazardous waste is, however, crucial to its classification, as it determines the protocols for its collection, transport, storage, and treatment. In addition, it will specify the type of organization responsible for it (local authority, waste producer, or specific agency).

3 Waste Management Policy

Waste management includes waste treatment, which must be carried out under the hierarchy of waste treatment methods. This is done by encouraging reuse, recycling, and avoiding disposal, saving natural resources. For example, local authorities have, over the past two decades, made significant and necessary investments in implementing a rigorous waste management policy. This management, coupled with strict regulation, encompasses three main aspects: the recovery of waste, consisting of its reuse and recycling with any other measures aimed at obtaining, from waste, reusable materials or energy; —the storage of special industrial waste recognized by its dangerous properties and appearing on a list fixed by decree; such special waste may not be deposited in storage facilities receiving other categories of waste; and finally—final waste, whether or not resulting from the treatment of waste which is no longer capable of being treated under the technical and economic conditions in particular by extraction of the recoverable part or by reduction of its polluting and dangerous nature. The principles of prevention, obligations, guidance, and prerogatives are well documented and explained in each country's legislation.

4 Recovery of Waste

Waste recovery is a set of processes whereby material waste is transformed into another material or energy product. Among these processes, composting and methanization may be mentioned, which make it possible to reuse green and organic waste. In addition, incineration and cogeneration can be used to recover energy during waste treatment. So, to recover one or more wastes, transform them into useful products with real benefits. A process in which wastes from extremely specific resources that have essentially harmful and problematic characteristics are transformed into useful sources of raw materials and energy that can be reused. Waste is no longer a problem to be treated; it is an emerging resource for economic and environmental order. In recent years, the recovery of waste resources has required inventive change, ecological transition, and economic linkage that adds value. Waste recovery includes reuse and recycling through composting and energy through incineration and biomethanization.

4.1 Reuse and Recycling

Reuse is defined as the operation by which residual substances, materials, products, or waste are reused. So, it defines any operation of control, cleaning or repair, or preparation of residual substances, materials, or products into a useful and reusable product.

Thus, reuse, recycling, and recovery of waste materials—similar words in the broad sense—play an essential role in combating environmental impacts associated with waste generation and disposal. This is also true for natural resources management. Both also limit greenhouse gas emissions and water consumption from industrial production. The strong demand for recycled raw materials has increased in recent years with the emergence of new economic powers such as Korea, China, and India. However, it has also been encouraged by the rising prices of many raw materials, including energy, and by the recent fight against climate change. Waste destined for recycling comes either from enterprises, following falls in production, scrap production, end-of-life equipment, and industrial or commercial packaging, or from households represented by household packaging, newspapers and magazines, and end-of-life equipment, such as end-of-life vehicles and household appliances. This waste is collected by municipal collection services, specialized companies, or recycled. But to develop the recycling of certain waste, it is necessary to collect it selectively following the establishment of a policy of selective collection and sustainable waste recycling. This policy is well organized and supported morally and financially in joint efforts by communities and public authorities.

Reuse or recycling is a waste or residue treatment operation distinguished by the reuse of residual substances, materials, or products. This is for the same use for which they were designed or prepared. Reuse operations are processes that, because of simple intervention, allow the reuse of the object, material, or substance, despite the nature of the waste. This is for the purpose for which the object, material, or substance was placed on the market.

Reuse or recycling is included in waste recovery in the broadest sense of the word. It plays an essential role in combating the environmental impacts associated with waste generation and disposal. It also plays a role in natural resources management. Both also limit and contribute to greenhouse gas emissions; water consumption is linked to industrial production; and above all, the growing demand for raw materials and their rising prices are linked mainly to energy prices. Thus, waste destined for recycling comes either from enterprises (production falls, scrap production, end-of-life equipment, and industrial or commercial packaging) or from households (household packaging, newspapers and magazines, and end-of-life equipment, such as end-of-life vehicles and household appliances). The term inert waste refers to rubble from construction and demolition work. Glass, breeze blocks, tiles, concrete: waste of this type is inert because it does not degrade naturally. It can be used as backfill and sub-base for road construction, for example. Much of this waste is buried in dedicated landfill sites.

Furthermore, recycling plastics, glassware, and metals is widely practiced worldwide. It is well governed by selective urban waste collection and recycling policies, i.e., sorting at source. This selective sorting guides future waste recovery operations.

4.2 Composting

Composting targets the transformation of any residual organic material, mainly green household waste, gardening and yard waste, agro-food waste, etc., into a product that fertilizes and restructures soils; it is the dominant form of residual material recovery. Two composting processes are distinguished: "industrial" composting and artisanal composting. "Industrial" composting is an essentially bio-physical–chemical process that consists of treating fermentable waste, namely fermentable waste such as green waste, sludge from sewage treatment plants, etc., to produce a solid, humus-rich residue like soil: compost. Industrial composting differs from natural composting by accelerating fermentation, which consists of, for household waste, disposing of non-fermentable materials by manual and mechanical sorting, including metals, glass, plastics, gravas, etc., crushing the fermentable material, thus freeing it of non-oxidizable materials, and sometimes even forced ventilation aided by the air compressor. Composting is an aerobic process in which organic matter rapidly oxidizes by emitting CO_2 and transforms into compost. This compost, rich in minerals and organic matter, is used as fertilizer to improve the chemical and physical characteristics of cultivated soils (Bouzaiane et al., 2000; Hachemi et al., 2018, 2020).

For this method of treatment, bio-waste or compostable or fermentable residual materials are introduced into a windrow or into a composter, where they ferment in the presence of air and water (Hassen et al., 2000; Hassen et al., 2021). Waste must be regularly ventilated and watered so fermentation is done properly. Composters can be practiced according to and of various natures: lombri-compostor (individual composter based on the digestion of bio-waste by earthworms or earthenware glasses) in an apartment; individual garden composter for a fireplace; collective composter, for example, at the foot of buildings; industrial composter in open-pit windrows or shelters.

The main goal of this operation of recycling fermentable materials is the production of compost. Compost can be used to fertilize cultivated or other soils. It would enrich the soil by returning organic material. This return to the ground must be carried out after a rigorous study presenting and ensuring the necessary and sufficient guarantees in terms of safety, compensation for the capacity of potentially toxic products such as heavy metals and emerging micro pollutants, etc., or of microorganisms pathogenic to man and plants, and above all, satisfactory agronomic quality to fertilize mineral elements. The latter quality refers to compost's richness in mineral elements such as nitrogen and phosphates. Research on the proper procedure for composting and its application to crop soils and its standardization has been carried out and updated. This research guides the safety and fertilizing efficiency of composts used in agriculture (Komolafe et al., 2015, 2021; Tian et al., 2001; Uzoh et al., 2015). Strict legislation and strict standards on compost and its applications, especially compost prepared from waste and derived sludge and based on unconventional organic amendments, have been prepared over the past 20 years, as have required applications setting out appropriate management methods and criteria for compost and its application in cropping soils.

4.3 Storage and Landfill of Waste

Storage is a waste disposal operation. This form of treatment of landfill waste has long been the preferred way of treating waste. This is because of its particularly low cost and the economic and environmental protection challenges associated with landfill operation. This landfill installation and operation generally require a significant right-of-way, whose immobilization lasts at least 30 years after waste reception, for non-inert waste, whether hazardous or not. The procedures and criteria for waste admission to landfills are generally well-defined and specified by well-established legislative decrees in each country. Three categories of waste storage facilities exist and are distinguished within the classification of facilities classified for environmental protection: inert waste storage facilities subject to site location conditions, safety, accident or pollution prevention measures, site operating rules, and emission monitoring measures. Given the inert nature of the waste to be received on-site, the specific provisions applicable to these installations are often less restrictive than those applicable to the storage of other categories of waste. Asbestos waste is considered inert waste, not eligible for disposal on-site. Non-hazardous waste storage facilities and hazardous waste storage facilities are classified as environmental protection facilities and require municipal permits in all cases. This authorization shall specify the maximum and annual capacity of the installation. It shall also specify the duration of the operation, the area of the installation in the area to be operated, and the technical requirements required. Ministerial orders regulate landfill sites. The ministerial decrees specify measures for the containment of waste, design and construction measures using several safety barriers to prevent risks of groundwater and soil pollution, but also measures for the prevention and monitoring of biogas emissions and provisions in terms of operating methods and monitoring procedures during the "post-operation" tracking phase after the last waste has been supplied.

They also set a minimum distance and require the operator to have control of the land in this "isolation strip" or an equivalent guarantee through public utility easements. Finally, financial guarantees are a prerequisite for the start of work on installations. Their purpose is to cover the costs of restoring the site if an environmental accident occurs. In addition, they are to restore the site in the event of operator failure.

4.4 *Incineration Without Energy Recovery*

Waste incineration with no energy recovery is carried out for waste with hardly any or no potential energy value or energy efficiency. A disposal operation is desired for this waste.

5 Electricity and Heat Recovery Potential

Energy recovery from waste is a valuable source of diverse renewable energy supplies. It makes it possible, under the hierarchy of waste management methods, to use waste not recycled or recovered into materials as a source of renewable energy. Over the last ten years and in Europe, renewable energy production from urban waste has been estimated at 2.5% of total European production, of which one-third is electricity and two-thirds are heating (source SOeS). Energy recovery also contributes to greenhouse gas reduction and limits fossil fuel use. By selling the energy produced, it is possible to reduce urban waste treatment costs by at least 20%.

This method of energy recovery is a treatment operation preferred to disposal but reserved for waste that cannot be reused or recycled. It recovers energy produced during waste treatment as heat, electricity, and fuel. Two types of energy recovery can be distinguished: recovery by heat treatment through incineration, co-incineration, pyrolysis, and gasification; and the recovery of biogas from non-hazardous waste storage facilities and the methanization of organic waste.

Paper, cardboard, putrescible waste, textiles, plastic, etc., account for about 70% of waste. In addition to household waste, many sectors of activity produce waste that can be recovered by incineration. First, there are many industrial wastes such as hydrocarbon residues, tars, used solvents, etc. There are also agricultural wastes such as wheat straw, corn, rice, etc., and agro-industrial wastes mainly from sugar mills and oil mills. In pulp mills, "black liquors," residuals from wood treatment after cellulose extraction, are burned to extract energy. It has recently been established and estimated that by exploiting its potential to recover energy from waste, Europe could supply 17 million households with electricity and 24 million with heating. Denmark is the European champion of electricity and heat production per capita.

Thus, the incineration operation allows a 70% reduction in the mass and a 90% reduction in the volume of household waste and its mineralization. This is because of machete and ash production. Today, it is one of the most cost-effective solutions for treating large quantities of waste on an industrial scale by recovering thermal energy and possibly producing electricity. However, incineration often appears to citizens as a process with a significant environmental impact. NIMBY, or "Not in My Back Yard" or "Not in This Near Me," is the result of these apprehensions and is translated into a reaction of rejection of any new incineration plant close to an agglomeration. The need to empower citizens and manage their modern way of life, in urban areas and industrialized countries is the ultimate means of eliminating them, and heavy consequences in terms of waste management policy are in sight with over 500 kg of waste generated per capita and year in urban areas. Incineration was the second method of treating household waste after controlled landfill.

In recent years, major efforts have been made to close polluting installations, often the smallest units, and the tendency to concentrate waste in large incineration units equipped with energy recovery systems with an hourly tonnage of treatment exceeding 20 tons and increased performance to meet the regulation and strengthen treatment and discharge standards.

5.1 *Heat Treatment Techniques*

Heat treatment of urban waste, and other waste, can be carried out using various processes and equipment. There are two main families of processes:

5.1.1 Thermolysis

It corresponds to a physico-chemical transformation of the organic part of the waste, carried out in a temperature range between about 450 and 600 C, and without oxygen. During thermolysis, the organic part of the waste is pyrolyzed and produces gaseous hydrocarbons, oils, and coal, which can be recovered. Several variants of the process have been developed for multinational companies, German and French.

5.1.2 Incineration

Incineration is a heat treatment by burning solid waste and certain liquid residues such as leachate from landfills, residues of the olive-growing industry main margins as colloidal residues with liquid dominance of water and fatty acids, and solid dominance of pomegranates consisting of hides and skins, residues of pulp, fragments of olive stones, etc., and by chemical oxidation with combustion gas production such as nitrogen compounds NOx, oxygen, CO_2, water vapor, etc., and solid residues such as ash and chips. Thus, incineration involves a reduction in ash and gas and certain liquid residues such as one of the modern techniques of waste management. It will be used for and/or generating electricity and heat for district heating. However, it is also a source of air pollution if it is poorly configured and implanted.

Before any urban waste recovery operation, whether composting, incineration, or biomethanization, it is necessary to characterize urban waste composition. This characterization operation is crucial for determining the most suitable methods of recovery and treatment. Hence, composting is the best technique to recover waste if more than half of it consists of putrescible materials and paper and paperboard associated with a prevailing high humidity of more than 40%. If the humidity is low and less than 30% with high calorific waste such as cardboard, wood, paper, plastic, paint residues, industrial and hydrocarbon sludge, glues and mastics, packaging, and soiled materials or filtering materials that come from the treatment of motor vehicle filters, etc., and with a fine-dimensional tendency of several tens of centimeters, the incineration operation proves highly advisable.

That means waste incineration is intended to be an effective way and a suitable procedure for disposing of a high percentage, 90% of the initial volume of waste transformed into gas and smoke and a valuable means of producing energy from these residues. It is a step forward in sanitation compared to huge landfills. For example, in some developed countries such as France, direct landfilling of recoverable waste has been banned since July 2002.

An incinerator is an industrial device for reducing or destroying objects by incineration, i.e., by burning them completely. It is generally presented as a furnace where the heat released by the materials during combustion ignites the materials added. Incineration plants are equipped with an oven and a post-combustion chamber. In the furnace, waste is decomposed by heat at temperatures between 800 and 900 C, which generates combustible gases. These gases will be recovered in energy at the furnace outlet, in the fumes, through a heat exchanger in which water or superheated steam circulates (Fig. 1).

5.1.3 Different Types of Incineration Kilns

There are different types of Solid Waste Incineration kilns on the market and incineration can be implemented in several major types of kilns such as the following:

A. Ovenswith Fixed or Mechanical Grids

These are the most commonly used ovens for household waste combustion. These fixed grid furnaces generally run batch-wise and treat small quantities of waste. Among ovens with mechanical grids, a distinction is made between ovens with rollers driven by a translational movement (pusher, recoil grid, alternating motion grid), ovens with rollers operating using rollers in rotation and arranged in stairs, and step ovens. All these systems ensure the mixing of the waste bed and its progression on the grid. These hardy but reliable fireplaces treat various types of heterogeneous waste such as wood, cardboard, and packaging, usually in bulk with high moisture levels. Inclined grids automate ash extraction. This furnace is simple to use and requires only a small investment. However, the grid operates intermittently, and its treatment capacity is reduced from 50 to 1000 kg per hour. This is a technique subject to large market fluctuations.

B. Mechanical or Rotary or Oscillating Grill Furnaces

These furnaces consist of a cylinder inclined horizontally in which the waste progresses under gravity and rotation. By turning over and stirring solids and oxidizing air, the process can be more efficient. Operating temperatures are around 1100–1300 C. These furnaces can treat a wide variety of wastes and are well suited to the treatment of industrial and toxic wastes because of the permanent stirring of the

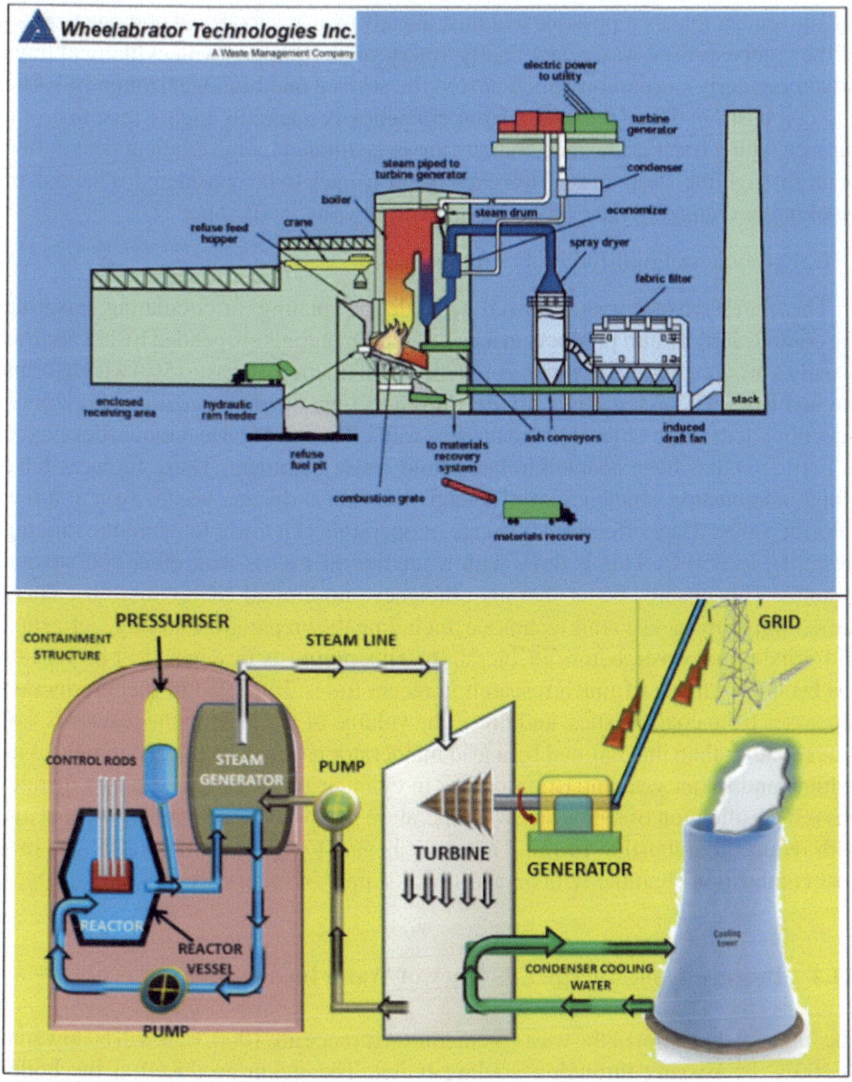

Fig. 1 Example diagram of a solid waste incinerator

fuel and many air injection orifices ensuring efficient combustion and the formation of good quality materials (low in unburned residual carbon). This furnace is derived from coal-fired boilers and treats coarse waste, such as household waste. The majority of municipal waste incinerators, with or without heat recovery, are designed on this model. Among the grill furnaces, one can distinguish the mobile carpet gratings used for coals, household garbage granules, and plant waste; tilting bar grids which make it possible to dispose of household waste in bulk, without past sorting, and the movement makes it possible to adjust the advance, reversal, and homogenization of the combustion of waste, and finally, rotary soles equipped with cylindrical trays driven regularly in rotation which ensure the stirring and homogenization of waste and combustion. Thus, the combustion efficiency is markedly higher than that of a furnace with a fixed grid, since combustion is facilitated by the efficient penetration of air through the waste mass. However, the drawback to be considered is the risk of blocking or damaging the grid during industrial waste incineration.

C. Pyrolytic or Fluidized Bed Furnaces

These are furnaces with fluidized beds, dense, rotating, or circulating, ensuring the combustion of solid products in a bed of inert materials suspended by hot air. It is usually a mixture of sand to which is added a small fraction of waste (5%) which forms the bed base. The assembly is fluidized by injecting air via vertical and horizontal injections at the base or level of the furnace walls. The fluid bed technique, developed to burn coal, has been adapted to household waste treatment. These furnaces have a high combustion efficiency of 99% and can receive diverse waste, even at a low calorific value. They offer the advantage of operating at a lower temperature ranging from 650 to 850 C. This is done with a smaller air excess than other combustion systems. This ensures better thermal efficiency and limited NOx emissions. There are several drawbacks to this technique, including the preparation, sorting, deferring, and grinding of waste before incineration. This pyrolysis or degasification reactor has been the subject of much research in recent times. It has led to fuel production processes from coal, plastics, and tires. The volume of gas leaving the reactor is 4–6 times smaller than that emitted by a grid incinerator as it can be filtered to reduce its volume and toxicity. Combustion occurs in cylinder-type chambers in two distinct phases. Gasification of solids at low temperature in a pyrolytic chamber, without air, with reduced combustion forming a pyrolysis gas. Re-combustion of this gas in a post-combustion chamber with an additional supply of secondary air.

5.1.4 Processes and Energy Efficiency of Waste Incineration

The fumes at the outlet of the waste incineration furnace are 1000°C; it will be lowered to 400°C by passing through a cooling boiler. The steam produced at the boiler during this cooling stage can then be recovered and recovered thermally, in three different forms of energy; thermal recovery is by feeding a district heating network or distributing it to enterprises and/or public establishments with a yield of up to 90%; (2.) thermal and electrical recovery or cogeneration with an average efficiency of up to 80%) and finally (3.) electrical recovery, following a transformation into electricity using a turbo-alternator and with relatively low efficiency of the order of 35%. In parallel, a flue gas treatment and refining procedure involving filtration at the boiler outlet will be carried out. This will rid the flue gases of potential, often harmful, residual pollutants such as dust, acid gases, heavy metals, and dioxins. The installation for treating fumes comprises several modules for dedusting by electro filter and/or bag filter, for neutralizing acid gases, by dry, semi-wet, or wet route, for treating dioxins and furans, by adsorption on activated carbon for example, or by selective catalytic reduction, for treating nitrogen oxides, by selective reduction, or by catalytic or non-catalytic route.

Flue gas treatment and filtration residues generally comprise fly ash, residues of the flue gas neutralization operation, and cakes for filtering flue gas washing water. Thus, the incineration of one ton of household waste can generate between 15 and 40 kg of residues depending on the smoke treatment. Residues are collected and removed from the site for disposal.

Finally, the removal and treatment of residual materials from solid waste incineration includes various handling operations, temporary storage, treatment, and maturation. Often, these operations produce a product that can be substituted for natural aggregate from quarries in road construction techniques. These machetes replace quarries' natural aggregates. They have been recognized as durable solid materials and do not cause significant environmental impacts. Statistics have shown, on average, that of the estimated annual production of household waste incinerators of 3 million tons, about 70% are processed at maturing plants. The remaining 30% is recovered or disposed of in a non-hazardous waste storage facility.

5.1.5 Waste Incineration Process

Complex phenomena and successive physico-chemical and thermal types of pyrolysis, transformation, and oxidation simultaneously govern waste combustion. The high heterogeneity of waste, due to its differences in size and composition, governs and correlates these phenomena. In addition, they are due to the relatively low calorific value. This calorific value can vary within a range of 5800–10,500 kJ kg-1 depending on the origin and quality of the waste. It can fluctuate significantly during the same day of operation, because of variations in humidity and temperature. As mentioned, the composition of household waste consists of an organic combustible part, moisture and constituent water, and an inert part composed of minerals and non-combustible materials such as glass, scrap metal, etc.; and waste incineration generates solid waste products such as machetes, ash and gases mainly CO_2 and H_2O.

In summary, incineration involves a succession of physico-chemical and thermal processes, some of which may occur simultaneously, resulting in the generation of energy, solid non-combustible, or chewy materials or residues, and ashes consisting of fine mineral particles collected under the grid, at the level of the boiler and in dust removal systems (Fig. 2). According to reports, machetes are often used as flaps and solid materials on construction roads, while ashes are buried in landfills. The water or moisture content factor of the waste to be incinerated plays a decisive role in waste incineration energy efficiency. The higher the moisture value, the more likely it is to reduce energy efficiency.

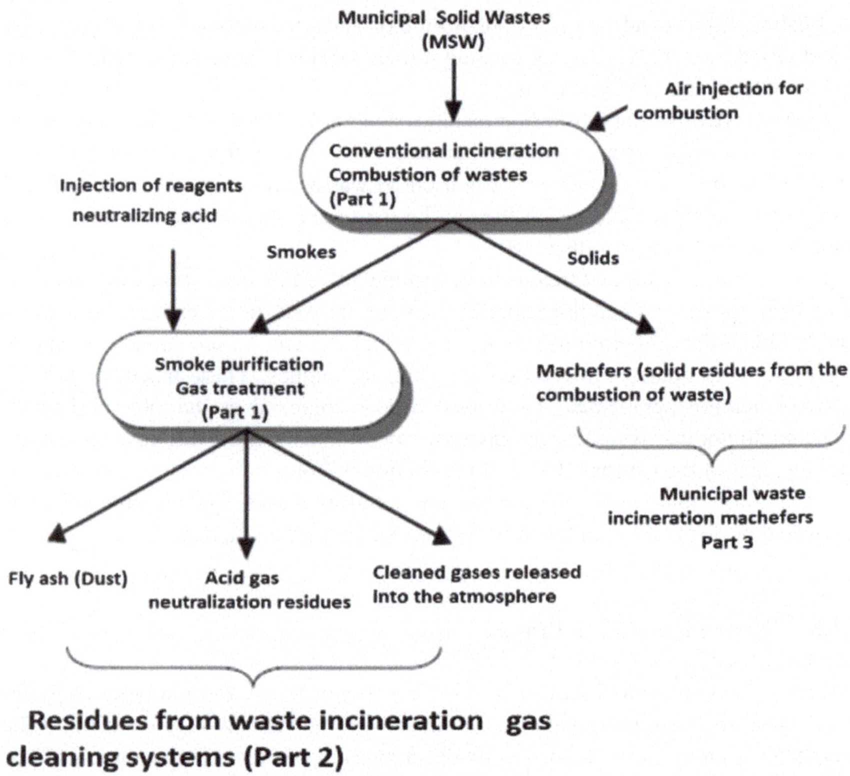

Fig. 2 Schematic procedures for municipal waste incineration

5.1.6 Advantages and Disadvantages

The incineration of waste greatly reduces the volume of our solid residues. However, if it is poorly installed and operated, it can pose enormous risks to man and his environment. Incineration allows a certain reduction in waste volumes deposited in public landfills, and it is well suited to many households and similar waste (ADEM, 2012). Incineration reduces waste mass and volume entering landfills by 70%. It also reduces landfill waste quantities. It also decreases waste harmfulness through the combustion of certain special hazardous wastes, such as medical waste. It contributes to the reduction of fossil fuel consumption. Thus, one ton of waste can produce 500 kWh of electricity. In addition, it can produce 1,700 kWh of heat alone.

Because of some past inappropriate practices, incineration may cause harmful emissions leading to air pollution, including dioxins, furans, heavy metals, acid gases, dust, etc. Which, in large doses, may affect health due to certain cancers or reproductive disorders, if these broadcasts are not properly filtered and processed. Thus, risk levels seem related to smoke treatment performance, although they seem low in recent installations. Organic pollutants recognized as persistent or (POPs), represented in particular by the dioxin/furan family, appear at about 800 C and are characterized by complex molecules defined by their more or less strong toxicity. Because of their high persistence and lifetime, which is confirmed by their thermal and chemical stability in the environment and by their high resistance to natural biological degradation, these residual components have been shown to have negative effects on people, animals, and plants. Additionally, they are transported over long distances and deposited away from their emission due to their strong bioaccumulation and concentrations in living tissues.

5.1.7 Health Centers and Incinerators

Today, incinerators are the most effective for health centers such as hospitals and clinics. UN classification considers medical waste the most hazardous waste after radioactive waste (Basel Convention). Medical waste is a by-product of health care. Although most wastes are harmless, some medical wastes pose a significant health risk. This is the case for contaminated waste (15–25% of all medical waste), where cutting waste accounts for 1%, organic waste for 1%, chemical or pharmaceutical waste for 3%, and radioactive, cytotoxic, or broken thermometers for less than 1%.

Today the waste produced is largely made of disposable materials, a result of the increasing use of plastic and composite materials. The percentage of hazardous materials such as mercury and other heavy metals has also increased. Medical waste developments are prompting us to reconsider our treatment methods. Ineffective management of medical waste puts the community at risk, patients, and health staff, both at risk due to inappropriate storage, transport and disposal of infected waste. In addition, there are environmental risks involved in the unprotected burning of waste in open pits or poorly maintained incinerators.

In developing countries where capacity is limited, a recent WHO policy paper notes that "small incinerators are considered an interim solution." 3 Incineration carried out under "Best Practices" may be an economical and less risky way to dispose of medical waste.

5.1.8 Waste Incineration's Energy Efficiency

Recent energy recovery policies aim to reduce greenhouse gas emissions in landfills as well as reduce energy bills and dependence on fossil fuels, among other things. Heat consumers such as iron and steel ovens (cement plants, lime ovens), paper mills, chemical platforms, as well as heat networks are likely to reduce their fuel consumption and energy bills.

Incineration plants produce heat and electricity. Each follows its own production method and their yields are disproportionate. Heat production by burning waste is efficient with values between 70 and 80%, or about 1500 kilowatts/thermal hour per ton of waste. This amount of energy generated would often be used in the surrounding agglomerations and near the incineration plant. Electricity generation will use steam produced after waste combustion directed to a turbine at the highest possible pressure. This produces electricity from an electric generator. This generated electricity can be supplied to the public electricity grid permanently throughout the year. Energy efficiency is 20–25%, or 300–400 kilowatt-hours. It has been estimated that it takes 5–7 tons of raw waste to obtain one ton of fuel oil. For example, the energy produced in the three incineration centers of the Syndicate of treatment of household waste in the Paris region (SYCTOM) heats 300,000 housing equivalents each year. Other data showed that 100 tons of raw MSW with 50–60% organic matter can generate about 1–1.5 Mega Watts of power, depending on the waste characteristics (Douard, 2021).

In France, for example, the values indicated that by recovering 50% of all household waste in France, one would obtain about 1% of the country's energy consumption. This solution is far from satisfying your needs and unprofitable. In addition, waste incineration fumes are toxic. They should be filtered and neutralized before discharge into the atmosphere. Standards and legislation on air pollution from incineration plants are particularly severe in most countries that have implemented this incineration practice.

Ultimately, energy from biomass is a renewable energy source that depends on the cycle of living plant and animal matter. Biomass chemical energy stored in chemical bonds is transformed into electricity and heat generated by combustion. This is mainly wood, plants, agricultural waste, and household and agro-food waste, or biogas from the fermentation of these materials in processing plants (Fig. 3).

Fig. 3 Conversion of biomass from combustion to energy

5.2 *Methanization or Biomethanization*

In parallel with the incineration of household or other solid waste, it can also be transformed into biogas, by fermentation. This is thanks to microorganisms. It can also be transformed into biomethane from anaerobic bacteria. Biogas is similar to natural gas and is composed of methane and CO_2. This biotransformation mainly concerns fermentable waste of household types such as green waste, paper, and cardboard, etc., agricultural waste such as manure, size waste, animal manure, etc., and urban sludge from sewage treatment plants. It is a natural process that occurs in natural ecosystems such as marshes, the digestive tract (rumen) of herbivorous mammals such as cattle and sheep, and soils, etc. (Egert et al., 2003). It is a biological process that digests organic matter into co-products, whether solid or liquid, without oxygen. During the process, products are obtained: biogas, mainly composed of methane (CH_4) and carbon dioxide (CO_2) and by-products such as digestate (with methanization of solid co-products), and/or methanization of colloidal-majority effluents such as manure, pots, sludge from purification plants, etc., which are stable at the pollutant level.

Thus, methanization is defined as the degradation of the waste in the absence of oxygen-producing biogas, a gaseous mixture composed mainly of methane and carbon dioxide as has been said, and transforms the initial organic material in these wastes or residues into a wet, partially stabilized product called digestate. All organic matter can be decomposed by methanization, besides stable compounds such as lignin (wood) which can, conversely, be easily composted. Methane-nizable materials may be organic waste of agricultural, agro-industrial, or municipal origin. There are thus small methanizers treating only livestock effluents on the farm, medium-sized methanizers mutually treating livestock effluents, or large methanizers treating all kinds of non-hazardous organic waste.

At present, the majority of methanized compounds are agriculture wastes; thus "on-farm methanization" makes it possible to diversify agriculture activities by producing exploitable energy from locally available deposits. This activity is predominant among poultry, cow, and cattle farmers in most countries worldwide. Besides the production of biogas which is the subject of energy recovery, the digestate or residue of the anaerobic digestion of fermentable materials, which is a waste with fertilizing characteristics, can be spread on agricultural land when it complies with rural code regulations. Agronomic recovery is possible in different ways. The compost can be used to fertilize and restructure agricultural soil after maturation through composting. Digestate can also be spread on agricultural land as part of an application plan, or registered to become a marketable product after a specific procedure. The short methanization or biomethanization of organic waste thus allows double recovery of organic matter and energy. This method can be used to treat greasy or wet organic waste, and which otherwise cannot be composted. Reducing greenhouse gas emissions by substituting fossil fuels for chemical fertilizers is also of importance. However, the technical stakes of methanization are not negligible. These installations lack flexibility, as they require a continuous supply of materials, which can be difficult to guarantee. Finally, the digestate obtained is relatively less rich in organic matter than compost; this is why composting characterized by digestates is sometimes practiced.

Also, the establishment of facilities to treat agri-food waste is becoming increasingly encouraged to limit on-site energy spending and costs in factories and contribute to the clean-up of the natural environment by industrial residues.

Biomethanization exploits the calorific value of certain wastes that cannot be recovered in material. This is done while ensuring control of the environmental impact of their treatment regarding their chemical composition. Physical characteristics such as moisture, the fine particle size of the materials to be methanized, and the composition of the waste must be well monitored and respected. In order to prepare waste for biomethanization, it is often necessary to sort and mix it first.

For example, animal methane production is estimated at 15% of total emissions, 75% of which are from cattle (Moss et al., 2000). Flooded soils are among the main natural sources of CH_4 (Le Mer & Roger, 2001). These sources account for 30% of CH_4 emissions.

A significant amount of greenhouse gas emissions is also generated by the degradation of organic waste assimilated into non-hazardous waste storage facilities. The treatment of these wastes through anaerobic digestion or methanization is of dual economic and ecological interest; essentially by obtaining an energy source via biogas and by reducing and stabilizing organic waste pollution. This bio-physical–chemical procedure reduces air pollution by limiting greenhouse gas emissions and avoiding soil pollution because of groundwater contamination by leachates overloaded with highly polluting materials.

Methane-transformable waste pollutants are often translated as dry matter (DM) and volatile matter (VM) for solid co-products and chemical oxygen demand (COD) for liquid co-products. Methanization encompasses two modes of working methanogenic bacteria responsible for the biochemical transformation of residues: (1) psychrophilic functioning bacteria, active in the temperature range 0–20 C and rarely used on the industrial scale of anaerobic digestion; and (2) mesophilic-functioning bacteria, active in the temperature range of 20–40 C, with an ideal around 37 C, and the most frequently found among the good number of existing methanization units. Waste recovery is well-established in many countries on an industrial or individual scale such as India, China, etc.

According to our previous discussion, the transformation of organic matter into energetic gaseous compounds, or biogas, involves a set of highly interdependent microorganisms that perform complementary activities (Tou et al., 2001). We can distinguish non-methanogenic flora, which includes facultative aero-anaerobic microorganisms and is subdivided into two groups: 1.) hydrolytic bacteria that transform complex organic matter into simpler compounds, and among these bacteria: hemicellulolytic bacteria like *Bacteroides tyminicola* and amylolytic bacteria like *Clostridium butiricum*; proteolytic bacteria like genera Clostridia; and at last, acid-producing bacteria that produce simple compounds like alcohols, volatile fatty acids, H_2, CO_2), which will be used as substrates for methanogenic bacteria; the second group is represented by methanogenic flora, including strict anaerobic, autotrophic microorganisms. Their action is conditioned by the prior action of the first two groups since they are stimulated by organic compounds such as acetate or cysteine to release methane and CO_2 through decarboxylation. Some of the CO_2 generated can be reduced to provide 30% of the final methane.

5.2.1 Biochemical Principles of Biomethanization

Figure 4 Illustrates the major steps of methanization and includes the main operating steps, namely hydrolysis, acidogenesis, acetogenesis, and finally methanogen

A. Hydrolysis

The hydrolysis phase corresponds to the decay of complex organic materials. During this phase, polymers (polysaccharides, lipids, proteins, nucleic acids) are transformed into monomers (monosaccharides [glucose], fatty acids and amino acids, and nitrogen bases). During the decay phase, microorganisms secrete extracellular enzymes which hydrolyze soluble polymers. This is ensured by extracellular enzymes. For particulate substrates, polymers are hydrolyzed in two stages. These stages comprise the colonization of the particle surfaces by microorganisms and the production of enzymes by these same organisms. Hydrolytic bacteria carry out transformation steps in anaerobic fermentation. A strict or optional anaerobic metabolism is characteristic of these bacteria. Many varieties of bacteria are available and functional during non-aerobic digestion, and this is conditioned according to temperature regimes, mesophilic or thermophilic regimes (Archer & Kirosp, 1990).

B. Acidogenesis:

This is the phase during which the monomers are transformed into volatile fatty acids (AGV) such as formic, acetic, propionic, butyric, isobutyric, valeric, and isovaleric acids, and alcohol. The gas produced during the fermentation phase is rich in carbon dioxide at 80% (CO_2) and partly in hydrogen at 20% (H_2). Acid-forming bacteria, such as Clostridium, are often strict or optionally anaerobic. Among the 4 steps of methanization, acidogenesis is the fastest due to rapid bacterial multiplication. Organic overload of a reactor often results in an accumulation of intermediates such as hydrogen and acetate during the acidogenesis stage.

Acetogenesis: Under the action of acetogenic bacteria, the products of the two preceding phases are transformed into acetate, hydrogen, and carbon dioxide (CO_2).

Different types of bacteria cause acetogenesis. They are essentially syntrophic bacteria represented by acetogenic-producing bacteria with a forced passage of hydrogen; these are homo-acetogenic bacteria divided into two groups according to the origin of the acetate produced. The first group is the acetate coming from a carbon substrate, and the second group is the acetate obtained by CO_2 reduction by H_2. Sulfate-reducing agents (BSR) are also grouped into two groups. The first group uses lactate as a carbon source and the incomplete oxide is acetate and CO_2. The bacteria in the second group oxidize acetate and some long-chain acids to CO_2.

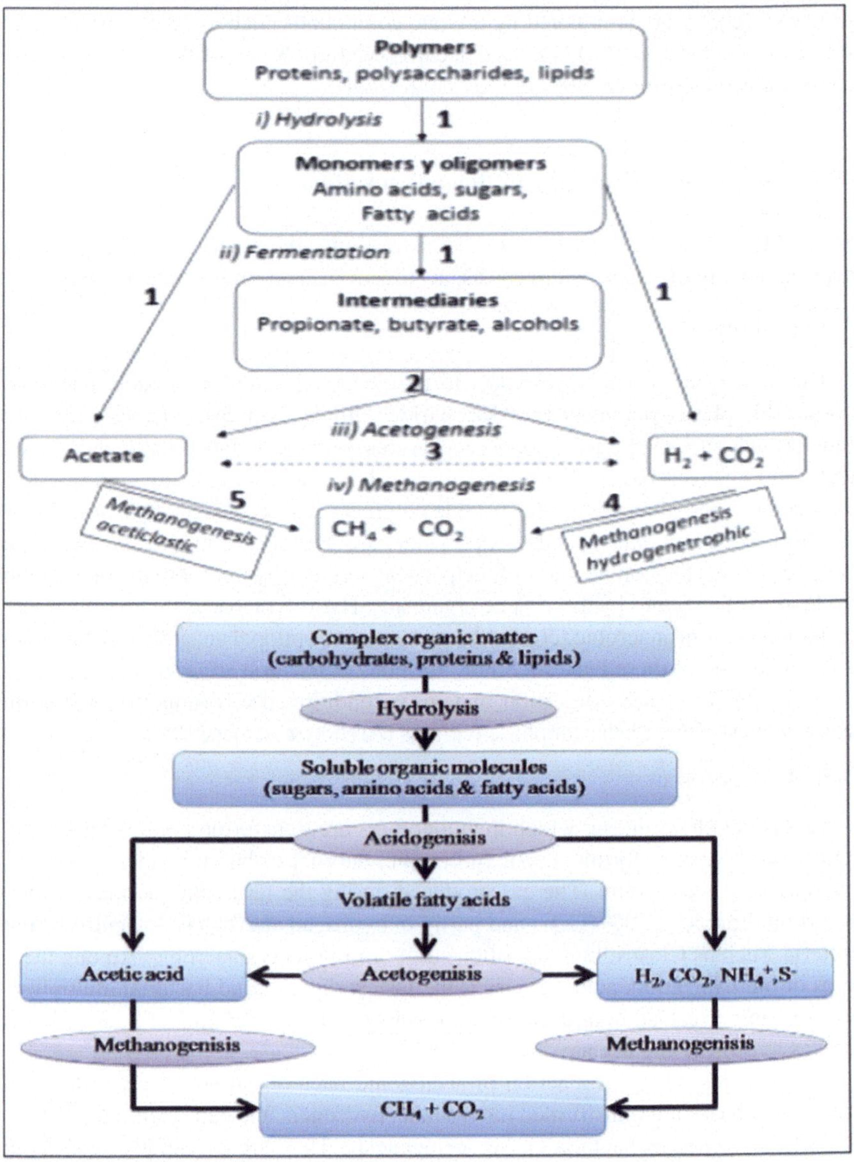

Fig. 4 Main biochemical steps of biogas production via organic waste residues

From the thermodynamic viewpoint, the oxidation reactions caused by the acetogenic bacteria forced on hydrogen are impossible unless $\Delta G > 0$ (the free enthalpy or the available energy ["free"] to do work), and the partial pressure in H_2 must be high.

Methanogen: The products obtained from the three preceding phases will be transformed into methane (CH_4). Hydrogenophilic (hydrogenotrophic) and acetoclastic (acetotrophic) methanogenic bacteria are responsible for this transformation.

Methanization reaction is summarized:

$$C_6H_{12}O_6 \ldots\ldots\ldots\ldots 3CO_2 + 3CH_4 \quad \text{(I–1)}$$

The stoichiometric relationship of the methanization reactsion varies from the location where the methanogen develops (Fig. 4).

In the rumen, the above relation will be transformed:

$$C_6H_{12}O_6 \ldots\ldots\ldots\ldots\ldots 65CH_3COOH + 20CH_3COOH + CH_3CH_2 + CH_3(CH_2)^2 + 35CH_4 + 60CO_2 + 25\,H_2O \quad \text{(I–2)}$$

Hydrogenotrophic or hydrogenophilic methanogens use the reduction of carbon dioxide or formic acid by hydrogen as an energy source, according to the following equation:

$$CO_2 + 4H_2 \rightarrow H_2O + CH_4 \quad \text{(I–3)}$$

For acetolactic or acetotrophic methanogens, organisms produce methane from acetic acid, methanol, and methylamine. Among the two types of bacteria often found in a reactor such as Methanosarcina and Methanosaeta (Morgan et al., 1991), the Archaea of the genus Methanosaeta use acetate as the sole source of carbon to produce methane according to the following reaction:

$$\text{Acetate} + H_2 \rightarrow CO_2 + CH_4 \quad \text{(I–4)}$$

5.2.2 Types of Biomethanization Procedures

There were two biomethanization procedures: that which caters to the physico-chemical nature of the waste matrix to be reacted, namely the matrix of liquid waste with colloidal consistency such as manure, cow cups, waste sludge, etc., most commonly found on agro-industrial and farm scales, along with household waste and other solid waste (Fig. 5).

Liquid waste can come from different sectors such as groups (stations of purification), agro-industry industries, waste from farms (animal excretions such as chicken slurry, cow dung, slurries), slaughterhouses (washing, blood).

The methanization of sludge from sewage treatment plants has evolved most rapidly in the European zone in the last 20 years. Sludge has a significant methanogenic power, as well as regulatory threshold values for certain pollutants. The technology of anaerobic digestion of sludge from sewage treatment plants is of

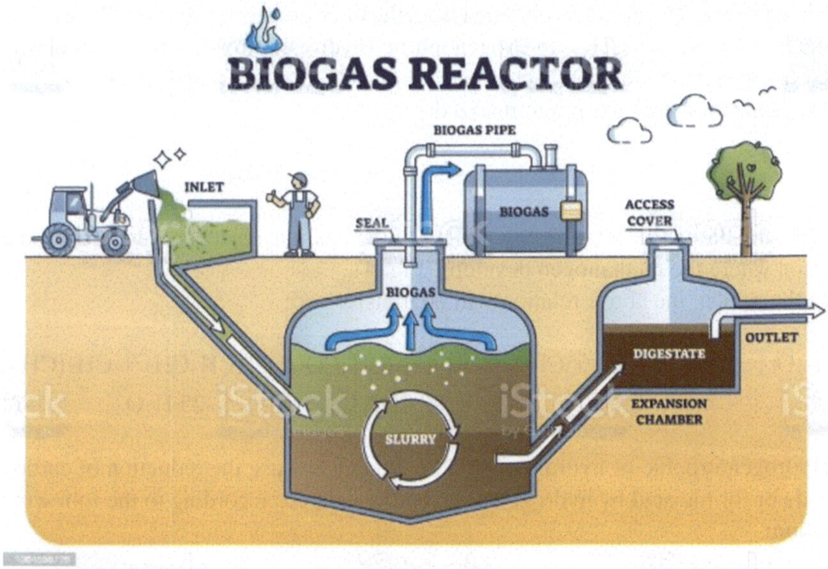

Fig. 5 Prototypical schematic of biogas reactors

dual interest: on the one hand, it reduces pollution of natural environments (soil, sea, lake) and produces bioenergy via biogas (Moletta & Verstraete, 2008).

Liquid waste's characteristics are primarily determined by its organic content. COD indirectly determines the concentration of organic matter in an effluent intended for liquid biomethanization. The COD of a liquid or paste effluent (sewage treatment plant sludge, pig manure, cow pots) can be split into soluble (biodegradable) and inert parts. By fractionating COD, it is possible to identify the biodegradable fraction of organic matter in the fluid effluent. Each fraction of COD, soluble or inert, is divided into highly and slowly biodegradable fractions and into soluble and particulate fractions. Figure I-2 shows Beline (2008)'s study of total COD fractionation on manure samples.

This fractionation of COD makes it possible to measure organic matter rapidly and slowly in the biomethanation reaction matrix (Fig. 6).

5.2.3 Key Factors Governing Biomethanization

Some significant factors govern the biomethanization process, such as:

The pH, or acidity, of the reaction medium: The waste matrix is variable according to its sources. Pig manure is one of the liquid effluents to be treated by biomethanation. This could ensure that pH equilibrium is maintained in the reaction medium. The pH limit values are between 6.8 and 7.4, and during the acidogenesis phase, the pH may

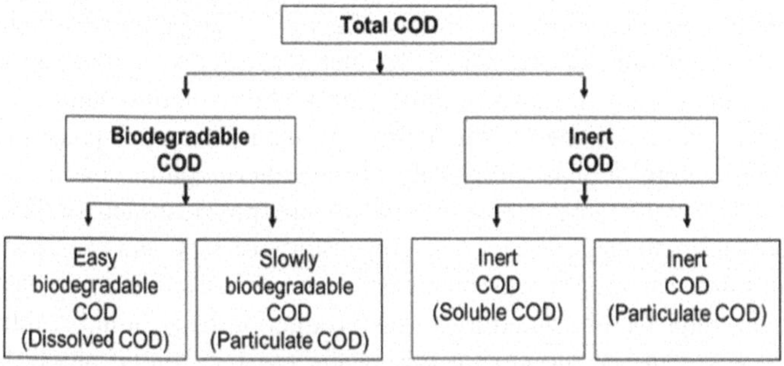

Fig. 6 Organic matter fractionation model

drop to 4.5. The appropriate pH during the methanogen phase of anaerobic digestion of distillery wastewater is around 6.0, according to Savant et al. (2002).

A. Methanogenic Potential

The methanogenic potential is defined as the volume (or quantity) of biogas produced per ton of dry volatile matter digested from waste during its methanization. The theoretical value of the potential can be estimated as a function of the composition of carbon, hydrogen, oxygen, nitrogen, and sulfur (CHONS) according to the Bus well equation (Buswell & Mueller, 1952). Moletta (2008) reported that for bovine manure with a percentage of dry matter of 5–12% and a ratio of volatile matter to dry matter of 75–80%, the methanogenic potential fluctuates between 200 and 350. Pig manure has 3–8%, 70–80%, and 350–500% methanogenic potential.

B. The Hydraulic Residence Time (TSH)

The TSH is the average time spent by the product, in a reactor, on a medium operating in continuous mode. The TSH is given by the ratio of the liquid volume in the reactor to the volume flow rate of the liquid discharge treated. The hydraulic residence time of effluents from urban wastewater stations varies from 4 to 8 h depending on the bioreactor operating at ambient temperature (20–25 C) (Lettinga et al., 1992). Sludge from sewage treatment plants is volatile (40–50%) in mesophilic mode (Moletta, 2008). The thermophilic regime accelerates product degradation. The TSH can be reduced by up to 12 days (Cecchi et al., 1992).

C. The C/N Ratio

As its name suggests, it is the ratio between organic carbon, COD, and total organic nitrogen. This report defines composting solid waste and the level of nutrients (nitrogen) present in the reaction matrix. The C/N ratio defines nitrogen uptake and availability. For liquid effluents from sewage treatment, the appropriate value can drop to 15 (Romano & Zhang, 2008).

D. The Hydraulic Residence Time

The hydraulic residence time is directly affected by temperature, which is also a physico-chemical parameter to consider. Temperature affects microorganism growth according to operation mode (psychrophilic, mesophilic, thermophilic).

Microorganism activity during the hydrolysis and methanogen phases is directly related to temperature. A phase of hydrolysis under thermophilic conditions accelerates the hydrolysis reaction. According to the texture (physical state) of the substrate, the two modes of methanogen have ideal temperatures of 35 °C and 55 C, respectively, which permit the growth of microorganisms.

Finally, the final factor related to biogas production from liquid residues is the inoculation or seeding of the reactor during the reactor's initial phase. Similarly, seeding can be translated into batch-fed, or fed-batch digestion, by the principle of dilution, by introducing a daily quantity of fresh matter. For a continuous reactor of the infinitely mixed type, this seeding results in sludge recirculation. Seeding a reactor optimizes the volume charge during the methanization start-up phase with a substrate low in organic matter. With leaching, inoculation compensates for biomass loss. Above all, it promotes the equilibrium of the reaction medium environment by adjusting the pH and the alkaline power.

Methanization is not only limited to the liquid substrate and most types of organic waste are methanizable, either liquid or solid. Anaerobic digestion of solid waste, called dry digestion, is difficult to implement and pretreatment with humidification is often necessary.

5.2.4 Types of Anaerobic Digestion Reactors

Liquid or solid digestion reactors are different. With liquid effluents, there is a free reactor (reactor of the infinitely mixed type) or with support (sludge bed, fixed biomass, or mechanical support).

The homogenization of the reaction medium of each type of reactor is ensured either by a mechanical agitator [dry digestion, semi-liquid medium, viscous] or by biogas recirculation or effluent recirculation (liquid digestion).

A. Types of Reactors Recommended for Liquid Methanization

Several types of bioreactors are available in laboratories and industries. There are mixed reactors for processes involving free microorganisms, with an anaerobic contact, a sludge bed, and a compartment. Microorganisms are free or floc. Microorganisms used to form a biofilm promote adhesion to a fixed or mobile support. These processes increase active biomass. The UASB (Up-Flow Anaerobic Sludge Blanket), the anaerobic filter, the fluidized bed, and the internal recirculation reactor are distinguished.

Types of reactors recommended for dry methanization of solids: The technology depends on the waste water content. There are limited mixed reactors or pistons. The methanization operation can operate in three modes, namely the continuous mode, the discontinuous mode, and finally the sequential discontinuous mode.

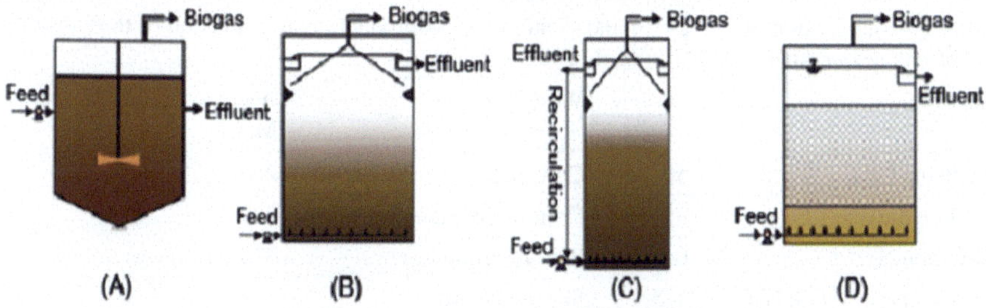

Fig. 7 Most commonly used anaerobic reactor types: [A] Completely mixed anaerobic digesters, [B] UASB reactors, [C] AFB reactors, [D] Up-flow AF reactors (Ersahin et al., 2011)

B. Continuous Operation

This mode of operation consists of continually supplying the biological reactor with a constant flow rate, also called nominal operation. The volume of effluent or solid waste in the reactor remains constant. A quantity equal to the daily incoming volume must be evacuated.

C. Operation in Batch Mode

The reactor operates without exchange with the external medium. The reactor is no longer supplied with organic materials once it is filled. When digestion is complete, the reactor is emptied for new fermentation.

D. Operation in Sequential Discontinuous Mode or Fed-Batch

This is the combination of continuous and discontinuous modes. The reactor operates in a cycle alternating filling, reaction, decantation, and emptying. By not completely emptying the reactor, some microorganisms are conserved in order to form a new cycle. In general, small-scaleanaerobic digestion functions sequentially and discontinuously (Fig. 7).

5.2.5 Biomethanization Yields

As mentioned above, digestate is the residual organic material not transformed into biogas during fermentation. This digestate can be an agricultural amendment because of its fertilizing qualities due to its nitrogen, phosphorus, potassium, and calcium content. It can also be used as a consolidated material for road installation instead of gravel and tiles.

Biogas or gas produced by methanogenic fermentation is known by different names. These include marsh gas, sewage gas in wet biomethanization, or biogas in dry biomethanization. Table 1 gives an example of the composition of suitable biogas.

Table 1 Composition of biogas (6) and yield of biogas production as a function of the substrate (according to Lagrange, 1979; Tou et al., 2001)

Gas	%	Waste	Average yield (Liters/kg M.S.)
Methane CH_4	50–70	Cow dung	200
Carbon dioxide CO_2	35–40	Crotons and dung of cows	200
Nitrogen N_2	0.5–3	Poultry droppings	310
Various gases: H_2S, NH_3, CnH_2n…	1–5	Sheep manure	135
Hydrogen H_2	1–3	Sewage	340
Oxygen O_2	0.1–1		

5.2.6 Properties and Characteristics of Biogas Produced

A. Properties of Biogas

Under the same temperature and pressure conditions and at equal volume, biogas is lighter than air; its density relative to air is about 0.7. So, in the event of a leak, biomethane will escape upwards unlike butane and propane, which avoids accidents.

Biogas production is closely related to the nature of the reaction substrate; thus, looking at the values reported in Table xxx, the ideal substrate consists of sewage, although for convenience cow cups and poultry droppings are also suitable substrates and materials.

B. Biogas' Calorific Value

Methane burns with a blue flame when combustion is complete. This comes with a strong heat release. Under these conditions, 1 m3 of methane can reach a temperature of 1400 C by burning and emit heat equivalent to 8500 and 9500 kcal. Comparing the calorific value of biogas with that of conventional fuels makes it possible to state these approximate equivalents: —1 m3 of biogas equals, from an energy viewpoint, 1.25 kilowatt-hours. —Its combustion produces energy equivalent to 0.7 L of oil or 0.6 L of diesel.

C. Biogas Utilization

Biogas has many uses, so it can produce heat. This can be used in cooking, drying, or heating or operating steam machines for electricity or mechanical energy production. Also, biogas can run an internal combustion engine to produce kinetic (mechanical) energy or electricity. There are equivalence standards:

- 1 m^3 of biogas lets a lamp illuminate for 6–7 h and diffuse a luminosity equivalent to 60 W of electric light;
- 1 m^3 of biogas can also produce 1.25 kWh of electricity. In addition, purified biogas can be used in the chemical industry for making many chemicals such as chloroform, acetylene, methanol, etc.

5.2.7 Main Benefits and Limitations of Biomethanization and Co-Biomethanization

However, the two processes of anaerobic digestion by liquid methanization and solid methanization of solid waste are both generally unstable when not tracked. Monitoring involves controlling the parameters that inhibit the reaction in the medium. Parameter controls are often done by adding chemicals to maintain anaerobic digestion balance and/or by heat treatment.

Anaerobic digestion of solid waste is complex to start with, sometimes manifesting itself in instability of the process. This is due to an accumulation of soluble compounds that may inhibit the process itself. On-line monitoring of control parameters such as pH, temperature, volatile fatty acids, etc., and homogenization of the reaction medium, is sometimes difficult or inaccessible. Accessibility to mechanical stirring depends on the reactor substrate size. When homogenization is impossible, temperature gradients may be formed in the reactor. This is while recalling that temperature is one of the keys and determining parameters of anaerobic digestion. To guarantee organic matter degradation in substrates, microorganism development conditions must be respected. This is not the case for waste with low dryness or water lack such as solid waste.

By adding sodium hydroxide to the acidic matrix or acid to the basic matrix, the reaction medium pH must be balanced. While an increase in the concentration of volatile fatty acids in the reaction medium, a major inhibitor of the anaerobic digestion process, may be manifested by an increase in soluble COD at the reactor outlet.

To deal with this problem, the technology of co-methanization of two or more combinations of solid and/or liquid waste may present a solution and a remedy. To avoid this risk of inhibition by the accumulation of soluble substrates, co-methanization of the co-products via meal residues and other carbohydrate-rich compounds such as manure and manure is an effective alternative to this inhibition problem. The digestion technique also reduces inhibition.

Finally, to track the evolution of transformation reactions in the digester and to better manage the evolution of the process and the risks of inhibitions, phenomenological models of anaerobic digestion have been developed.

Biomethanization benefits are multiple and can be categorized into 4 main plans:

- Waste treatment plant by a natural process requiring less space than aerobic composting or landfilling and therebyreducing the volume and weight of waste to be landfilled;
- Energy plan by net energy production, generation of high-quality renewable fuel that can be used in several end applications;
- Environmental plan through significant reduction of CO2 and CH_4 emissions, removal of odors, production of clean (sterile) compost and nutrient-rich liquid fertilizer for agriculture;
- And finally, the economic plan is the most profitable among other treatment forms.

6 General Conclusion

Solid household waste management remains a major challenge for our authorities. In today's environment, diversity, enormous household waste, and landfills that ignore regulations encourage recycling. Incineration does not "disappear" waste, it transforms it into:

- Recyclable material (everything that does not burn): about 30% of waste weight. They can be used as building or earthmoving materials.
- Ash and REFIOM (residues from household waste incineration and fume cleaning): about 3% of waste. Their toxicity, which depends on the fume treatment chosen, is measured, which determines the class of discharge to which the ash is directed.
- Smoke.

The treatment costs of household waste incineration plants currently in operation vary considerably in treatment costs. This depends on the date of commissioning, the choice of equipment, energy recovery, recovery and disposal of by-products, etc.

Biomethanization or short methanization is an effective procedure for treating liquid effluents and solid waste. This includes household waste landfill leachates, wastewater, sewage treatment plant sludge, effluent from the agri-food industries and agricultural waste, manure and similar animal waste. The treatment of waste through methanization decreases the pollutants in the by-product or residue at the end of a process by decreasing organic matter such as dry matter, volatile matter, etc., and reducing or limiting olfactory pollution and air. The process of methanization of waste makes it possible to obtain energy via biogas produced. This can effectively reduce fossil energy consumption and the negative effects on human well-being and the environment.

Different technologies for recovering useful energy from Municipal Solid Wastes already exist and are extensively used in different countries for their multiple benefits. It is necessary for the success of these technologies in India to evolve an Integrated Waste Management system, coupled with necessary legislative and control measures. A detailed feasibility study needs to be conducted in each case, taking into account the available waste quantities and characteristics and the local conditions. In addition, a relative assessment of the different waste disposal options. Suitable safeguards and pollution control measures also need to be incorporated into the design of each facility to fully comply with environmental regulations and safeguard public health. Waste-to-Energy facilities, when set up with such consideration, can effectively bridge the gap between waste recycling, composting and landfilling, for tackling the increasing problems of waste disposal in urban areas, in an environmentally benign manner, besides augmenting power generation in the country. The Ministry of Non-Conventional Energy Sources provides financial assistance for energy recovery projects.

References

ADEM. (2012, December). L'incinération des déchets ménagers et assimilés. Les Avis de l'ADEME.

Archer, D. B., Kirsop, B. H. (1990). The microbiology and control of anaerobic digestion. In Wheatley. Anaerobic digestion, a waste treatment technology, Critical reports an applied chemistry (pp. 43–91). London, Elsevier.

Beline, F. Fractionnement des lisiers. (2008). Le CEMAGREF, Unité de recherche de gestion des effluents d'élevage et des déchets municipaux.

Ben Ayed, L., Hassen, A., Jedidi, N., Saidi, N., Bouzaiane, O., & Murano, F. (2005). Caractérisation des paramètres physico-chimiques et microbiologiques au cours d'un cycle de compostage d'ordures ménagères. Environnement, Ingénierie and Développement, Epi-sciences, 2005, N 40 - 4ème Trimestre, (pp. 4–11). https://doi.org/10.4267/déchets-sciences-techniques.2001.hal-03175434.

Bouzaiane, O., Jedidi, N., Hassen, A. (2000). Microbial biomass improvement following municipal solid waste compost application in agricultural soil. October 2014, in the book: *Composting for sustainable agriculture* (pp. 199–208). https://doi.org/10.1007/978-3-319-08004-8_10.

Buswell, A. M., & Mueller, H. F. (1952). Mechanism of fermentation. *Industrial and Engineering Chemistry., 1952*(44), 2442–2460.

Cecchi, F., Pavan, P., Musacco, A., Mata-Alvarez, J. & et Sans, C. (1992). Comparison between thermophilic and mesophilic anaerobic digestion of sewage sludge coming from urban waste watertreatment plants. *Water Science and Technology, 26*(9–11), 2409–2412.

Douard, F. (2021). La grande métamorphose de l'unité de valorisation énergétique de Saint-Ouen-sur-Seine. L'agence Métropolitaine des Déchets Ménagers (SYCTOM). Bioénergie International n 68 d'octobre, 2020.

Egert, M., Wagner, B., Lemke, T., Brune, A., & Fridrich, M. W. (2003). Microbial community structure in midgut and hindgut of the humus-feedind larva of Pachnoda epphipiata (Coloeptera: Scarabaeidae. *Applied and Environmental Microbiology, 69*, 6659–6668.

El Ouaer, M., Turki, N., Ibrahim, C., Hussain, Chaaben, S., Kallel, A., Trabelsi, I., & Hassen, A. (2022). Insights into the use of landfill leachate to grow *Chlorella sp.* For lipid and fatty acids production. *Clean Technologies and Environmental Policy*. https://doi.org/10.1007/s10098-022-02460-5.

Ersahin, M. E., Ozgun, H., Dereli, R. K. & Ozturk, I. (2011). Anaerobic treatment of industrial effluents: An overview of applications. In F. S. García (Ed.), *Wastewater—treatment and reuse*. In Tech.

Hassen, A., Belguith, K., Jedidi, N., Cherif, A., Cherif, M., & Boudabous, A. (2002, January). Microbial characterization during composting of municipal solid waste. *Bioresource Technology, 80*(3), 217–225. https://doi.org/10.1016/S0960-8524(01)00065-7.

Hassen, W., Hassen, B., Ammeri Werheni, R., Hidri, Y., Hassen, A. (2021, May). Co-Composting of Various residual organic waste and olive mill wastewater for organic soil amendments, In the book: *Humic Substance* Publisher: IntechOpen License CC BY 3.0, Humic Substances, *47*. https://doi.org/10.5772/intechopen.97050.

Hechmi, S., Hamdi, H., Mokni-Tlili, S., Ghorbel, M., Khelil M. N., Zoghlami, I. R., Benzarti, S., Jellali, S., Hassen, A., & Jedidi, N. (2020). Carbon mineralization, biological indicators, and phytotoxicity to assess the impact of urban sewage sludge on two light-textured soils in a microcosm. *Journal of Environmental Quality*, Mars *49*(2), 460–471. PMID: 33,016,418. https://doi.org/10.1002/jeq2.20011.

Hechmi, S., Hamdi, H., Mokni-Tlili, S., Zoghlami, I. R., Khelil, M. N., Jellali, S., Hassen, A., & Jedidi, N. (2018, August). Repetitive land application of urban sewage sludge: Effect of amendment rates and soil texture on fertility and degradation parameters. *Catena, 172* (JAN 2019), 11–20. https://doi.org/10.1016/j.catena.2018.08.015.

Komolafe, S. E., Akangbe, J. A., Ajiboye, B. O., & Dairo, O. D. (2015). Influence of developmental strategies on agricultural development in Ilorin south local government area of Kwara State,

Nigeria. *Ghana Journal of Science, Technology and Development, 2*(1), 45–53. https://gjstd.org/index.php/GJSTD. ISSN: 2343-6727.

Komolafe, A. F., Adejuyigbe, C. O., Babalola, O. A., Soretire, A. A., & Kayode, C. O. (2021). Fertilizer values of composts as affected by plant materials and composting duration on maize (Zea mays) performance. *Agro-Science, 20*(1), 87–94. https://doi.org/10.4314/as.v20i1.14

Lagrange, B. (1979). *Biométhane, Principes–Techniques–Utilisations'*, (Ed.), Edi-Sud (p. 246).

Le Mer, J., & Roger, P. A. (2001). Production, oxidation, emission and consumption of methane by soils a review. *European Journal of Soil Biology, 2001*(32), 25–50.

Lettinga, G., & Hulshoff, L. W. (1992). UASB Process design for various of wastewaters. In J. F. Malina, & F. G. Pohland (Eds.), *Design of anaerobic process for treatment and industrial and municipal wastes.* (Vol. 7, pp. 119–145). Water Quality Management Library, Technomic Publishing Co.

Moletta, R., & Verstraete, W. (2008). La méthanisation dans la problématique énergétique et environnementale. In Moletta, R. (Ed.), La méthanisation. Technique et documentation (pp. 3–8). Éditions Lavoisier.

Morgan, J. W., Evison, L. M., & Forster, C. F. (1991). Change to microbial ecology in anaerobic digestion treating.

Moss, A. R., Jouany, J. P., & Newbold, J. (2000). Methane production by ruminants: Its contribution to global warming. *Annales Zootechnie, 2000*(49), 231–253.

Romano, R. T., & Zhang, R. H. (2008). Co-digestion of onion juice and wastewater sludge using an anaerobic mixed biofilm reactor. *Bioresource Technology, 99*(3), 631–637.

Savant, D. V., Souche, Y. S., Prakash, S., & Ranade, D. R. (2002). Methano*brevibacter acididurans sp* nov., a novel methanogen from a sour anaerobic digester. *International Journal of Systematic and Evolutionary Microbiology, 55*(4), 1081–1087.

Tian, G., Salako, F. K., & Ishda, F. (2001). Replenishment of C, N, and P in a degraded alfisol under humid tropical conditions: Effects of fellow species and litter polyphenols. *Soil Science, 166*, 614–621.

Tou, I., Igoud, S., & Touzi, A. (2001). Production de Biométhane à Partir des Déjections Animales Rev. Energ. Ren.: Production et Valorisation – Biomasse, 103–108.

Uzoh, I. M., Obalum, S. E., & Ene, J. (2015). Mineralization rate constants, half-lives and effects of two organic amendments on maize yield and carbon-nitrogen status of loamy ultisol in South-Eastern Nigeria. *Agro-Science, 14*(3), 35–40. https://doi.org/10.4314/as.v14i3.7

Business Model of Resident Participation Based Sustainable Bioreactor Landfill Gas: Case Study of the City of Depok

F. Farizal, James Marisi, Deevantianto Rafif Wibisono, Rahmat Nurcahyo, and Misri Gozan

Abstract Municipal solid waste (MSW) is a problem in many big cities worldwide. Treating the MSW with better waste management, i.e., bioreactor plant, can turn the waste into a useful thing, i.e., energy; waste to energy (WtE): Converting the methane gas produced during the decomposition of the organic waste compounds to electricity. In addition, the bioreactor also saves landfill space, and minimizes spoiling the environment. However, benefits come with costs, i.e., the magnitude of the investment involved, and the complexity of running the bioreactor project compared to conventional landfill. This study proposes a business model for sustainable bioreactor landfills based on resident participation (RSLFG), and shows the model's applicability. The study was carried out on the basis of regulations, and mechanisms to implement the model. As a result, a limited liability company is proposed as the best suit entity type to run the project. Even though all types of funding collection mechanisms are feasible, direct-personal collection by collector officers is better than indirect-machine collection, virtual account whether using ATM, internet banking, or mobile banking.

F. Farizal (✉) · J. Marisi · D. R. Wibisono · R. Nurcahyo
Department of Industrial Engineering, Universitas Indonesia, Depok 16424, Indonesia
e-mail: farizal@eng.ui.ac.id

J. Marisi
e-mail: marisijames@gmail.com

D. R. Wibisono
e-mail: deevanwibisono@gmail.com

R. Nurcahyo
e-mail: rahmat@eng.ui.ac.id

M. Gozan
Department of Chemical Engineering, Universitas Indonesia, Depok 16424, Indonesia

F. Farizal · M. Gozan (✉)
Research Center for Biomass Valorization (RCBV), Universitas Indonesia, Depok, Indonesia
e-mail: mgozan@ui.ac.id

A. Anouzla and S. Souabi (eds.), *Generation of Energy from Municipal Solid Waste*,
https://doi.org/10.1007/978-3-031-74334-4_4

1 Introduction

The level of human growth and consumerism lifestyle is high for citizens living in urban areas, which in turn causes the amount of trash generated. Municipal solid waste (MSW) is the waste produced by a household. Aside from the waste problem, other problems occur in the waste dumping site, i.e., landfills. The issues include demographic concerns, environmental conditions (smell and pollution), working environment (social, health, and environmental), and income level (Sasaki et al., 2014). The negative impact resulting from dumped waste is climate change, greenhouse gas emission, and pollution of the ecosystem with heavy metal in the air, land, and water (Albores et al., 2016). Among other factors, family size, income level, and education level highly influence the amount of MSW generated (Kolekar et al., 2016). In addition, the weather conditions may also affect the magnitude of the amount of trash and garbage composition formed (Jarallah & Aleisa, 2013).

MSW generally consists of organic biodegradable materials by 65.68% (Farizal et al., 2014; Kwarteng et al., 2016). The biodegraded compounds can be used as waste to energy (WtE) because the energy contained is more than 15,000 kJ/Kg (Bawaain et al., 2017). The waste can produce methane gas from bacterial decomposition of the organic material (Aich & Ghosh, 2016). This type of energy is renewable since human beings always produce MSW. It is also green since it can save coal consumption by about 3.04 million MT and reduce 35.42 million MT of CO_2 emissions per year (Aich & Ghosh, 2016). MSW-generated methane can produce electrical energy as a feedstock (Scarlat et al., 2015). When a landfill can produce 371,000 tons of gas, it will generate 2,650 GWh of electricity (Fazeli et al., 2016).

MSW processing techniques can be selected by identifying the waste composition. For example, when the garbage contains predominantly organic compounds, processing with an anaerobic digestion system is appropriate (Sefouhi et al., 2014). There are four ways to do the process: sanitary landfill (engineered landfill), open dumping landfill, incineration, and composting. System processing with composting produces carbon emissions slightly less than the incineration system. The bond proceeds derive from sanitary landfill system processing (Wang & Liwei, 2015). Landfill gas (LFG) mainly contains greenhouse gases such as 50% methane and 45% carbon dioxide. WtE processing facility has lower emissions than fossil fuels facilities producing electricity. Utilizing WtE sources will reduce dependency on the electricity produced from fossil fuels and simultaneously reduce carbon dioxide emissions (Chaliki et al., 2016). Methane gas is a renewable energy source (Ahmed et al., 2015).

In Indonesia, it is estimated that the amount of MSW generated daily is 200,000 tons (Kurniawan et al., 2021). However, in many cities, at most only 69% of MSW are transported and dumped to a landfill, leaving the rest of 10% to be buried, 7% to be composted and recycled, 5% to be burned, and the rest of 7% is not managed at all. Farizal and Tammarar (2018) showed that the tipping fee set by many local governments is optimal already. Raising the tipping fee to get better MSW management is difficult. Fortunately, Farizal et al. (2019) proposed an alternative scheme of funding based on resident participation to manage the waste better. Their study showed that the funding generated is sufficient to implement a better MSW management, i.e., develop a resident participation-based sustainable bioreactor landfill gas plant (RSLFG). In this scheme, the resident funding will be used not only to construct the RSLFG facility but also to operate and maintain the project during its economic lifetime, which is assumed to be 20 years. The issue is managing the residents' willingness to fund the project over time. What is the meaningful way to collect the funding so the residents are encouraged to pay? Running the possible funding collecting way, is the RSLFG still able to sustain its operation? This study will highlight these issues.

2 Theoretical Aspect

2.1 *Sustainable Bioreactor Landfill Gas Plant*

US EPA (2017) introduced a bioreactor landfill, where liquids are injected into the MSW landfill to accelerate the bacteria degrading the waste (Thampan & Chandel, 2015). By so doing, the methane gas produced will have a more stable rate, and the waste will stabilize faster. There are two types of bioreactors, namely aerobic and anaerobic. In an aerobic bioreactor landfill, the air is added to enhance the aerobic process. Meanwhile, waste degradation in an anaerobic bioreactor landfill does not need oxygen input to produce the gas. Figure 1 shows the schematic of an anaerobic bioreactor landfill plant.

In a landfill, methane gas is generated due to the existence of water, nutrients, and micro-organism through the following mechanism: Organic waste + H_2O + nutrient $\xrightarrow{microbes}$ CH_4 + CO_2 + NH_3 + H_2s + Biomass + heat (Tchobanoglous et al., 1993).

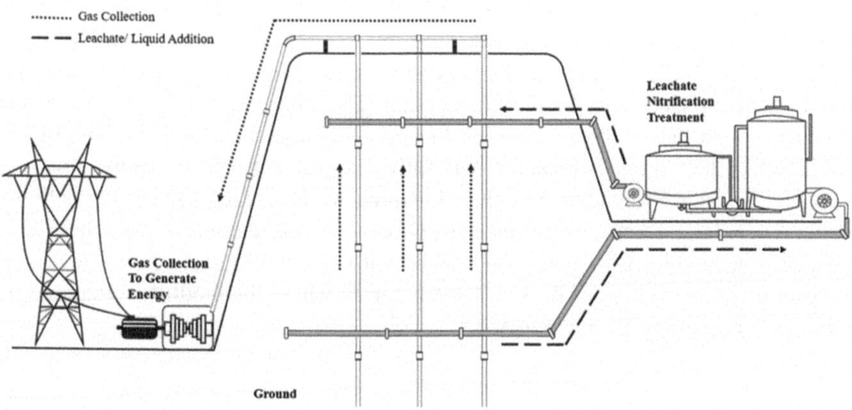

Fig. 1 Anaerobic bioreactor landfill plant schematic

Compared to non-bioreactor landfills, waste decomposition using bioreactor landfills is accelerated to minimize long-term environmental risks, reduce operating costs, and generate a more stable volume of methane gas. This later stabilizes the electricity generated (Sang et al., 2012). However, bioreactor is more challenging to construct and operate. Its operation is more complex than the conventional landfills (sanitary or engineered landfills). The landfill must be monitored regularly to check the gas emissions, odors, and landfill fires to ensure the waste moisture (US EPA, 2017). Bioreactor landfill also consumes more capital, operation, and maintenance costs. The combustion of methane gas will still produce carbon dioxide, but its effect on global warming is lesser than releasing pure methane gas into the atmosphere (Kardono & Purwanta, 2017).

The methane gas is collected and converted to electricity in a bioreactor plant. The amount of electricity generated is estimated using Landgem (landfill gas emissions model) developed by US EPA through Eq. (1).

$$Q_{CH_4} = \sum_{i=1}^{n} \sum_{j=0.1}^{1} kL_o \left(\frac{M_i}{10} \right) e^{-kt_{i,j}} \tag{1}$$

where Q_{CH_4} is annual methane generation in the year of the calculation (m^3/year), i is the time in 1 year increment, n is (year of the calculation)—(waste acceptance initial year), j is 0.1 year time increment, k is methane generation rate (year$-^1$), L_o is potential methane generation capacity (m^3/Mg), M_i is mass of waste accepted in the ith year (Mg), $t_{i,j}$ is the age of the jth section of waste mass M_i accepted in the ith year (decimal years, e.g., 3.2 years).

However, Eq. (1) says that the methane generated will diminish as time evolves. So, the bioreactor plant should be closed permanently after operating within a certain period since harvesting its electricity is not economical anymore. At this point, a new landfill should be opened. Ardiansyah (2014) developed a sustainable bioreactor landfill model (SLFG) that generates relatively stable electricity over time. Given the rate of MSW dumped into the landfill, the model can estimate the size of the bioreactor landfills to be constructed, the time to open and close the landfill, and the amount of electricity that can be harvested. Table 1 shows a conceptual activity plan to generate electricity by the SLFG plant. The table indicates that to make the bioreactor plant sustainable, during the 24 years, 8 bioreactors should be built on the site selected. Just about Bioreactor A is fully dumped with MSW, Bioreactor B is prepared. So once Bioreactor A is closed, Bioreactor B is ready to be filled with the waste. A year later, electricity produced by Bioreactor A is ready to be harvested. This setting is repeated for Bioreactor C, D. So on until the 8th bioreactor, i.e., Bioreactor H. After this, the cycle goes back to Bioreactor A, where the landfill is prepared by mining the remaining Bioreactor A.

Table 1 Activities' Plan of Sustainable Bioreactor Landfill Gas Plant, Conceptual (Ardiansyah, 2014)

Activity	Time (Year)																													
	-4	-3	-2	-1	0	1	2	3	4	5	6	7	8	9	10	11	12	13	14	15	16	17	18	19	20	21	22	23	24	25
Landfill Preparation	A			B			C			D			E			F			G			H			A					
Piling Up Waste																														
Landfill Gas Construction																														
LFG Production																														

2.2 *Project Funding*

In financing renewable energy infrastructure, the main problem public authorities face is the lack of the necessary capital. Conversely, private investors are generally reluctant to fund this track due to its high transaction costs and high-risk turnover (Yildiz, 2014). The SLFG project needs more funding than the regular LFG due to the type of technology used and the repetitive nature of the project (Ardiyansyah, 2014; US EPA, 2017). Instead of involving private investment, involving the community in this sector is another option. In Germany, community participation is combined with public institutions. Group of citizens and local government share ownership in the energy supply sector (Hall et al., 2016). Citizen's funding can be an important tool for energy projects at the local level. Citizens' willingness to engage in community-based renewable energy projects is due to their beliefs and social norms (Kalkbrenner & Roosen, 2016; Taurina et al., 2021). Some funding concepts through private contributions of individuals were described by Yildiz (2014). The funding can be made in several ways: capitalization (equity finance), such as cooperative (Briggeman et al., 2016), Mutual Funds (Close-end Fund), other equity finance in the form of Company Limited (Limited Liability Company), Company Stock, Scheme Based Debt, and Mezzanine (Bobinaite & Tarvydas, 2014), crowdfunding (Zhang et al., 2016, 2019).

An alternative source of funding for WtE based on resident participation has been proposed and investigated by Farizal et al. (2019b). Even though both funding schemes rely on the participants' willingness, resident-based funding differs from the existing crowdfunding. Crowdfunding is an umbrella term that uses small amounts of money from many individuals or organizations to finance a project, business or personal loan, or other needs through an online web-based platform (Kirby & Worner, 2014). Table 2 summarizes the different aspects of Crowdfunding and Resident-Based Funding.

2.3 *Business Model*

A business model involves exchanging value among the stakeholders (Gordijn et al., 2000). Osterwalder (2004) defined a business model as "a conceptual tool that contains a set of elements, and their relationships, and allows expressing a company's

Table 2 Different aspects of crowdfunding, and resident-based funding

No	Aspect	Crowdfunding	Resident-based participation funding
1	Payment frequency	One time	Repeated time, and regular (monthly)
2	Funding purpose	Capital spending	Capital, and OM spending
3	Timing	The initial time of the project	The whole duration of the project (from the initial to the end of the project)
4	Participant	Can be sparse in many areas	Residents (ones who live surrounding the project)
5	Mechanism	Through the internet	Cash deposit and/or internet

logic of earning money." It illustrates how the company produces, delivers, and captures value (Amit & Zott, 2001; Osterwalder & Pigneur, 2010). A business model is a tool for managers to develop their strategy (Baden-Fuller & Morgan, 2010). A good business model generates a compelling value proposition to customers, a significant value in producing and delivering products and services (Teece, 2010). Implementing digital business models improves the retail industry's productivity, services, and business competitiveness (Harting et al., 2019). Creating a business model that forms an abstract representation, conceptual, operational, and financial inter-related, is crucial for an entity to achieve its goal (Al-Debei et al., 2008). This business model is expected to make a business sustainable.

A sustainable business model encourages robust value relationships among customers, the company, and the environment (Abdelkafi & Tauscher, 2015; Dembek et al., 2018). The sustainable models should include environmental impact, energy recovery, resident participation, and profit sharing (Cucchiella et al., 2017). A company adopts sustainable business models succeed in creating value for both the firm and their customers in a logistic company (Sxoinaraki & Panou, 2017), the fashion industry (Todeschini et al., 2017), and the banking industry (Yip & Bocken, 2018). Redesigning business models with sustainable concern requires a holistic understanding of the entire system, for instance, in the automotive industry (Wells, 2013).

In 2010, Osterwalder & Pigneur developed a tool, Business Model Canvas (BMC), to assist in developing a business model. BMC has nine pillars: value proposition, key partners, key activities, key resources, customer relationship, channels, customer segments, cost structure, and revenue streams. The value proposition is the benefit the customers will receive from the products or services offered. Key activities are the main activity carried out to deliver the benefit. Key resources are the means the business owns to carry out the activities. Channels are the media used to deliver the value proposition. Customer segments mean customer classification involved in the business activities. A company should also clearly and completely identify income sources and possible expenditures. The quantity of the income and expenditures should be estimated using sound methods and approaches.

The main barrier to designing and structuring a business model is the lack of systems and guidance to support and strengthen sustainable development (França et al., 2017). In addition, the barriers are the imbalance between the cost and the benefit, not the well-maintained environment, the application of an incorrect business model, the unfit between value proposition and customer segments, financial infeasibility, the partially integrated environment, and execution errors (Batocchio et al., 2016).

3 Methodology

The methodology of this study is depicted in Fig. 2. This study is the last part of our study on managing MSW. Based on the resident-based sustainable bioreactor landfill gas plant (RSLFG) model developed by Farizal et al., (2019a & b), this study investigates and proposes the best-fit business model and its mechanism to run the plant. Regulations investigation and BMC are used to determine the best-fit business model. Regulations study is important since it will ensure that the proposed business model is applicable (do not conflict with existing rules and regulation). BMC is used to structure the model logically and draw the process. The study continued with determining the applicable business model's processes and mechanisms to run the model. Since residents' funding is a critical variable for the success of RSLFG implementation (Farizal et al., 2019b), this study is concentrated on determining the best-fit residents' donations collection mechanism that encourages the donors to pay their commitment. Four existing payment mechanisms are studied based on the resident's payment common practices and parameters leading the residents to use the mechanism. To ensure that the proposed business model and its best-fit funding collection mechanism is a good-to-go project, this study is closed by executing the feasibility study on the proposed model and mechanism. The feasibility study is conducted using the net present value (NPV) approach and followed by running sensitivity analysis (SA) on several critical components of a project funding.

4 Results, and Discussion

4.1 Case Study: Municipal Solid Waste at the City of Depok

The City of Depok is a relatively new municipality in Indonesia, established in 1999. The city is one of Jakarta's favorite suburbs that attracts many Indonesians to stay. Starting from having 6 districts when it was established, the city has grown up to 11 districts. Figure 3 gives a glance at information on the City of Depok and its population density (Kota Depok, 2019).

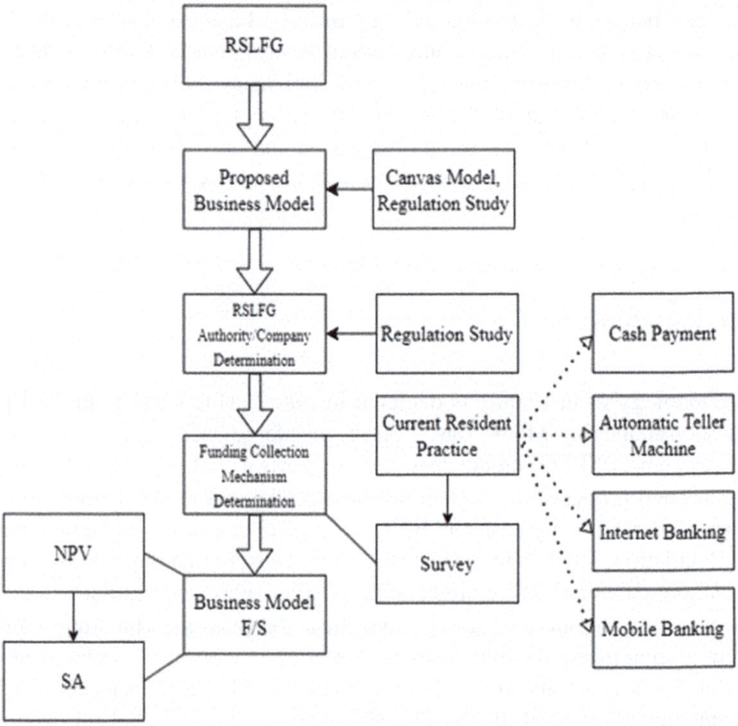

Fig. 2 Flowchart of the methodology

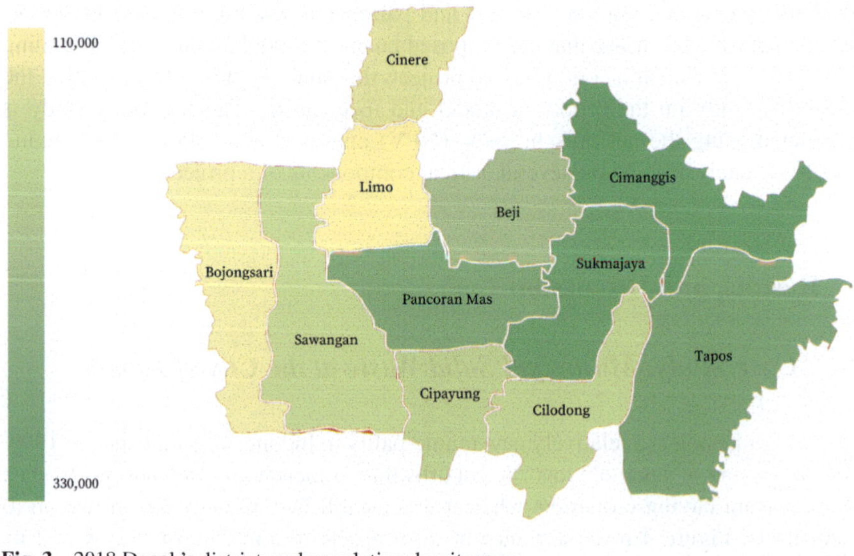

Fig. 3. 2018 Depok's district, and population density map

The population of Depok in 2022 is 2,123,350, while its average population growth rate per year from 2020 to 2022 is 1.79% (BPS Kota Depok, 2023). Depok has one landfill located in Cipayung District to handle its MSW. The landfill occupies an 11.2-hectare area with 3 pools. On average, 900–1,000 tons/day of MSW are dumped into this landfill. The landfill, opened in 1984, actually has reached its full capacity. However, it is still operating since it is difficult to find and open a new one to replace the existing Cipayung landfill. People are unwilling to face the bad impact of an open-dumping landfill. An open-dumping landfill creates a bad smell and an awful appearance for the surroundings. It is the home of disgusted pests such as insects, rodents, and flies. The open-dumping landfill also attracts scavengers to come.

The good thing for the City of Depok is utilizing its MSW as a source of electrical energy per Presidential Regulation 18, 2016. Based on the Minister of Energy and Mineral Resources Regulation Number 44 Year of 2015, the state electricity company (PLN) must purchase electricity from an LFG project with a flat rate for 20 years.

4.2 *Proposed RSLFG Business Model*

The proposed business model of RSLFG for the City of Depok is shown in Fig. 4. Figure 4 shows the 6 main players (indicated by the double-line rectangle), and their interrelation. The following describes the business model viewed from the 9 pillars of BMC.

1. Value proposition

Among many types of landfills, bioreactor landfill gives a more stable and greater methane gas (US EPA, 2017; Kaszubska & Wzorek, 2017). It also has a shorter waste decomposition time, which is beneficial to reduce the waste volume. Sustainable bioreactor landfill gas (RSLFG) provides a good alternative to MSW management that fits with the Indonesia setting, where the country has quite a large population but a limited landfill area. However, the bioreactor takes bigger investment, operation, and maintenance costs than a conventional landfill. Farizal et al. (2019b) showed that the Depok residents are willing to donate some funding for better MSW handling. The funding is used to finance the sustainable bioreactor plant's construction and operation. Resident-based sustainable bioreactor landfill gas (RSLFG) is an environmentally friendly energy source (Ahmed et al., 2015; Chaliki et al., 2016).

2. Key Partners and Resources

To run the RSLFG plant, six partners are involved, i.e., central government, local government, the office of cleaning and gardening (OCG), MSW authority/company, PLN, and the residents of Depok. In Fig. 4, the six partners are indicated by the double-line rectangle. Among partners involved in managing MSW, key partners in the proposed business model are the residents of Depok and the OCG of the City of Depok. As a key partner, the resident of the City of Depok has an important role. Aside from being the MSW producer, the community may participate in financing

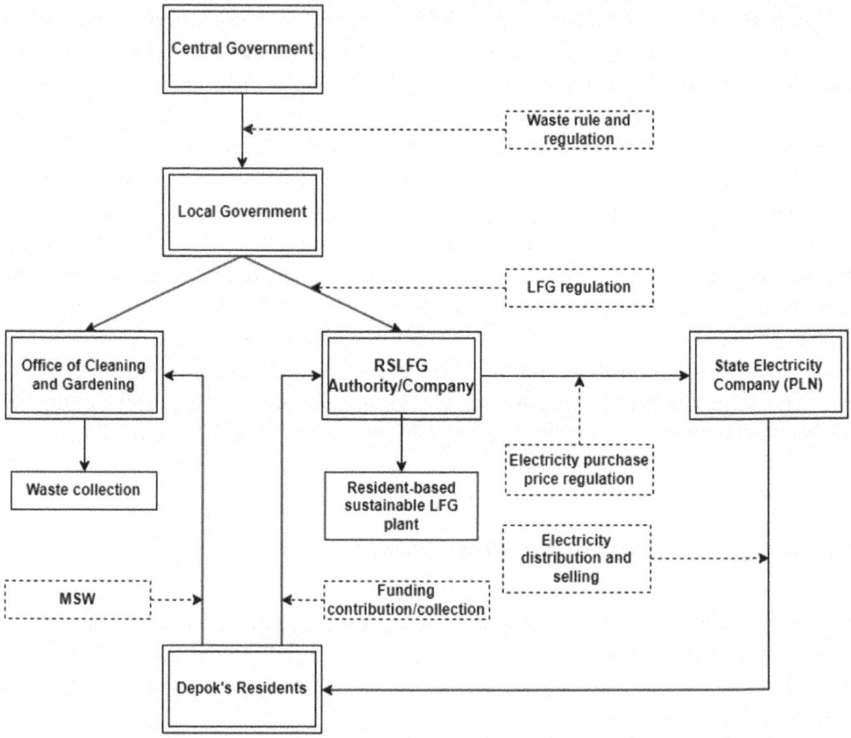

Fig. 4 The proposed business model of resident-based sustainable LFG (RSLFG)

the RSLFG project. The residents may also be involved in MSW processing and recycling. Another resident's important role is serving as the buyer of the electricity produced by the RSLFG plant.

The OCG's role is collecting MSW from the residents' houses (or a designated collecting point), shipping, and dumping the garbage at the landfill. In the landfill site, the MSW is managed by the authority of the RSLFG plant. The authority is given a license and assigned responsibilities to run the bioreactor plant, harvesting the methane gas, converting the gas to electricity, and then transmitting the electricity to PLN through the grid. The central government has a role in issuing regulations and legislation on waste management, which then authorizes the local governments to identify the agency or organization to run the RSLFG project.

Based on Depok's Local Regulation Number 05 Year of 2014 on Waste Management, Chapter II, Section 2, MSW management can be done by the city government, the community, and the agency. In Chapter 6 of Article 27, the forms of cooperation and partnership of the waste management system can be done through:

a. Cooperation between the city government office, provincial government office, and other city district governments
b. City Government cooperation with the private sector (PPP)

c. Provision of the initial investment for the construction of a waste management system of government or
d. Licensing of private investment

Government Regulation Number 81 Year of 2012 Article 26 says a local/city government can partner with a company, society, or business entities in accordance with local regulation, i.e., Depok Rule Number 05 Year of 2014 on Chapter 6 of Article 27 about the form of cooperation, and partnership on the implementation of the waste management system. It says the cooperation between the city government and the private sector (PPP).

This study proposes a kind of limited liability entity (enterprise) to handle the operation of the RSLFG plant. The limited liability company is supervised by an enterprise with a structure following the City of Depok Rule Number 10 Year of 2011. The following structure, i.e., the structure of MSW authority/company:

1. Board of Directors
2. Board of Supervisory that consists of a chairman, secretary, and members
3. Manager of Operations
4. Manager of Finance
5. Supervisor of RSLFG
6. Staff of RSLFG as many as 43 people

3. Value Stream and Key Activity

RSLFG's income is obtained from selling the electricity to PLN, funding from the residents, and funding credit from the Clean Development Mechanism (CDM). The electricity produced by the RSLFG plant is sold to the PLN with a purchase scheme as stated in the Regulation of the President of the Republic of Indonesia Number 18 Year of 2016. This regulation gives an authority to PLN to purchase electricity from assigned regionally owned business entities. Chapter IV Article 8 explains that profit from selling the electricity is the right of the RSLFG business entity. In addition, the Regulation of the Minister of Energy and Mineral Resources of the Republic of Indonesia Number 44 Year of 2015 also regulates PLN to purchase the RSLFG's electricity.

Another source of funding for RSLFG comes from residents' participation. The residents fund the RSLFG by paying regular tipping fees and participating in resident-based donations. A study by Farizal et al. (2019b) for Depok shows that 86% of Depok's population is willing to pay more to overhaul Depok's current waste management system to be a bioreactor landfill. The amount of donations is shown in Fig. 5. The donations nominated in IDR are 20.000, IDR 50.000, IDR 80.000, IDR 100.000, and IDR 200.000. Based on that scheme, the amount of donations gathered is estimated as IDR 328.394.427.337.

Farizal et al. (2019a) also studied the impact of declining the number of residents willing to donate. The result showed that, declining by 90% of the donors, the NPV is still positive. This shows that despite only having 10% of the donors, the RSLFG project can still be operated and be feasible since the NPV is still positive. But, if there

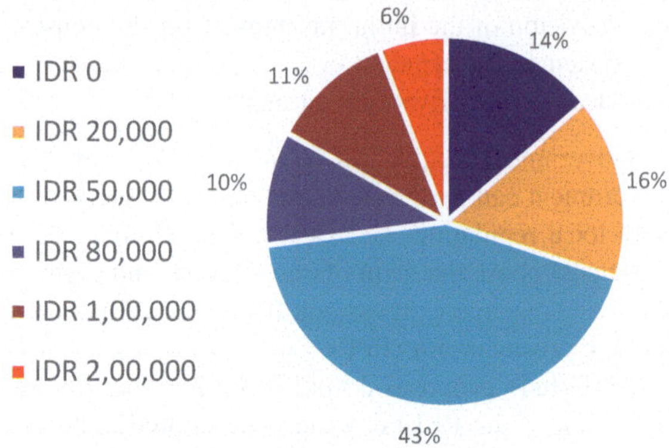

Fig. 5 Distribution of Depok's residents by donation category

are no donors, the NPV is negative. If the RSLFG scheme is constructed, the project needs residents' donations to operate. The study indicates that residents' donations are critical toward the operational sustainability of the RSLFG plant compared to other funding sources such as CDM and electricity sales.

4. Cost Structure

Cost, an important business model element, is a success factor in a new company (Chammassian & Sabatier, 2020). Cost should be enumerated clearly and completely. Operating the RSLFG includes direct costs such as investment, operation, and maintenance costs, employee salary and labor wage, indirect costs such as tools and equipment, and overhead costs such as insurance, bonuses, and taxes. In addition to regular activities to operate and maintain the RSLFG to produce electricity, the RSLFG authority will also have expenditures related to resident donations' collection. The expenditure also includes costs to utilize the banking systems.

The number of salaries and wages for the RSLFG personnel are regulated through Local Regulation of Depok City Number 10 in the Year 2011 regarding wage weight of Regional Owned Enterprises (BUMD), as follows:

1. Salary of the Board of Directors wages are accounted as 100%
2. Salary of the Chairman of the Supervisory Board amounted to 45% of the CEO's wage
3. Salary of the Secretary of the Supervisory Board amounted to 40% of the CEO's wage
4. Salary of the Supervisory Board Member amounted to 35% of the CEO's wage

Company Obligations Towards Employees In addition to the salaries and wages, the RSLFG authority has additional obligations toward their employees that are dictated by Law No. 24 Year of 2011 Article 15 Paragraph (2), the law on the implementation of social security for workers. The article says that every company (employer) must register all employees, without any exception, as BPJS Health participants. Even though the company has already enrolled with a private health insurance provider, it still must register its employees with BPJS.

As a limited liability corporation, the company also must provide financing for senior resident security programs (Regulation of the Government of the Republic of Indonesia Number 46 the Year of 2015) regarding health insurance (Regulation of the President of the Republic of Indonesia Number 28 the Year of 2016 regarding the third amendment to Presidential Regulation Number 12 the Year of 2015). Implementing a pension guarantee program (Regulation of the Government of the Republic of Indonesia Number 45 Year of 2015) and providing religious holiday allowance (based on the Regulation of the Minister of Labor of the Republic of Indonesia Number 6 Year of 2016). The assessment of the amount of financing that is the responsibility of the company is as follows:

a. Senior Resident Guarantee (JHT): This payment is charged to the company for 3.7% of the individual income each month.
b. Pension Guarantee: This payment is charged to the company at 2% of the individual monthly income.
c. Death Benefit (JKM): This payment is charged to the company for 0.3% of the individual monthly income.
d. Work Accident Benefit (JKK): This payment is charged to the company for 1.27% of the individual monthly income.
e. BPJS Health: The payment fee charged to the company is 4% of the cost of employee wages.
f. Holiday Festival Allowances (THR). Based on the Regulation of the Minister of Manpower of the Republic of Indonesia Number 6 Year of 2016 regarding religious holiday allowances for workers/laborers in companies.

5. Channel

Farizal et al. (2019b) showed that the residents of Depok are willing to make donations for the RSLFG project. Since the project duration is quite long, it makes sense to investigate funding collection mechanisms and payment modes encouraging residents to pay their donations. An appropriate payment mechanism can avoid inefficiencies and unnecessary expenses (Kim & Iorio, 2017). Furthermore, payment will facilitate and encourage people to participate in contributing. Therefore, payment has to meet people's expectations. In payment transactions, privacy is the most important issue since it protects user's information from being misused by others (Sreejesh et al., 2016).

Other payment system concerns are accessibility and affordability (Iqbal & Sami, 2017). Conventional bank services have low accessibility since users are required to come to the bank to conduct the transactions. This procedure makes the payment system ineffective for bank customers (Deb & Agrawal, 2017). Affordability is related to the fee users pay to use the service. In this case, the fee will be paid by the RSLFG authority, which collects the donation instead of the donors. The fee will be a big consideration since it will reduce the donation amount. Moreover, customer behavior also affects the choice of payment method (Chen et al., 2017). Knowing payment standards will make it easier to determine which payment systems are important to develop (Sun et al., 2017).

4.3 Survey

This survey aims to disclose residents of Depok opinions on several parameters in transaction mechanisms, including transaction convenience (Alalwan et al., 2016), transaction security (Kovács & David, 2016), transaction notification, transaction fee, transaction time speed, and pride. The survey questionnaire was developed using a Likert scale where 1: Very Not Good, 2: Not Good, 3: Pretty Good, 4: Good, and 5: Very Good. The scale evaluated four payment methods: cash payment, ATM card usage, Internet banking, and mobile application (m-banking). The cash payment system refers to transactions by depositing cash either through a teller at a bank or the officer who comes to collect the payment. ATM payment transactions refer to payment by visiting or through ATM machines placed at gas stations, shopping centers, or grocery stores. Internet payment refers to payment at the website receiver company using computer/laptop media. Internet payment can be made at home or office. M-banking payment refers to using an application software installed on a mobile phone.

Using an appropriate payment mechanism that fits with Depok residents' characteristics will encourage the donors to fulfill their commitment by paying their donations.

4.3.1. Survey on the characteristics of Depok Resident

This survey was conducted to get information regarding common transaction methods used by Depok residents when they pay their electricity bill, water, telephone, or house installment payments, and donations. The objective is to disclose residents' experience using the transaction system in many ways, and accommodate convenience and relief when conducting the transactions. Moreover, this survey also identifies users' knowledge levels when performing existing transaction methods. The questionnaire also asked about respondents' bank accounts to minimize transfer fees caused by using different accounts.

Based on the data supplied by the respondents, the results showed that 40 to 60% of transactions conducted by 55.94% were carried out by non-cash payment. Conversely, 45.45% of the respondents pay their bills in cash. The results are summarized in Table 3.

Another finding showed that the cash payment system dominates payments for electricity, water, and house installments (45.45%) and donations (56.64%). In addition, most respondents have 2 different bank accounts, 1 account with a state-owned bank and another with a private bank. These findings indicate that the respondents are representative since they have enough experience doing cash and non-cash transactions.

Figures 6, 7, and 8 show demographic information of the survey respondents. Figure 6 is related to the age of the respondents, Fig. 7 is to their education, and Fig. 8 is to their monthly income.

Figure 6 shows that many of the respondents belong to the group between 40–60 years old. While Fig. 7 shows that more than 80% of the respondents have a senior high school or higher education.

Table 3 Respondent's transaction experience survey results

No	Respondents' experience with payment methods	Percentage (%)
1	The payment system used for bill payments	
	(Electricity, water, etc.)	
	(*can choose more than one)	
	Cash	45.45
	Internet Banking	34.27
	Smartphone app (*m-banking*)	29.37
	ATM cards	35.66
2	The payment system used for donation (*can choose more than one)	
	Cash	56.64
	Internet Banking	23.08
	Smartphone app (*m-banking*)	18.18
	ATM cards	38.04
3	Type of Savings Account (Bank) owned	
	(*can choose more than one)	
	State Owned Banks	67.1
	Private Banks	67.1

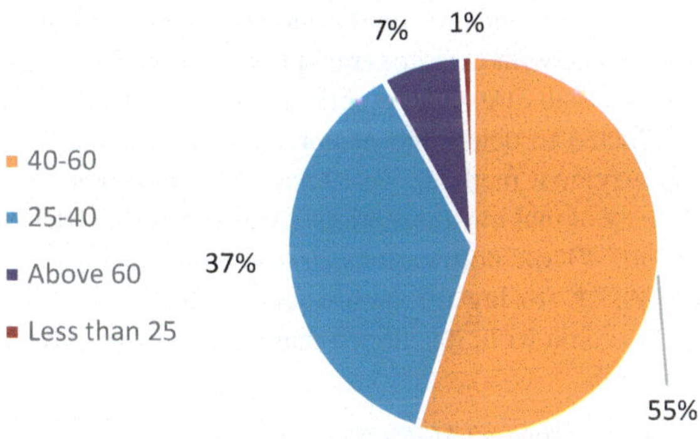

Fig. 6 Classification of the respondents based on their age

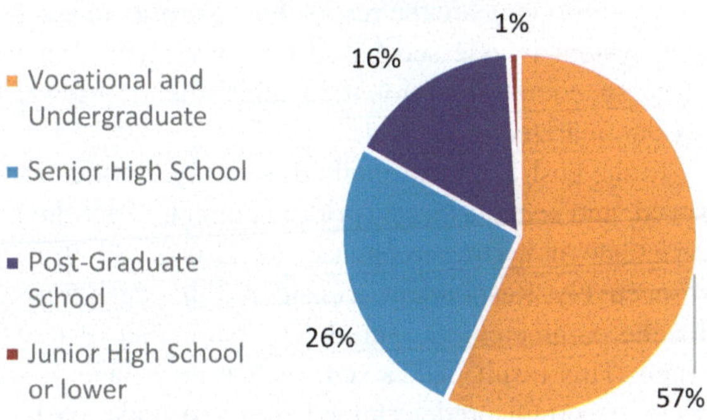

Fig. 7 Classification of the respondents based on their education level

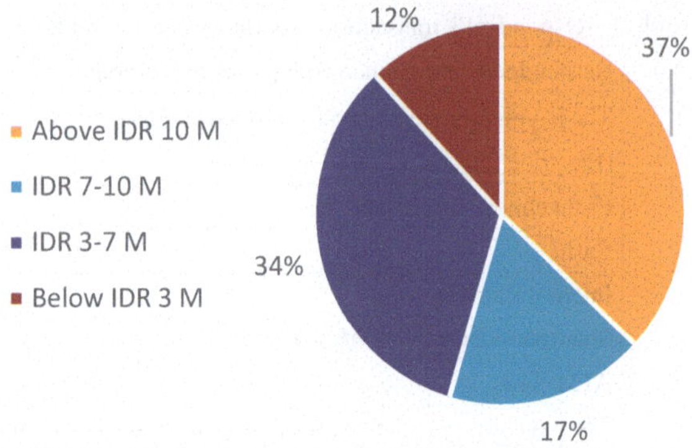

Fig. 8 Income level distribution of the respondents

Figure 8 shows that more than 85% of Depok residents have income above the minimum salary. Figures 7 and 8 reveal that most Depok residents are educated with more than enough income. Checking the figures, the respondents' demographic results of this survey are very much like the results of the previous study conducted by Farizal et al. (2019b). This finding justified using the resident-based participation funding model they proposed as the base calculation for this study.

4.3.2. Survey on the Usage of Payment System

The questionnaires on this survey are focused on disclosing residents of Depok's experience with many payment transaction systems, including cash payment, ATM, internet banking, and mobile phone (m-banking) applications. The survey was conducted to determine respondents' experience in making payment transactions. The payment methods were tested based on several transaction parameters such as the ease of making transactions (Alalwan et al., 2016), transaction security (Kovács & David, 2016), notifications, transaction fees, speed of time, and social or pride aspects. Knowing the parameters of the respondents' payment system they used and suitable with, it is hoped that the donations payment for the RSLFG project can run well.

1. *Cash Payment Method*

The following are the respondents' results in assessing the cash payment system. The system is assessed based on convenience, security, notifications, transaction costs, speed of transaction time, and social or pride aspects. Respondents' responses are summarized in Table 4.

In the cash payment method, security, notification, transaction fees, transaction speed, and social aspects (pride) parameters get the highest rating even though it is on a scale of 3. The percentages are 41.60%, 43.36%, 37.06%, 40.56%, and 53.85%, respectively. Respondents considered that cash transactions have a moderate value for the parameters measured. The parameter of transaction easiness is considered small. This result makes sense since they have to spend extra effort going to the bank at certain predetermined times to make cash payments. It can be said that the convenience of a cash payment transaction system is not good enough. Among

Table 4 Cash transaction parameter assessment results

Parameter/Scale	Score (%)				
	1	2	3	4	5
Convenience	4.20	6.29	34.27	37.06	16.18
Security	2.80	13.29	41.60	25.87	16.78
Notification	4.90	15.38	43.36	25.17	11.19
Transaction fees	9.09	9.79	37.06	33.57	10.49
Transaction speed	5.59	9.09	40.56	33.57	11.19
Social aspect	4.20	11.19	53.85	24.48	6.29

four parameters with high scores (Security, Notification, Transaction Speed, and Social Aspects), the Social Aspect has the highest value, 53.85%. The Social Aspect is the strongest point of cash transactions. Respondents considered that they may mingle with acquaintances/friends and socialize with bank officials by making cash transactions.

2. *Automated Teller Machine (ATM)*

ATM is one of the most frequently used payment channels (Wan et al., 2005). However, the weakness of using an ATM is if the customers have to wait in a queue. This especially occurs if the ATM is strategically located at the corner of a busy market or a melting point. This long queue makes customers uncomfortable (Mahmood & Shaikh, 2013). In this case, the distribution of ATM locations must be balanced (Bilginol et al., 2015). However, the distribution of the number of ATMs is constrained by the investment amount (Ou et al., 2009). Other than ATM distribution around the town, another hindrance impeding ATM utilization is the lack of a security system (Suhr et al., 2012).

Depok residents' assessment results on ATM utilization end up with high grades at scale 4 on aspects of easiness, security, notification, and transaction time with values of 39.86%, 35.66%, 48.25%, and 44.06%, respectively (see Table 5). However, since ATM transaction incurs some fees, this matter is considered as the weakness of ATM. ATM transaction also removes social aspect. The highest value of transaction through ATM is Notification. After using an ATM, users receive a transaction receipt of their payment or transfer. The users will also receive the balance of their accounts. The next strength point of the ATM transaction is the transaction time. The time is relatively faster than the time through cash payment. The ATM transaction can be performed 24/7.

3. *Internet Banking*

Technology and good information systems for transactions using Internet banking contribute to the high satisfaction of its users (Zhou, 2013). However, this mode of transaction will be hampered during a bad internet connection or server breakdown (Suryotrisongko et al., 2012).

Table 5 ATM parameter assessment results

Parameter/Scale	Score (%)				
	1	2	3	4	5
Convenience	2.10	4.90	34.97	39.86	18.18
Security	4.20	9.79	34.97	35.66	15.38
Notifications	3.50	9.09	28.67	48.25	10.49
Transaction fees	10.49	8.39	41.96	31.47	7.69
Transaction speed	3.50	6.29	33.57	44.06	12.59
Social aspect	4.90	6.99	46.15	35.66	6.29

Products and services offered by banking provide an opportunity to manage finance via the Internet (Kolodinsky et al., 2004). Satisfaction, trust, simplicity of use, accessibility, variety of products and services, and relatively inexpensive fees are the reasons to use Internet banking for payment (Daniel, 1999; Rexha et al., 2003). In addition, customers' education level and income are also determinant factors (Juwaheer et al., 2012). Higher-income customers are more willing to use online payment (Black, 2005). The size and reputation of a bank also influence customers to use this service (Malhotra & Singh, 2010).

Residents of Depok survey results on Internet banking payment indicate that convenience and transaction speed are factors with high grades (Table 6). The table shows that the values are quite high for scales 4 and 5. Parameter Convenience has 35.66% and 33.57% values for scales 4 and 5, respectively. At the same time, the parameter Transaction Speed has values of 34.97% and 33.58%, respectively. Surprisingly, the parameter Social Aspect of Internet banking is high, i.e., 41.94% (even at scale 3). The reason is that Internet banking transactions can be performed simultaneously with internet browsing and social networking.

4. *Mobile Banking*

Study shows that residents are attracted to using mobile banking (m-banking) due to its easiness, risk, trust, and lifestyle (Alalwan, 2016; Koksal, 2016; Shaikh & Karjaluoto, 2015; Veríssimo, 2016). Characters of m-banking include convenience,

Table 6 Internet banking parameter assessment results

Parameter/Scale	Score (%)				
	1	2	3	4	5
Convenience	6.29	2.80	21.68	35.66	33.57
Security	9.79	6.29	25.17	37.06	25.17
Notifications	1.40	2.10	34.27	37.06	25.17
Transaction fees	9.09	12.59	38.46	33.57	6.29
Transaction speed	4.20	3.50	23.78	34.97	33.57
Social aspect	4.20	7.69	41.96	31.47	14.69

transaction accuracy, variety of product services, and non-stop open-hour satisfaction (Jun & Palacios, 2016). For financial institutions, m-banking allows them to get long-distance services (Leiva et al., 2017).

With the emergence of social media networking, this new technology can be integrated with m-banking so that payments can be made using a smartphone. Blending Internet banking and m-banking allows users to make payments anytime and anywhere without leaving their houses or offices (Laukkanen, 2007). Mobile banking payments provide convenient and convenient services (Nguyen et al., 2016). This feature becomes more popular with the growth of mobile commerce (Yang et al., 2015).

Key factors of m-banking utilization are the users' satisfaction and low risk (Safeena et al., 2012). The demographic factor, i.e., the age of the customers, also influences m-banking utilization. Laukkanen (2016) showed that young customers dominate m-banking customers. However, m-banking has weaknesses, such as a lack of customer protection against fraud (Purwanegara et al., 2014) and high transaction fees (Au & Kauffman, 2008).

Table 7 shows the resident of Depok survey results on m-banking payment using a handphone. The table indicates that almost all parameters except the Social Aspect score a high scale of 4. These results align with the findings mentioned in the previous paragraph.

Important Parameter to Encourage Payment Transaction The collected respondent data indicates that the most important parameter in the transaction is convenience, security, and transaction time (Table 8). The table shows that these three parameters have a high level of customer satisfaction on a scale of 5, with percentages of 51%, 59%, and 59% for convenience, security, and time transactions, respectively. These three parameters should be attached to the type of payments to be selected for collecting donations from the residents of Depok.

Using an appropriate type of payment method will encourage the residents of Depok to fulfill their pledge to donate their money to funding the RSLFG. Hard-to-access and unreliable payment mechanisms will prevent residents' participation (Iqbal & Sami, 2017).

Table 7 M-banking parameter assessment results

Parameter/Scale	Score (%)				
	1	2	3	4	5
Convenience	5.59	5.59	24.48	33.57	30.77
Security	4.20	8.39	38.46	35.66	13.29
Notifications	2.10	7.69	32.17	42.66	15.38
Transaction fees	7.69	10.49	41.96	30.07	9.79
Transaction speed	2.80	6.29	27.97	34.97	27.97
Social aspect	3.50	6.99	57.34	17.48	14.69

Table 8 Assessment of important payment parameter

Parameter/Scale	Score (%)				
	1	2	3	4	5
Convenience	6	3	29	54	51
Security	7	4	26	47	59
Notifications	2	5	40	62	34
Transaction fees	10	12	47	38	36
Transaction speed	3	3	28	50	59
Social aspect	11	14	53	44	21

Based on this finding, all four payment methods discussed in this section fulfill the requirements. They are suitable to be implemented.

4.4 Feasibility Study on the Proposed Business Model and Funding Mechanism

To evaluate RSLFG's worthiness, a feasibility study was conducted on the model and its mechanism applicability to run the business. Learning that the setting conducted on the payment mechanism survey for this study is very much the same as the setting on the study conducted by Farizal et al. (2019b), Farizal et al. (2019A) NPV calculation result was used as the basis for calculating NPV on this study (which takes into consideration of salary, and wage of the RSLFG operators, and funding collection activity). In this study, the tax rate is assumed to be 25%.

4.4.1 Net present value (NPV) with Officers' Donation Collection Scheme

If residents' funding is collected directly by the collector officers, the officers collect the donations directly from house to house every month. Once the money is collected, the officers deposit the funding into the company's account. To carry out the job, the number of officers is estimated based on the number of clusters in the corresponding sub-district (note: Cluster in Indonesia is called Rukun Tetangga, RT). Depok City has 4,856 RTs spreading across its 11 sub-districts. In total, 15 collector officers are needed to collect the donations. Table 9 summarizes the distribution of the collector officers.

For this collection scheme, each officer is equipped with a motorcycle to support collectors' job traveling from donors' house to house. Table 10 shows the estimated expenses for carrying out this scheme.

Table 9 Distribution of donation collection officers

Sub-District	Number of RTs	Number of Collectors
Sawangan	355	1
Bojong	300	1
Pancoran Mas	608	2
Cipayung	321	1
Sukma Jaya	875	2
Cilodong	340	1
Cimanggis	640	2
Tapos	620	2
Beji	372	1
Limo	218	1
Cinere	207	1
Total		15

Table 10 Donation collection cost components for officers

Component	Quantity	Cost* (Million IDR)
Officer wage	15	593.55/year*
Transportation (Motorcycle)	15	240**

* Data based on Regional Minimum Wage (UMK) Depok C, i.e., IDR 3,297,489
** Data based on the average price of a motorcycle in 2017

Based on RSLFG's structure described in Sect. 4.2, the operational costs (salaries and wages) incurred due to the company's structure are summarized in Table 11. The standard salary amount is based on the salaries of similar companies or services. The nominal salaries were edited from the Kelly salary guide (2016). While wages for staff and operators are based on the minimum wage (UMK) standard for Depok in 2017, IDR 3,297,489.

Table 11 Monthly salary of the company's boards and staff

Position	Salary (Million IDR)
Board of directors	120
Chairman of supervisory board	54
Secretary of supervisory board	52.8
Supervisory board member	42
Operational manager	40
Financial manager	40
Supervisor	10
Operators and staff (43 people)	141.79
Total	500.59

Table 12 Social security and health costs for 1 year

Number of workers	Health, and social security costs (Million IDR)
50	267.49

Table 13 The company pays THR for 1 Year

Number of workers	THR costs (Million IDR)
50	500.59

Table 14 Additional first-year operational costs incurred by employing collector officers

Component	Quantity	Cost (Million IDR)	Total (Million IDR)
Motorcycle cost	15	16	240
Administrative fee (1 year)	–	0.48	0.48
Initial cost + collection card	–	0.14	0.14
Grand Total			240.62

Table 12 lists estimated expenditures related to BPJS employment and health insurance costs that are obligated by Law Number 24, the Year of 2011 Article 15 paragraph (2), while Table 13 lists estimated THR allowance as governed by the Regulation of the Minister of Manpower of the Republic of Indonesia Number 6 the Year of 2016. The minimum value of BPJS employment and health insurance is calculated according to the respective regional minimum wage (UMK), with a maximum value of IDR 8,000,000.

However, under this scheme (residents of Depok donations are collected directly by collector officers), additional operational costs are due to employing 15 additional staff. Table 14 shows the additional costs under this scheme, including motorcycle spending and related fees.

Considering all additional costs described above to the NPV calculation, those costs do not drag the NPV to be negative. This result means that operating the RSLFG plant under this scheme (collecting donations directly by collector officers) is worthy (profitable). NPV calculation result under this scheme is IDR 1,850,151,964,718. Complete NPV calculation results for some discount rate values are displayed in the second column of Table 16, and indicated with red ink in Fig. 9.

4.4.2 NPV Using a Virtual Account Scheme

Collecting residents funding using a virtual account payment mechanism will let Depok residents pay their donations directly at any time. To exercise this scheme, the RSLFG authority is assumed to open 3 accounts with the state bank and 2 with the private bank. By so doing, it is expected that the system will be able to provide a convenient experience and reduce transaction costs between banks. Under this

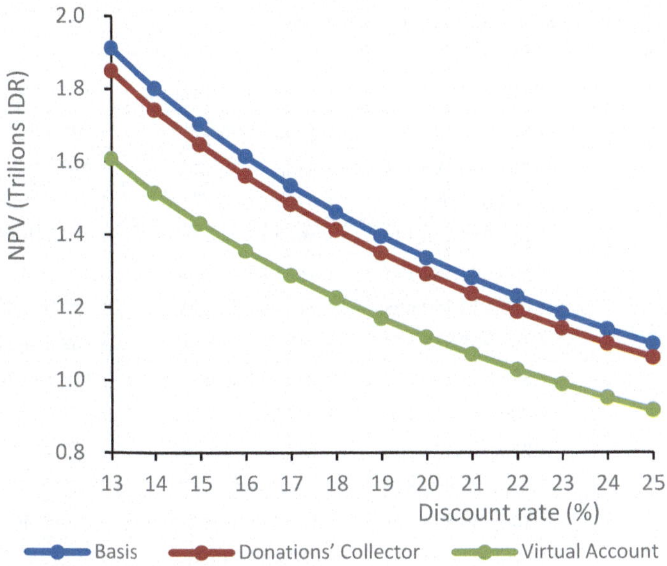

Fig. 9 NPV versus discount rates for three different schemes

scheme, each donor will be given a virtual account to pay for their pledge. Virtual accounts can also reconcile real-time payments. With a virtual account, transaction data such as payer identification is already coded so that donors do not need to provide proof of payment transactions. The fees charged to billers or companies that use virtual account services differ from bank to bank.

The costs incurred using virtual accounts are listed in Table 15. This study assumes that the transfer fee is IDR 5,000 per transaction. The bank charges the fee to the biller, i.e., RSLFG authority. Details of the total administration cost are explained in Table 15.

Based on data in Table 15, the calculated cost of using virtual accounts is quite high compared to the cost of the donations collection scheme. Fortunately, taking this cost to the NPV calculation, the NPV comes up with a positive value. This

Table 15 The costs of using a virtual account

Types of Costs	Costs (Thousand IDR)
Total Administration Costs to open 5 accounts	12,900
Administration fees/transaction	5
Initial card purchase	–
The initial cost of a virtual account	–
Total transaction costs charged to the biller per year *	29,789,940

*Total transaction costs/year = adm. fees/transaction × 12 months x number of donors

Table 16 NPV comparison results versus discount rate

Discount Rate (%)	NPV (IDR)		
	Basis*	Donations' collector	Virtual account
13	1,912,746,819,572	1,850,151,964,718	1,608,129,341,570
14	1,802,650,108,554	1,743,476,591,521	1,514,746,096,031
15	1,704,001,083,369	1,647,887,506,737	1,431,045,085,617
16	1,615,253,286,149	1,561,887,342,517	1,355,719,660,494
17	1,535,104,382,102	1,484,215,103,080	1,287,669,429,721
18	1,462,453,359,668	1,413,804,720,599	1,225,964,084,236
19	1,396,365,835,568	1,349,751,460,146	1,169,814,074,936
20	1,336,045,834,415	1,291,284,594,599	1,118,546,765,921
21	1,280,812,757,146	1,237,745,104,291	1,071,586,974,922
22	1,230,082,521,260	1,188,567,416,387	1,028,441,040,403
23	1,183,352,065,819	1,143,264,402,430	988,683,732,663
24	1,140,186,578,901	1,101,415,012,009	951,947,465,694
25	1,100,208,934,699	1,062,654,045,962	917,913,376,152

* (Farizal et al., 2019a)

means collecting residents' funding with a virtual account is also good to go, even though it gives less profit than direct collection. Another benefit of the virtual account payment method is company orderly administration and user satisfaction. The NPV for this scheme is IDR 1,608,129,341,570. Complete NPV calculation results for some discount rate values with virtual account payment mechanisms are displayed in the third column of Table 16 and indicated with green ink in Fig. 9.

4.5 *Sensitivity Analysis*

Sensitivity analysis (SA) examines the impact of data input uncertainty on the NPV value (Saltelli, 2002). In a feasibility study (FS), all data inputs are not known for sure, but they are estimated. SA is carried out to give a sense of caution to the stakeholder as well as to identify which parameters are more sensitive (give more impact to the NPV). Based on the FS result, SA will go in two directions. If the NPV is positive, SA goes by increasing the costs, and decreasing the revenue (income). On the other hand, if the NPV is negative, SA is performed by increasing the revenue (income), and reducing the costs. SA calculation is performed by changing the input within 10 to 50%.

1. *Sensitivity Analysis for decreasing revenue*

Since the NPV for this scheme is positive, SA is conducted by reducing the revenue. As mentioned, RSLFG's revenue comes from 3 sources: selling the electricity to PLN, claiming CDM, and residents' donations. For instance, a revenue decline is possible when leakage in the bioreactor plant causes the amount of methane gas to be harvested. Thus, the electricity produced is reduced. Another possible event that causes revenue to decline is the residents' movement (migration); residents who pledged for donations are moving out and then replaced by new residents unwilling to donate or any other events. Based on a discount rate of 13%, SA for declining income is given in Fig. 10. The figure shows that declining income reduces the NPV for donations collection and virtual account schemes. However, the figure shows that the NPVs are still positive (which means the RSLFG project is still profitable) when income declines up to 50%. The NPV when the revenue drops 50% is IDR 893,778,554,932 and IDR 651,755,931,784 for the donations collecting scheme and virtual account scheme, respectively. Figure 10 also shows that even though the reduction rate of both NPVs is very much the same, the virtual account scheme is more sensitive than the donations collection scheme. The NPV of the virtual account scheme becomes null faster than the donations collection scheme. The NPV of the virtual account will reach zero when revenue decrement is around 52%, while the NPV of donation collection is around 70's percent. From this perspective, selecting a donations collection mechanism is more benevolent than a virtual account scheme.

2. *Sensitivity Analysis for increasing salaries/wages*

Figure 11 shows SA due to salaries and wage increments based on a 13% discount rate. The phenomenon that salaries and wages go up is not impossible. In many cases, salaries and wages are intentionally raised by a company to help its employees balance the inflation. Figure 11 shows that as the salaries and wages increase, the NPVs decrease. For the case of salaries/wage increments, both schemes have the same sensitivity as the gradients of the lines are the same.

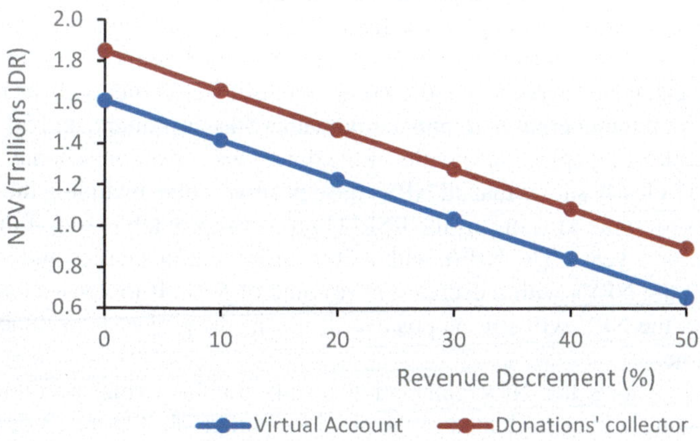

Fig. 10 Sensitivity analysis on the impact of revenue declining toward NPV

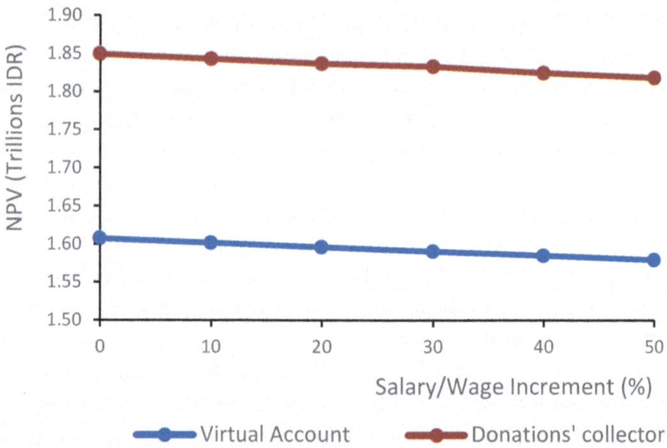

Fig. 11 Sensitivity analysis on the impact of salaries/wages increment toward NPV

Comparing the lines in Figs. 10 and 11, it can be inferred that RSLFG authority should be more cautious regarding the reduction of the revenue than the increment of salaries/wages of its employees since the NPV is reduced larger due to losing some income than spending more on salaries/wages.

3. *SA due to project duration time changing*

The third sensitivity analysis is performed toward project duration time. Every project has a definite lifetime that is set beforehand. However, a project might be terminated earlier or extended. The project is terminated early, for instance, in the case of force majeure. If suffering a big loss, the project should be terminated earlier to avoid more loss. On the other side, a project is extended, for instance, due to the absence of new equipment. Since we cannot guarantee the lifetime, SA is conducted to describe the impact of the project lifetime on the investment return.

Like many energy-related projects, the RSLFG project lifetime is assumed to be 25 years. SA will be conducted for 3, 5, 10, and 20 years. Based on a discount rate of 13%, SA is calculated according to the 2 schemes discussed earlier with the following setting: normal remuneration, remuneration with a 50% increment, and 50% lack of income. Figure 12 displays the results of the donations collection scheme.

Figure 12 clearly shows that all NPVs have positive value, including the one with just 3 years lifetime. This means the RSLFG project is profitable even if the project is running for 3 years. The NPVs with a 50% increment in salaries and wages are bigger than the NPVs with a decrease in revenue of 50%. If the project is stopped at any time, the NPV will still be positive. It means the project is profitable at any duration time.

Figure 13 shows the NPV calculation results for the virtual account scheme under the same setting explained in the previous paragraph. Figure 13 displays the same pattern as Fig. 12 but with smaller NPV values. The results indicate that the RSLFG project is worthy regardless of the funding collection mechanism used and

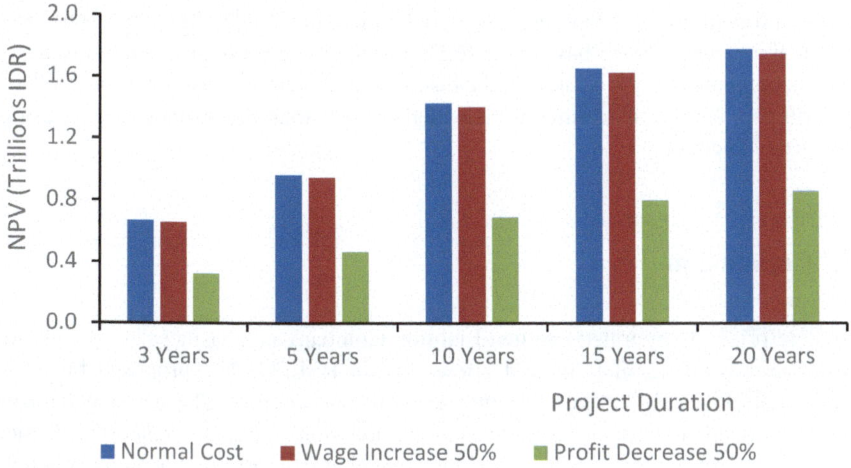

Fig. 12 SA for RSLFG for some project durations for the donations collection scheme

the project's lifetime. The NPV for 3 3-year project lifetime under 50% salaries/ wages increment is IDR 559,846,749,265, while the NPV for the one with 50% revenue decrement is IDR 217,083,190,728. All NPVs have positive values, which means the project is profitable.

Fig. 13 SA for RSLFG for some project durations for the virtual account scheme

Even though the RSLFG business model with a donators' collection scheme and a virtual account scheme have positive NPV (which means both schemes are profitable), comparing Figs. 12 and 13, the business model employing donations' collection officers is preferable since it gives higher NPV than the business model using the virtual account.

5 Conclusion

An enterprise-incorporated Limited Liability Company is a business entity that fits with the existing regulations and allows to run RSLFG. The proposed business model comprises six entities that interact with one another. The best mechanism to collect residents' donations is through the donations' collector officers' scheme. The scheme gives better profit and is less sensitive than a virtual account scheme. Sensitivity analysis on scenarios of the decrement revenue, increment salaries, wages, and changing project lifetime indicates the RSLFG with the proposed business model and donations' collection officers' scheme is a profitable project whether run within normal lifetime duration or terminated early. The RSLFG business model can be an alternative solution to cities with MSW problems.

Acknowledgement This work has been supported partially by a PUTI Q2 2020 Grant funded by DRP Universitas Indonesia under Contract No. NKB-1727/UN2.RST/HKP.05.00/2020

References

Abdelkafi, N., & Tauscher, K. (2015). Business models for sustainability from a system dynamics perspective. *Organization, and Environment, 29*(1), 74–96.

Ahmed, S. I., Johari, A., Hashim, H., Lim, J., Jusoh, M., Mat, R., & Alkali, H. (2015). Economic, and environmental evaluation of landfill gas utilization: A multi-period optimization approach for low carbon regions. *International Biodeterioration & Biodegradation, 102*, 191–201.

Aich., A. & Ghosh, S. K. (2016). Organic fraction of municipal solid waste—A valuable source of green energy in India. *International Journal of Energy Sector Management, 10*, 526–545

Alalwan, A. A., Dwivedi, Y. K., Rana, N. P. P., & Williams, M. D. (2016). Consumer adoption of mobile banking in Jordan: Examining the role of usefulness, ease of use, perceived risk, and self-efficacy. *Journal of Enterprise Information Management, 29*, 118–139.

Albores, P., Petridis, K., & Dey, P. K. (2016). Analyzing efficiency of waste to energy systems: Using data envelopment analysis in municipal solid waste management. *Procedia Environmental Sciences, 35*, 265–278.

Al-Debei, M M., El-Haddadeh, R. & Avison, D. (2008). Defining the business model in the new world of digital business. *Proceedings of the Fourteenth Americas Conference on Information Systems*. Toronto.

Amit, R., & Zott, C. (2001). Value creation in e-business. *Strategic Management Journal, 22*, 493–520.

Ardiansyah, S. (2014). Analisis Model Pembangunan Pembangkit Listrik Tenaga Sampah Berkelanjutan. Unprinted Bachelor Thesis. Universitas Indonesia, Fakultas Teknik, Jakarta. In Indonesian language.

Au, Y. A., & Kauffman, R. J. (2008). The economics of mobile payments: Understanding stakeholder issues for an emerging financial technology application. *Electronic Commerce Research, and Applications, 7*, 141–164.

Baden-Fuller, C. & Morgan, M.S. (2010). Business models as models. *Long Range Planning, 43*, 156–171.

Batocchio, A., Ghezzi, A., & Rangone, A. (2016). A method for evaluating business models implementation process. *Business Process Management Journal, 22*, 712–735.

Bawaain, M., Mamun, A. A., & Omidvarborna, H. (2017). Ultimate composition analysis of municipal solid waste in Muscat. *Journal of Cleaner Production, 48*, 355–362.

Bilginol, K., Denli, H. K., & Seker, D. Z. (2015). Ordinary least squares regression method approach for site selection of Automated Teller Machines (ATMs). *Procedia Environmental Sciences, 26*, 66–69.

Black, G. S. (2005). Predictors of consumer trust: Likelihood to pay online. *Marketing Intelligence & Planning, 23*, 648–658.

Bobinaite, V., & Tarvydas, D. (2014). Financing instruments, and channels for the increasing production, and consumption of renewable energy: Lithuanian case. *Renewable, and Sustainable Energy Reviews, 38*, 259–276.

BPS Kota Depok. (2023). *Depok Municipality in Figures*. Badan Pusat Statistik Kota Depok. https://depokkota.bps.go.id/publication/download.html?nrbvfeve=NTUyNDJhYzA2MmRhMDQ4YWE3NmFhZmZk&xzmn=aHR0cHM6Ly9kZXBva2tvdGEuYnBzLmdvLmlkL3B1YmxpY2F0aW9uLzIwMjMvMDIvMjgvNTUyNDJhYzA2MmRhMDQ4YWE3NmFhZmZkL2tvdGEtZGVwb2stZGFsYW0tYW5na2EtMjAyMy5odG1s&twoadfnoarfeauf=MjAyMy0wOS0xMyAxMzowNzoyMg%3D%3D. Retrieved on June 2023.

Briggeman, B. C., Jacobs, K. L., Kenkel, P., & Mckee, G. (2016). Current trends in cooperative finance. *Agricultural Finance Review, 76*, 402–410.

Chaliki, P., Psomopoulos, C. S., & Themelis, N. J. (2016). WtE plants installed in European cities: A review of success stories. *Management of Environmental Quality: An International Journal, 27*, 606–620.

Chammassian, R. G., & Sabatier, V. (2020). The role of costs in business model design for early-stage technology startups. *Technological Forecasting & Social Exchange, 157*, 120090.

Chen, R., Xu, X., & Shen, H. (2017). Go beyond just paying: Effects of payment method on level of construal. *Journal of Consumer Psychology, 27*(2), 207–217.

Cucchiella, F., D'Adamo, I., & Gastaldi, M. (2017). Sustainable waste management: Waste to energy plant as an alternative to landfill. *Energy Conservation, and Management, 131*, 18–31.

Daniel, E. (1999). Provision of electronic banking in the UK, and the Republic of Ireland. *International Journal of Bank Marketing, 17*, 72–83.

Deb, M., & Agrawal, A. (2017). Factors impacting the adoption of m-banking: Understanding brand India's potential for financial inclusion. *Journal of Asia Business Studies, 11*, 22–40.

Dembek, K., York, J., & Singh, P. J. (2018). Creating value for multiple stakeholders: Sustainable business models at the base of the pyramid. *Journal of Cleaner Production, 196*, 1600–1612.

Farizal, Zulys, A., Ihwan, Kh., Prihastuti, I., Yulianda, S., & Dissa, E. (2014). *Sanitary Landfill Gas for Riau Rural Electricity*. Project report to EEP (Energy and Environment Partnership) with Indonesia.

Fazeli, A., Bakhtvar, F., Jahanshaloo, L., Sidik, N. A., & Bayat, A. E. (2016). Malaysia's stand on municipal solid waste conversion to energy: A review. *Renewable and Sustainable Energy Reviews, 58*, 1007–1016.

França, C. L., Broman, G., Robert, K. H., Basile, G., & Trygg, L. (2017). An approach to business model innovation and design for strategic sustainable development. *Journal of Cleaner Production, 140*, 155–166.

Farizal, & Tammarar, E. (2018, November). Tipping fee determination to support the waste to energy concept at the city of Depok, Indonesia. *The Proceeding of Conference on Emerging Energy and Process Technology* (CONCEPT) 7, UTM. https://doi.org/10.1051/e3sconf/20199001007

Farizal, Amalin, N., & Rachman, A. (2019a). Waste to energy financial model design based on resident participation. *European Journal of Sustainable Development, 8*(4), 391–400. https://doi.org/10.14207/ejsd.2019.v8n4p391

Farizal, Amalin, N., & Rachman, A. (2019b, June). A new funding scheme for MSW management to support WtE program. *The Proceeding of International Conference on Advances in Energy Systems and Environment* (ASEE19). PWR, Poland. http://www.asee.pwr.edu.pl/

Gordijn, J., Akkermans, M., & Vliet, H. V. (2000). *Business modelling is not process modelling.* ECOMO 2000, Salt Lake City, USA, Springer.

Hall, S., Foxon, T. J., & Bolton, R. (2016). Financing the civic energy sector: How financial institutions affect ownership models in Germany, and the United Kingdom. *Energy Research & Social Science, 12*, 5–15.

Harting, R., Reichstein, C., Laemmle, P., & Sprengel, A. (2019). Potentials of digital business models in the retail industry–empirical results from European experts. *Procedia Computer Science, 159*, 1053–1062.

Iqbal, B., & Sami, S. (2017). Role of banks in financial inclusion in India. *Contaduria y Administracion, 62*(2), 644–656.

Jarallah, R. A., & Aleisa, E. (2013). Investigating cause contributing to increased municipal solid waste in Kuwait: A national survey. *Journal of Engineering Research, 1*, 123–143.

Jun, M., & Palacios, S. (2016). Examining the key dimensions of mobile banking service quality: An exploratory study. *International Journal of Bank Marketing, 34*, 307–326.

Juwaheer, T. D., Pudaruth, S., & Ramdin, P. (2012). Factors influencing the adoption of Internet banking: A case study of commercial banks in Mauritius. *World Journal of Science, Technology, and Sustainable Development, 9*, 204–234.

Kalkbrenner, B. J., & Roosen, J. (2016). Citizens' willingness to participate in local renewable energy projects: The role of community, and trust in Germany. *Energy Research & Social Science, 13*, 60–70.

Kardono & Purwanta, W. (2017). Landfill gas for energy: Its status, and prospect in Indonesia. *Proceedings of International Symposium* on Ecotopia Science, ISETS07.

Kaszubska, M., & Wzorek, M. (2017). The bioreactor–an innovative method of disposal of solid waste. *Economic and Environmental Studies, 17*(2), 347–361.

Kim, K., & Iorio, R. (2017). The 5 clinical pillars of value for total joint arthroplasty in a bundled payment paradigm. *The Journal of Arthroplasty, 36*, 1712–1716.

Kirby, E., & Worner, S. (2014). *Crowdfunding: An Infant Industry Growing Fast.* Staff working paper of the International Organization of Securities Commission (IOSCO) Research Department, Staff Working paper: [SWP3/2014], p. 8.

Kolekar, K. A., Hazra, T., & Chakrabarty, S. N. (2016). A review on prediction of municipal solid waste generation models. *Procedia Environmental Sciences, 35*, 238–244.

Kolodinsky, J. M., Hogarth, J. M., & Hilgert, M. (2004). The adoption of electronic banking technologies by US consumers. *International Journal of Bank Marketing, 22*, 238–259.

Koksal, M. H. (2016). The intentions of Lebanese consumers to adopt mobile banking. *International Journal of Bank Marketing, 34*, 327–346.

Kota Depok. (2019). *Depok Municipality in figures.* BPS-Statistics of Depok Municipality. Depok. ISSN: 978-602-0925-77-6

Kovács, L., & David, S. (2016). Fraud risk in electronic payment transactions. *Journal of Money Laundering Control, 19*, 148–157.

Kurniawan, T. A., Avtar, R., Singh, D., Xue, W., Othman, M. H. D., Hwang, G. H., Iswanto, I., Albadarin, A. B., & Kern, A. O. (2021). Reforming MSWM in Sukunan (Yogyakarta, Indonesia): A case-study of applying a zero-waste approach based on circular economy paradigm. *Journal of Cleaner Production, 284*, 124775. https://doi.org/10.1016/j.jclepro.2020.124775

Kwarteng, S. O., Anarfi, K. P., & Essandoh, H. M. K. (2016). Source separation and recycling potential of municipal solid waste in Ghana. *Management of Environmental Quality, 27*, 210–226.

Laukkanen, T. (2016). Consumer adoption versus rejection decisions in seemingly similar service innovations: The case of the Internet, and mobile banking. *Journal of Business Research, 69*, 2432–2439.

Laukkanen, T. (2007). Internet vs mobile banking: Comparing customer value perceptions. *Business Process Management Journal, 13*, 788–797.

Leiva, F. M., Climent, S. C., & Cabanillas, F. L. (2017). Determinants of intention to use the mobile banking apps: An extension of the classic TAMmodel Spanish. *Journal of Marketing, 21*, 25–38.

Mahmood, T., & Shaikh, G. M. (2013). Adaptive automated teller machines. *Expert Systems with Applications, 40*, 1152–1169.

Malhotra, P., & Singh, B. (2010). An analysis of internet banking offerings and its determinants in India. *Internet Research, 20*, 87–106.

Nguyen, T. N., Cao, T. K., Dang, P. L., & Nguyen, H. A. (2016). Predicting consumer intention to use mobile payment services. *International Journal of Marketing Studies, 8*, 117–124.

Osterwalder, A. (2004), The Model Ontology–A proposition in a Design Science Approach, *These pour l'obtention du grade de Docteur en Informatique de Gestion. Ecole des Hautes Etudes Commerciales de l'Université de Lausanne.*

Osterwalder, A., & Pigneur, Y. (2010). *Business model generation: A handbook for visionaries, game changers, and challengers*. John Wiley & Sons Inc.

Ou, C,S., Yen, D.C., & Hung, C.S. (2009). Determinants of information technology investments: The case of ATM in an emerging economy. *Advances in Accounting, incorporating Advances in International Accounting, 25*, 278–283.

Purwanegara, M., Apriningsih, A., & Andika, F. (2014). Snapshot on Indonesia regulation in mobile internet banking users attitudes. *Procedia—Social, and Behavioral Sciences, 115*, 147–155.

Rexha, N., Kingshott, R. P. J., & Aw, A. S. S. (2003). The impact of the relational plan on adoption of electronic banking. *Journal of Services Marketing, 17*, 53–67.

Safeena, R., Date, H., Kammani, A., & Hundewale, N. (2012). Technology adoption, and Indian consumers: Study on mobile banking. *International Journal of Computer Theory and Engineering, 4*(6), 1020–1024.

Sang, N. N., Soda, S., Ishigaki, T., & Ike, M. (2012). Microorganisms in landfill bioreactors for accelerated stabilization of solid wastes. *Journal of Bioscience and Bioengineering, 114*, 243–250.

Sasaki, S., Araki, T., Tambunan, A. H., & Prasadja, H. (2014). Household income, living, and working conditions of dumpsite waste pickers in Bantar Gebang: Toward integrated waste management in Indonesia. *Resources, Conservation, and Recycling, 89*, 11–21.

Saltelli, A. (2002). Sensitivity analysis for importance assessment. *Risk Analysis, 22*(3), 579–590.

Scarlat, N., Motola, V., Dallemand, J. F., Ferrario, F. M., & Mofor, L. (2015). Evaluation of energy potential of municipal solid waste from African urban areas. *Renewable and Sustainable Energy Reviews, 50*, 1269–1286.

Sefouhi, L., Kalla, M., & Bahmed, L. (2014). Assessment of different methods of treatment for an integrated municipal waste management for an Algerian city. *Management of Environmental Quality, 25*, 493–504.

Shaikh, A. A., & Karjaluoto, H. (2015). Mobile banking adoption: A literature review. *Telematics, and Informatics, 32*, 129–142.

Sreejesh, S. R., Anusree, M., & Mitra, A. (2016). Effect of information content and form on customers' attitude and transaction intention in mobile banking: Moderating role of perceived privacy concern. *International Journal of Bank Marketing, 34*, 1092–1113.

Suhr, J. K., Eum, S., Jung, H. G., Li, G., Kim, G., & Kim, J. (2012). Recognizability assessment of facial images for automated teller machine applications. *Pattern Recognition, 45*, 1899–1914.

Sun, J., Dang, Z., & Zheng, S. (2017). Development of payment standards for ecosystem services in the largest inter basin water transfer projects in the world. *Agricultural Water Management, 182*, 158–164.

Suryotrisongko, H., Sugiharsono, Setiawan, B. (2012). A novel mobile payment scheme based on secure quick response payment with minimal infrastructure for cooperative enterprise in developing countries. *Procedia—Social, and Behavioral Sciences, 65,* 906–912.

Sxoinaraki, E., & Panou, K. (2017). Innovative business models for exploiting green vehicle potential in urban logistics. *Transportation Research Procedia, 25,* 954–964.

Taurina, Z., Dachyar, M., & Farizal, F. (2021). Factors influencing residents' intention toward source separation of plastic noodle wrapper waste: A case study of Jabodetabek, Indonesia. *Proceeding 4th Asia Pacific Conference on Research in Industrial, and Systems Engineering* (UPCORISE) 2021, Depok, pp. 412–418, ISBN: 978-1-4503-9038-5.

Tchobanoglous, G., Theisen, H., & Vigil, S. A. (1993). *Integrated Solid Waste Management.* McGraw-Hill International, New York, NY.

Teece, D. J. (2010). Business models, business strategy, and innovation. *Long Range Planning, 43,* 172–194.

Thampan, A., & Chandel, M. K. (2015). Bioreactor landfill technology. *International Journal of Science and Research, 4*(6), 256–260.

Todeschini, B. V., Cortimiglia, M. N., Menezes, D. C., & Ghezzi, A. (2017). Innovative, and sustainable business models in the fashion industry: Entrepreneurial drivers, opportunities, and challenges. *Business Horizons, 60,* 759–770.

US EPA. (2017). Bioreactor landfills. https://www.epa.gov/landfills/bioreactor-landfills. Accessed on April 27, 2017.

Veríssimo, J. M. C. (2016). Enablers, and restrictors of mobile banking app use: A fuzzy set qualitative comparative analysis (fsQCA). *Journal of Business Research, 6,* 5456–5460.

Wan, W. W. N., Luk, C.-L., & Chow, C. W. C. (2005). Customers' adoption of banking channels in Hong Kong. *International Journal of Bank Marketing, 23*(3), 255–272.

Wang, Z., & Liwei, G. (2015). Carbon emissions calculation from municipal solid waste, and the influencing factors analysis in China. *Journal of Cleaner Production, 104,* 177–184.

Wells, P. (2013). Sustainable business models, and the automotive industry: A commentary. *IIMB Management Review, 25,* 228–239.

Yang, Y., Liu, Y., Li, H., & Yu, B. (2015). Understanding perceived risks in mobile payment acceptance. *Industrial Management & Data System, 115,* 253–269.

Yildiz, O. (2014). Financing renewable energy infrastructures via financial citizen participation-the case of Germany. *Renewable Energy, 68,* 677–685.

Yip, A. W. H., & Bocken, N. M. P. (2018). Sustainable business model archetype for the banking industry. *Journal of Cleaner Production, 174,* 150–169.

Zhang, C., Campana, P. E., Yang, J., & Yan, J. (2016). Analysis of distributed photovoltaic financing: A case study approach of crowdfunding with photovoltaic water pumping system in microgrids. *Energy Procedia, 103,* 387–393. https://doi.org/10.1016/j.egypro.2016.11.304

Zhang, C., Campana, P. E., Liu, C., Wang, K., Zhang, Y., & Yan, J. (2019). Purchase intention for crowdfunded milk products with integrated photovoltaic water pumping systems in China. *Energy Procedia, 159,* 503–508. https://doi.org/10.1016/j.egypro.2018.12.008

Zhou, T. (2013). Understanding continuance usage of mobile sites Industrial. *Management & Data Systems, 113,* 1286–1299.

Municipal Solid Waste Management: Waste to Energy Technologies Versus Composting—Two Sustainable Development Models

Wadii Abid, Fatma Masmoudi, and Emna Ammar

Abstract In the world, the fast growing urbanization and industrialization occasioned increases in municipal solid waste (MSW) generation and energy consumption. According to the World Bank projection, global MSW generation will increase to 2.2 billion tons per year by 2025 and the MSW management has become a major concern which aims to transform the linear economy model into a circular one, minimizing consequently environmental impacts and hence public health risks. Depending on the countries socio-economic framework, to manage MSW, several procedures have been established and practiced; these include landfilling, thermal, chemical and biological treatments. However, open dumping and landfill remain the most applied methods for MSW disposal in spite of their detrimental environmental impacts on air pollution, water and soil contaminations, and climate changes induction. At present, in developing countries, energy demand and consumption is in continuous increase. Generally, all over the world, the high-energy demands are covered through fossil fuels resources i.e. natural gas and oil. Moreover, Waste-to-Energy (WTE) technologies, specifically biodrying, incineration, pyrolysis, and gasification could be considered as sustainable processes leading to products that could be fully recovered, recycled and reused, and will be an alternative energy sources, contributing to solve the energy crisis.

In developed countries, the application of the WTE project would moderate the informal landfilling as well as greenhouse gas (GHG) releases, promoting renewable energy production. The MSW management through WTE technologies may have impacts on both of the environment and human life, categorized in direct and indirect methodologies, taking into account the circular economy perspectives.

W. Abid · F. Masmoudi · E. Ammar (✉)
Laboratory of Environmental Sciences and Sustainable Development, University of Sfax, Preparatory Institute of Engineering Studies of Sfax, Sfax, Tunisia
e-mail: ammarenis@yahoo.fr

E. Ammar
National Engineering School of Sfax, Sfax, Tunisia

A. Anouzla and S. Souabi (eds.), *Generation of Energy from Municipal Solid Waste*,
https://doi.org/10.1007/978-3-031-74334-4_5

The direct approach involves MSW incineration for heat recovery, while the indirect approach, including thermochemical and biochemical processes, is more complicated but attractive because of the valorized products variety—such as biochar, digestate, humus, syngas and bio-oil. The development of these technologies can lead to suitable eco-friendly approaches, and WTE technologies are important components of circular economy that will assist to achieve the demand of clean energy in the future.

The present chapter focuses on the municipal solid waste management strategies and waste-to-energy conversion technologies as a model for circular economy and sustainable MSW option development. After sorting the MSW, their disposal and the technologies used to valorize this waste will be developed, presenting first the common management technologies practiced, with composting agronomic benefits, then emphasizing on waste-to-energy technology, its advantages and its challenges' as a sustainable and clean bioresources issue, enhancing environmental sustainability, facing energetic crisis. A comparison between composting process and WTE will be considered.

1 Introduction

The global population increase, expected to reach 9.7 billion in 2050, will raise some concerns considering the global biomass consumption, fossil fuels depletion, metals and minerals needs, with an annual waste production increase that will rise by 70% in the next 40 years, following these prospects (Chiocchio et al., 2021; United Nations, 2023). Indeed, it was reported that in developed countries, food waste produced by households is estimated to 42%, while losses occurring in the food industries are of 39%, food service sector generates 14% of waste and food remaining in retail and distribution represents 5% (Mirabella et al., 2014).

Certainly, the human needs and issues may incite to move toward a sustainable development. This approach is gaining worldwide attention, since at present wastes and by-products, especially those from the food and agricultural industries, are associated to pollution with a dilemma of hunger and malnutrition in some parts of the world, and food over-consumption with its related diseases in other parts, both of the phenomena leading to food waste production increase (FAO, 2019).

Furthermore, the Food and Agriculture Organization of the United Nations (FAO, 2019) revealed that more than 30% of the world's agricultural lands are used to produce food that is eventually lost or wasted, responsible of additional pollution and greenhouse gases emissions. These have a negative impact on the economy, with financial loss at the supply chain' different levels (Mirabella et al., 2014). Actually, the food loss reduction associated to waste performing management may potentially generate considerable economic value.

Actually, the human population increase with global industrial evolution have been responsible for the determined rise of the waste generated since decades. In 2010, the global solid waste discharge was 10.4 billion tons that will reach about

148 billion tons by 2025. Municipal solid waste (MSW) is an important component of solid waste generated in all over the countries (Souza Lima Cano et al., 2022). Consequently, on the global scale, the MSW generated amount is in an unceasing increase. The World Bank announced that in 2018, the global annual MSW was of 2.01 billion tons while it was of 1.3 billion tons (0.74 kg/capita/day) in 2012. By 2030, the expected increase of the global annual MSW will reach 2.59 billion tons, and in 2050, it will be of 3.40 billion tons/year (1.42 kg/capita/day) (Bello et al. 2022; Farooq et al., 2021; Maniruzzaman et al., 2022; Roy et al., 2022). This expected increase will be of 50% from the current level (Tiseo, 2022), with almost the half (53%) of this growth will come from developing countries (Fernondo & Zutshi, 2023).

Furthermore, the MSW generated by capita may vary according to regions and countries, and in the same country, it depends on urban or rural location (Stunžėnas & Kliopova, 2021). The European Union alone discards daily 88 million tons of edible food, and its daily MSW is estimated to be ranged from 0.3–1.4 kg/capita/day while in some cities in Africa, the registered average is 0.78 kg/capita/day. In Asia, the continent that represents 53% of the world's population, and houses urban areas, the MSW daily generated is about 760,000 tons, this production is assumed to increase to 1.8 million tons/day in 2025 (Maniruzzaman et al., 2022; Stunžėnas & Kliopova, 2021).

The developing countries contribute about 56% of the world's total solid waste, which is composed by essentially organic matter then paper, and plastic (Alam & Qiao, 2019; Das et al., 2019; Sharma et al., 2020). Therefore, waste production tends to increase with the population rise as well as with the economic growth. This generates waste management problems affecting not only environment, but also public health (Bello et al. 2022; Kumar & Agrawal, 2020; Sharma et al., 2020). However, statistics expected that MSW would reach 6.1 million tons/day by 2025 (The Sharma et al., 2020; Traven, 2023; World Bank, 2019). This speedy growing in waste progress is also accompanied by an increase in energy demand due to rapid economic growth, urbanization, living standards and industrialization (Kaur et al. 2023).

These organic wastes are a major contributor to greenhouse gas emission, and present remarkable health risks like gastro-intestinal disorders, dermatitis, malaria, asthma and bronchitis for the nearby residents of waste dumping yards (Al-Ghouti et al., 2021; Das et al., 2019; Deus et al., 2020; Kumar & Agrawal, 2020; Sharma et al., 2020). In addition to health risks, wastes dumping poses environmental problem through the groundwater contamination with the generated leachates, rich in heavy metals, fluoride, chloride, ammonia and moreover large content of chloride (Ferronato & Torretta, 2019; Kumar & Agrawal, 2020; Sharma et al., 2020). It should be mentioned that 80% to 90% of the municipal waste is disposed-off in landfills without any proper management practice (Chen, 2018; Kumar & Agrawal, 2020; Raharjo et al., 2017). Moreover, solid wastes require a long period for being degraded, which calls for an urgent need to find out other performing solid wastes management solutions (Al-Ghouti et al., 2021; Ashani et al., 2020; Wang et al., 2017a). However, MSW global average cost is 205.4 USD/ton that will reach a five folds rise by 2025 and 2030. In fact, the production of solid waste is involvedly concomitant with air, soil, and water pollution (Das et al., 2019).

In this regard, waste management appears to be a particularly interesting theme insofar as it is precisely the "non-appropriation" and rejection that underpin the status of waste. It may be underlined that as mentioned by Zouaoui (2021), waste does not only refer to zero or negative economic value, but rather to a complete, complex set or a system of socio-cultural values. Authentically, an appropriate solid waste management including the 3R activities consisting in reduction reuse and recycling, will contribute to new jobs creation and environmental burdens reduction that are the main challenge for developing countries (Alam & Qiao, 2019; Raharjo et al. 2017; Suthar et al., 2016). Therefore, MSW incineration agrees its energy potential and that solid waste can be converted into useful products (Al-Ghouti et al., 2021; Ashani et al., 2020; Dou et al., 2017; Wang et al., 2017b).

At present, in developed countries advanced technologies such as thermal treatment, landfill bioreactors, composting and anaerobic co-digestion are practiced to overcome environmental and various other concerns. These technologies need skilled and trained personnel and substantial funds (Al-Ghouti et al., 2021; Dou et al., 2017). However, 46 billion USD is annually spent by developed countries to manage the solid waste, which most likely will double if sustainability is not incorporated (Das et al., 2019; The World Bank, 2019).

Waste to Energy (WTE) project would moderate the informal landfill as well as greenhouse gas (GHG) releases, promoting renewable energy production. The MSW management through WTE technologies may have impacts on the environment and human life, categorized in direct and indirect methodologies taking into account the circular economy perspectives.

In this chapter, we aim to discuss the MSW composition and the importance of its management technologies with a specific focus on the composting and waste-to-energy technologies as sustainable development models promoting agriculture and energy production to satisfy food security and energetic demand by renewable energy sources use respectively, because of the poverty of soils in organic matter, and the fossil fuel depletion as well as environmental issues. A comparison between MSW composting and its WTE process will be investigated.

2 Municipal Solid Waste Classification

Since the 1950s, many developed countries have implemented policies for waste sorting. Indeed, solid waste management may reduce or eliminate the waste hostile impact on human health and environment. Consequently, municipalities adopt some performing processes to manage effectively waste; these are monitoring, collection, transport, processing, recycling and disposal. However, people lifestyle' and seasons are among the main factors affecting the waste amount generated. For example, a recent study held in India showed that the chemical characterization of MSW revealed that when based on seasons, moisture content varies between 43.4% and 58.3%, pH

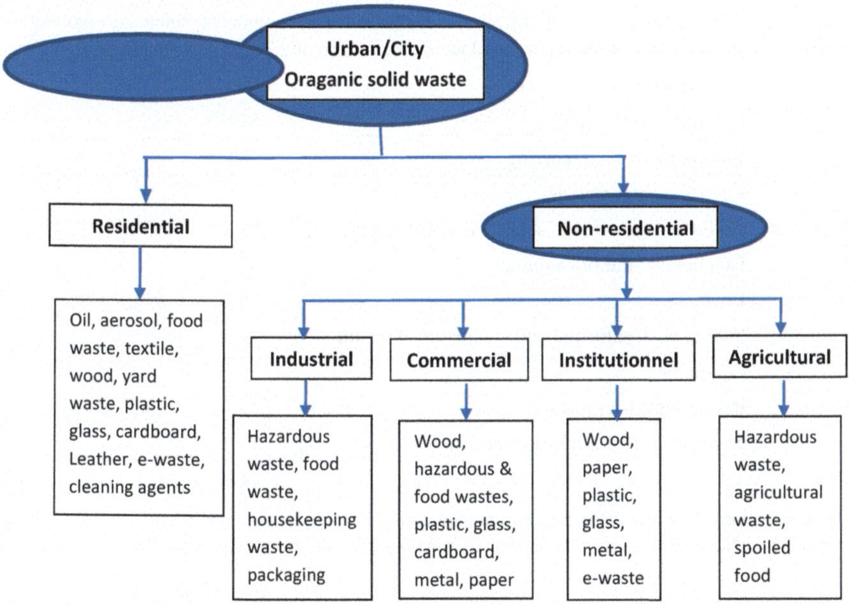

Fig. 1 Sources and types of organic solid wastes generated (Bello et al., 2022; Kumar & Agrawell, 2020; Soni et al., 2022)

between 5.5 and 6.5, volatile solid content from 32.9 to 58.9%, and the calorific value between 1203 and 3015 kcal/kg.

In reality, each entity generating waste shall separate its waste and sort it in different parts; these are biodegradable, non-biodegradable and hazardous household waste, they are then stored differently, and finally passed to the approved rag pickers or collectors (Singhal et al., 2022). Earlier investigation of Kumar and Agrawal (2020) classified solid organic wastes into residential and non-residential types (Fig. 1).

More recently, a waste categorization investigation, based on Web of Science, was carried out and distinguished 106 categories issued from consulting 1925 published papers (Zhao & Zhang, 2023). Over the 10 top, the first fourth categories (1 to 4) deal with environmental problems analysis and suggested solutions, while the fourth category included a variety of disciplines where technological studies reducing negative human activities impact on the environment were involved to promote sustainability (Table 1).

Since 2017, the number of annually published papers on waste sorting and recycling behavior status and trends has increased. The majority of these publications are emerging from Asia, then Europe followed by North America, because of their high population densities requiring waste sorting and recycling. Actually, many analyses of waste sorting behavior have been conducted by psychologists (Zhao & Zhang, 2023).

Table 1 The 10 top categories of publications dealing with environmental sciences, engineering as well as green and sustainable science and technology Zaho and Zhang (2023)

N°	Category (Web of Science)	Publications number*
1	Environmental sciences	512
2	Environmental engineering	284
3	Environmental studies	213
4	Green and sustainable science and technology	188
5	Psychology, multidisciplinary	81
6	Business	55
7	Public, environmental and occupational health	39
8	Economics	38
9	Psychology, sociology	33
10	Materials science, multidisciplinary	28
Total		1471

* Total countries: 77 in the different continents with publication percentages repartition: Asia 40%, Europe 37%, North America 16%, Austria 3%, Africa 2%, and South America 2%

3 Composition and Characteristics of Municipal Solid Waste

The MSW refers to the waste generated through daily consumers' activities, that are collected from houses, schools, offices, some industries, and may include packaging waste, food waste, paper/cardboard, green waste, batteries, clothes; these are largely produced in households and non-renewable items like plastic containers. It also includes waste generated in public places such as parks and streets, stations, schools, and small businesses that are similar to households (Souza Lima Cano et al., 2022; Kaur et al., 2023). The MSW is also regarded as unwanted and discarded materials issued from society and normal daily activities (Kaur et al. 2023). It is assumed principally as the solid debris, refuse, or produced waste where it was generated (the boundary or the district, or the municipality), without according interest to the production location (Bello et al., 2022).

Consequently, various MSW classifications were suggested, their hierarchy according to their origin (urban/rural), industrial activity issue, and rural sources is well used. Its first class is subdivided into residential and non-residential sources for solid waste from human domiciles, while the rural waste division includes agricultural waste and animal husbandry activities (Fig. 2). All of the classifications represent different groups, specifically based on the raw material used whatever is the activity (Bello et al., 2022; Zhao & Zhang, 2023).

These kinds of waste could be arranged into different categories, such as food wastes (Stunžėnas & Kliopova, 2021), paper and cardboards, metals, plastics and glass, differing from one country to another one (Table 2). The MSW composition varies all around the world, depending on country economic status, geographical

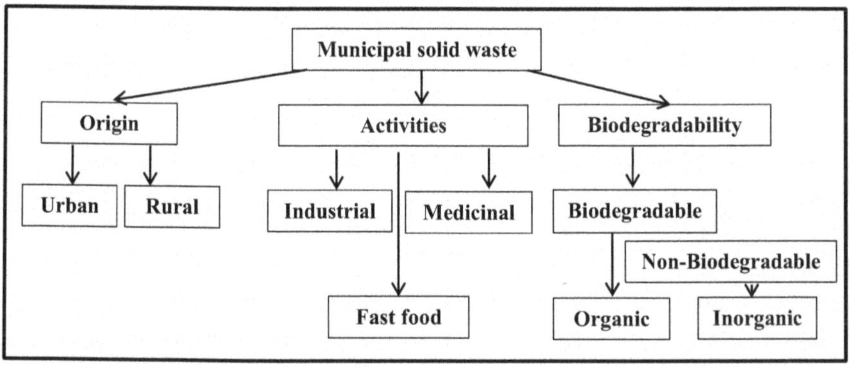

Fig. 2 Municipal solid waste classification

locations, consumption habits, literacy rate, oldness distribution, and variation in seasonality as well as climatic conditions (Das et al., 2019; Farooq et al., 2021).

While the organic fraction can be composted, the others waste categories such as plastics, metal and clothes can be recycled and reused to produce new products, contributing substantially to a strategy of a neutral carbon, providing secondary materials cost-effective and of high quality (Das et al., 2019; Kaur et al., 2023).

Generally, the MSW is made up of unsegregated degradable, partially degradable, or completely non-degradable materials (Soni et al., 2022). To recover energy, the non-recyclable combustible materials could be used as they can substitute the fossil fuels use, contributing to decreasing the global warming (Omari et al., 2014). It could be noticed that developing countries waste a lot of food, while the USA use a considerable amount of paper/cardboards and metal, Poland and Algeria are high consumers of plastics (Table 2).

The MSW treatment and processing representing an available resource with an interesting potential in terms of material and energy recovery, could be characterized by its chemical ultimate stream (Table 3), although the analyzes results are approximate and my differ according to countries because of the waste heterogeneity and variability.

It could be pointed out that food and yard are relatively rich in minerals, while paper and yard have high contents in carbon and oxygen but low nitrogen rate.

Considering physicochemical composition, few studies dealt with the MSW composition, this is due to the difficulty of getting a representative sample during all the year. Indeed, sampling is among the important factors that should be considered prior to waste characterization, and formulation of waste generation map. Actually, solid waste characterization is a crucial feature to develop and design an efficient management system. A German study investigated the MSW composition in rural and urban areas, and published the results found after a year-sampling exploration (Sailer et al., 2021). Some MSW physicochemical properties are presented in (Table 4) based on available data. These wastes are generated in some medium to large cities of a few developing countries. Generally, two main issues are considered namely the

Table 2 Municipal solid wastes composition in different countries of the world

Country	Organic (%)		Inorganic (%)				Reference, Date
	Food wastes	Papers/ Cardboards	Plastics	Metal	Glass	Others	
China	47.7	9.2	10.3	1.2	2.9	25	A, 2021
USA	21.59	23.05	12.20	8.76	4.19	18	B, 2018
Russia	34	19	14	4	12	17	C, 2019
Kazakhstan	46.3	12.8	15.2	–	–	5.9	D, 2020
India	50	5	4	5		29	E, 2018
Nepal	43.6	22.7	13.8	2.7	6.4	6.6	F, 2020
Thailand	64	8	17.6	2	–	–	G, 2019
Malaysia	45	7	24	6	18		H, 2017
Bangladesh	67.7	9.7	5.1	0.3	1.2	8.6	I, 2019
Indonesia	60	9	14	4.3	1.7	2.4	J, 2021
Poland	38	15	18	1	26	2	K, 2020
Germany	52.50	17.64	13.07	–	5.44	13.72	L, 2018
Greece	44.3	22.2	13.9	3.9	4.3	11.4	M, 2019
Turkey	52.3	8.48	10.74	1.37	4.29	18.85	N, 2021
Egypt	56	10	13	2	4	15	O, 2022
Tunisia	63.2	8.6	9.4	1.6	1.1	7.4	P, 2020
Algeria	54	8	18	2	2	5.2	O, 2022
Morocco	65	10	10	4	3	8	Q, 2020
Tanzania	67	11	7	1	4	10	R, 2021

A: Zhu et al. (2021); B: EPA (2018); C: Fedotkina et al. (2019); D: Abylkhani et al., 2020); E: Ahluwalia and Patel (2018); F: Khatoon (2020); G: Funatsou, 2019; H: Zainu and Songip (2017); I: Alam and Qiao (2019); J: Tetra Tech (2021); K: Przydatek (2020); L: Federal Ministry for the Environment, Nature Conservation and Nuclear Safety (2018); M: Sikalidis and Emmanouil, (2019); N: Demirarslan and Yener (2021); O: Hemidat et al. (2022); P: Chaabane (2020); Q: Hasib et al. (2020); R: Richard et al. (2021)

Table 3 Municipal solid waste streams ultimate analysis expressed in mass percentage (Arafat et al., 2015)

Waste category	C	H	O	N	Ash	Chemical formula
Food	44.99	6.43	28.76	3.3	16.0	$C_{3.7}H_{6.4}O_{1.8}N_{0.2}$
Paper	43.41	5.28	44.32	0.25	6.0	$C_{3.6}H_{5.3}O_{2.8}N_{0.02}$
Plastics	60.0	7.2	22.8	0	10.0	$C_{5.0}H_{7.2}O_{1.4}$
Wood	49.4	6.1	43.7	0.1	0.6	$C_{4.1}H_{6.1}O_{2.7}N_{0.007}$
Yard	40.31	5.64	39.0	2.0	13.0	$C_{3.4}H_{5.6}O_{2.4}N_{0.14}$
Textiles	55.0	6.6	31.2	4.6	2.4	$C_{4.6}H_{6.6}O_{1.9}N_{0.33}$

C: carbon; H: hydrogen; O: oxygen; N: nitrogen

Table 4 MSW Physicochemical properties generated in some areas in the word

City (Country)	Tübingen (Germany)	Kilchberg Weilheim, Kreßbach, Bühl (rural, Germany)	Osogbo (Nigeria)	Dehli (India)	Dhaka (Bengladesh)	Kuala Lumpur (Malaysia)
Reference (Date)	Sailer et al. (2021)		Abedoye et al. (2022)	Das et al. (2019)	Alam et Qiao (2019)	Das et al. (2019)
pH	ND	ND	ND	8.20	8.69	9.08
C/N	ND	ND	10.01/1.56	ND	10.2/1	ND
Moist. (%)	36.31	32.11	48.74	40.1	ND	55.0
Ash	ND	ND	6.86	5.49	29.4	9.26
VS (%)	ND	ND	ND	28.2	71	31.9
OC (%)	38.55	46.82	50.09	8.35	ND	ND
N_{Tot} (%)	1.82	2.46	1.56	0.94	0.89	1.26
H (%)	4.99	6.33	ND	1.16	ND	ND
P (%)	ND	ND	ND	0.62	0.31	ND
S (%)	ND	ND	1.23	0.01	ND	0.23
Cd (ppm)	ND	ND	ND	0.22	0.66	8.84
Hg (ppm)	ND	ND	ND	ND	0.19	0.99
Pb (ppm)	8.109	7.212	ND	0.04	41.05	26.3
Cr (ppm)	ND	ND	ND	ND	34.49	14.4

ND: Not determined; Moist.: moisture; VS: Volatile solids; OC: Organic Carbon; N_{Tot}: Total nitrogen

toxic effects and reactivity of pollutants like heavy metals, and organic substances such as hydrocarbons, polycyclic hydrocarbons and chlorinated substances. These organic compounds could be evidenced by relied and sophisticated techniques i.e. SEM, XRD and FTIR.

4 MSW Disposal and Management Technologies

In the context of circular economy, to promote environmental and human health protection and to boost the economic productivity of a region via resource recovery, MSW sustainable management is indispensable and constitutes a prerequisite to achieve the sustainable development of a country (Souza Lima Cano et al., 2022). However, MSW composition and qualities are important factors that determine the complexity level of waste management in addition to economic, social, cultural, legal, political, and environmental variables. Moreover, available resources and geography location are significant features that should be considered while managing waste

(Roy et al., 2022; Soni et al., 2022). When considering the waste composition, the developed countries generate rather dry waste made up of paper, cardboards, plastic, metal and glass, which are relatively easy to recycle, while in developing economies countries more than the half of total SMW produced, is organic waste and presents difficulties in its management (Farooq et al., 2021). Indeed, in developing countries, adequate MSW remediation is dependent and influenced by several factors like population density, socio-economic aspects, cultural facets, average living standards, unplanned management and lack of stringent environmental policies (Nanda & Berruti, 2021). Consequently, the MSW management practiced currently used in most developing countries is rather limited, ineffective or insufficient; these facts are impeding sustainable development (Fermando & Zutshi, 2023).

According to Nanda and Berruti (2021), MSW management is composed of three stages: The first is the waste generation, the second concerns the waste collection, handling and transfer and the third includes waste disposal, processing and treatment. The waste management hierarchy is presented in (Fig. 3).

In developed countries, the municipal solid waste conversion to fuels, heat and power as waste-to-energy technologies are well established, but municipalities are more concentrated on waste collection than the treatment (Moya et al., 2017; Nanda

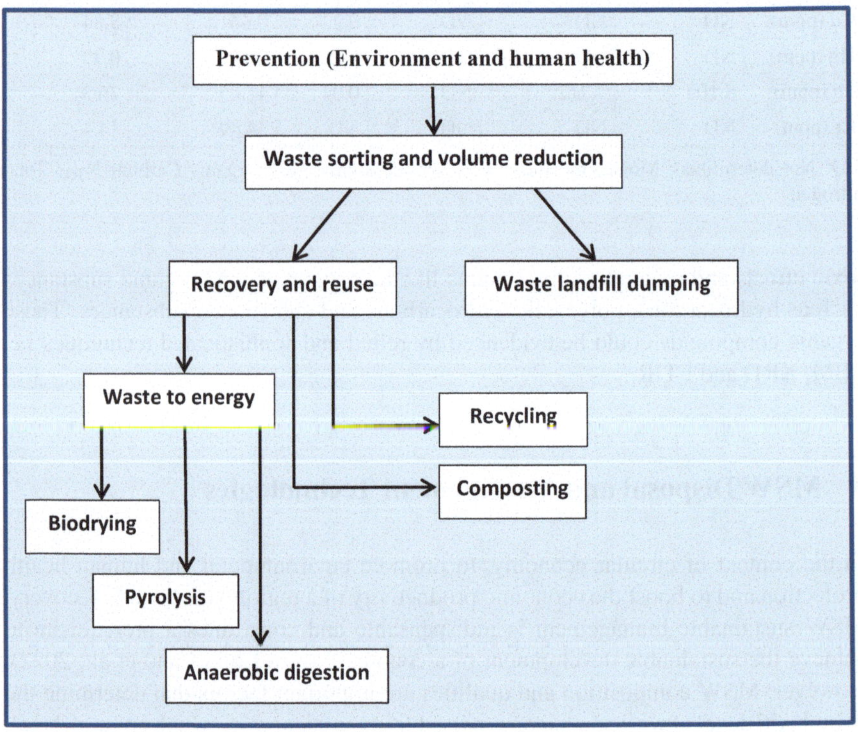

Fig. 3 Municipal solid waste management technologies hierarchy

and Berruti 2020, 2021; Soni et al., 2022). In the context of sustainability and environment protection, several others waste management techniques were innovated and modernized such as microbial conversion techniques i.e. composting and anaerobic digestion (AD). These processes produce renewable sources of fertile compost and biogas; while thermochemical processes like gasification and pyrolysis give syngas and bio-oil (Cucchiella et al., 2017; Lu et al., 2020; Nanda & Berruti, 2021; Roy et al., 2022).

4.1 Landfilling

Landfilling is the most lucrative waste disposal procedure practiced in many countries, in which biodegradable and non-biodegradable waste is buried in a designated terrestrial burial site for further treatment and utilization (Nanda & Berruti, 2021; Roy et al., 2022). In regard to the MSW incineration and recycling technologies, landfilling presents many advantages like its cost-effectiveness and the green energy production by the investigation on its landfill gas and leachate, promoting a passage from "waste storehouses" to "energy powerhouses" (Bello et al., 2022; Nanda & Berruti, 2021). On the other hand, MSW landfill site leads to enormous negative consequences and pollution in all the ecosystem notably the soil, water ground and surfaces, exposing human health to severe diseases (Bello et al., 2022). These consequences are related to the direct MSW dumping, without any previous treatment in most countries of the world. This leads to an acidic liquid "leachate" release, composed by non-desirable substances, with high concentrations of organic and inorganic chemicals, organic matter (both dissolved and suspended) as well as heavy metals and the emission of gas into the atmosphere resulting from microbial organic matter degradation (Bello et al., 2022). Furthermore, over the global methane emission, 17% is generated by landfills and open dumps, causing health hazards, especially respiratory diseases and smog, besides fires and explosion risk, contributing to global warming (Abubakar et al., 2022; Kumar et al., 2017). Moreover, problems concomitant with odor and leachates migration to receiving waters especially during the summer, are persistent, as well as discarded tires at dumps, which collect water-allowing mosquitoes to breed, increasing the risk of infectious diseases (Kumar et al., 2017).

In developing countries, due to high population density, MSW landfilling has the limitations of open space, contrarily to most developed countries where it is a common organized procedure (Nanda & Berruti, 2021). Landfills can be classified into open dump landfills, semi-controlled landfills and sanitary landfills, or according to its functionality such as secure landfill, reusable landfill (stabilization of waste for long time followed by excavation to recover metals, glass, plastics and compost), and bioreactor landfill that is the best by its engineering sanitary. Indeed, the latest technique allows the leachate collection and the microbial activity optimization for waste biodegradation without affecting environmental and human health (Jayasinghe et al., 2014; Nanda & Berruti, 2021).

4.2 *Composting*

Composting is another reliable biological method of waste disposal, particularly for domestic waste, rich in organic matter that is broken down and decomposed into simpler forms by saprophytic microorganisms, such as bacteria and fungi under controlled factors (principally moisture aeration, and C/N) (Lin et al., 2023). Composting consists of bioconverting various biodegradable wastes into safely products that are principally beneficial for agriculture field: Bio-fertilizers and soil amendment (Mbachu et al., 2022). It is the result of mixed microbial communities activities' notably bacteria, actinomycetes, molds and yeasts, and globally occurs during three phases: Mesophilic, thermophilic then the last is mesophilic, ending by insuring a mature and stable product, rich in humus including macro elements i.e. nitrogen, phosphorus, potassium and some trace elements (Magdich et al., 2022). During this process, temperature, moisture and aeration should be adequate for successful composting operation. This technology contributes to the sustainable agriculture, and presents an opportunity to promote the material recover from MSW organic component to be reused in the form of a biofertilizer enhancing soil fertility, in addition to its low cost operation and social impact with new green jobs creation (Abid et al., 2018, 2020; Abubakar et al., 2022; Bello et al., 2022). The composting process could be defined according to the following simplified reaction:

$C_6H_{10}O_5 + 6\ O_2\ 6\ CO_2 + 5\ H_2O + \text{Heat}$

The MSW organic fraction composed of food, agricultural, paper and yard debris wastes, ranges from 46% to more than 68% in some developed countries like Tunisia and Bangledesch (Table 2). However, compost monitoring and specific standards should be considered to lead to a safe and qualified product that could be used in the soil as a biofertilizer (Bello et al., 2022; Cao et al., 2023; Fouguira et al., 2023). In developing countries, the MSW organic fraction has high moisture content because of the presence of an important food waste proportion. Across the world, 1.3 billion tons of food waste is yearly generated, affecting the environment by the carbon footprint of which CO_2-equivalent estimation is 3.3 billion tons (FAO, 2020; Nanda & Berruti, 2021).

Considering its agronomic diversified uses as a fertilizer or as a material for plant nursery substrates production, the compost production and its quality performance has gained researchers interest (Fouguira et al., 2023; Jodar et al., 2017; Mbachu et al., 2022). Indeed, in compost, the presence of plant growth-promoting organisms and nutrients, improves agricultural yields and soil organic matter content, helping in ensuring food security (Magdich et al., 2022). Besides organic fertilizer, compost is applicable in bioremediation, plant disease control, weed control, erosion control, landscaping and wetland restoration (Abid et al., 2020; Mbachu et al., 2022). Moreover, the compost use as organic fertilizer decreases the cost, and reduces chemical fertilizer' health and environmental risk.

The implementation of MSW compost sorting and adequate quality control, would offer preferment applications of this soil amendment (Roy et al., 2022). Since MSW compost includes important amounts of organic matter and both organic and inorganic nitrogen, its maturity degree and the amended soil characteristics are important factors contributing to defining the required amount to be provided for plant nitrogen and phosphorus nutritional intakes while using the composted waste. Actually, MSW composts provides the soil with equivalent phosphorus amount as mineral fertilizers. In saline soils, MSW compost, with high organic matter content and low inorganic and organic pollutants concentrations, allows an improvement of physical, chemical, and biochemical characteristics and constitutes low-cost soil recovery (Lakhdar et al., 2009).

Despite all the merits that composting technology presents as biological treatment, reducing waste and generating new organic agricultural products, some disadvantageous may be attributed to composing, such as nitrogen release through ammonia gas (NH_3) by volatilization emission of greenhouse gases (GHGs), methane (CH_4), and di nitrogen oxide (N_2O), (Abubakar et al., 2022; Bello et al., 2022). In addition, MSW presents several environmental risks as soil amendment, like the nutrients/pollutants/organic-soluble compounds excess that could affect waters as well as plants, and the persistence of non-biodegradable and undesirable matters. In fact, MSW compost long-term application may result in accumulation of toxic metals in amended soil (Almendro-Candel et al., 2019).

5 Waste to Energy Technologies

5.1 Incineration

At present, to supplement the increasing energy demands, municipal solid waste conversion into solid, liquid and gaseous fuels using different waste-to-energy technologies including thermochemical and biological methods are available solutions. Recently, with the continuously increase in MSW generation, incineration practiced as a waste management technology has tremendously amplified, in spite of its high energy consumption (Bello et al., 2022; Ouda et al., 2016). The recent projects dealing with WTE offer a substitute to fossil fuels. Indeed in the word, 765 WTE plants are in operation based on WSM; these use annually 83 million tons over the 2.01 billion tons of the MSW produced (Roy et al., 2022). Last statistics exhibited that worldwide, more than 1.100 WTE plants are currently in operation, with a daily combined treatment capacity exceeding 700.000 tons of MSW (Traven, 2023).

The incineration process advantages are that it utilizes almost all types of waste fractions, reduces the MSW volume by up to 80–95% and mass up to 70–85%. In addition, an incinerator can manage at least 50,000 metric tons of MSW annually. However, this technology presents some disadvantages like the carcinogenic chemicals production such as dioxins and air pollution generation, in addition to its relatively high cost (Farooq et al., 2021; Mukherjee et al., 2020; Nanda & Berruti, 2021; Roy et al., 2022). Incineration technology can generate a substantial amount of power by transforming around 65–80% of the energy stored in organic materials into heat. Nevertheless, 1179 incineration plants are able to generate a total volume of over 700,000 Mg/day (Mg stands for Megagram which equals a metric ton) (Bello et al., 2022; Ouda et al., 2016).

In Europe, the most prevalent waste management practice is landfilling with 58% in 2014 (Jodar et al., 2017). Recently, attention has been moved from landfilling (−10.31%) to incineration and composting, more adapted methods to waste management (+11.78% and + 22.43%, respectively) than landfilling (Mbachu et al., 2022).

Incineration plants that use thermal power and electricity plants cogeneration have more efficiency reaching up to 80%, than those generating electricity (only 20%). The incineration process efficiency was increased when solid waste pre-processing of through advanced separation, biological and thermal waste treatment, and other methods (roasting, shredding and size reduction) adapted. Moreover, the produced fly ash can be used as raw material in road maintenance, cement production and construction (Roy et al., 2022).

5.2 *Biomethanization*

Biomethanization is the microbial anaerobic organic materials digestion process, using agricultural, vegetable market and restaurant waste, producing energy and a digestate, an organic fertilizer by specific methanogenic microbial flora (Qin et al., 2017). This process is considered as the oldest one used to stabilize sludge from wastewater treatment plant, producing renewable energy (Xu et al., 2020).

In a first stage, the organic materials like carbohydrates, proteins and fats are bioconverted into soluble organic materials such as sugars, amino and fatty acids by primary fermenting bacteria (hydrolytic-acidogenic) producing acetic and propionic acids. Then, acetogenesis will occur allowing the organic molecules breakdown into acetic acid, H_2 and CO_2 by anaerobic oxidizing bacteria (syntrophic-acidogenic). In the final stage methanogenesis occurs, producing methane (CH_4) from acetic acid and H_2 by methanogenic *Archaea* (Fig. 4) (Farooq et al., 2021; Ouda et al., 2016; Roy et al., 2022). This process is dependent on pH and temperature, defining the favorable conditions for microbial activity (Farooq et al., 2021; Qazi et al., 2018). The biogas, which is the biomethanization main product, contains 50–80% methane (CH_4), 20–50% carbon dioxide (CO_2), and traces of sulfide (<1% H_2S) and ammonia. Two products are cogenerated with the biogas; these are fiber and liquid digestate. The latter can be used as compost/biofertilizer, allowing nutrients recycling (nitrogen, phosphorus) (Farooq et al., 2021; Roy et al., 2022).

When compared to thermochemical technologies, biomethanization presents the advantages that it reduces leachate production, decreasing the risk of land and water contamination, with less amount of GHG emitted in the environment. In addition, its efficiency is in treating and managing organic waste, allowing the use of nutrient-rich digestate as organic fertilizer (Farooq et al., 2021; Roy et al., 2022).

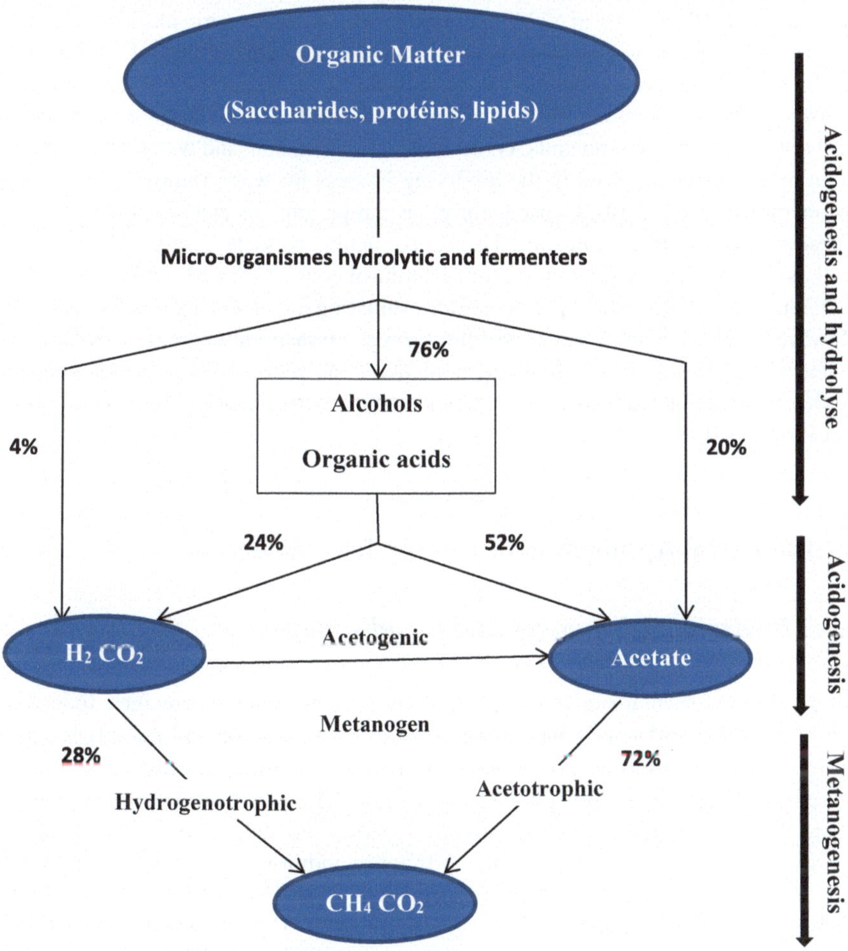

Fig. 4 Anaerobic digestion principle (Moletta, 2002)

5.3 *Biodrying Process*

The WTE technology is able to reduce 90% of MSW' volume. The recovered energy may depends on some waste physical properties such as density, composition and moisture percentage. As a result, biodrying or a biological drying, is a smart alternative for high moisture content waste treatment (Kechaou & Ammar, 2018; Tom et al., 2016; Zaman et al., 2018). Indeed, biodrying is an automatic and natural heating technology process, in which the drying process is reinforced by biological heat released at the on-site organic matter decomposition. This is used to evaporate water at the lowest possible residence time. This fact aims to improve the solid recovered fuel (SRF) quality by reducing the MSW moisture, preserving most of the waste gross calorific value (Pecorini et al., 2020; Sutthasil et al., 2022; Zaman et al., 2018). In fact, it is a process similar to composting, with an investigation in the heat released during the biodegradable waste decomposition, in order to stabilize and reduce the waste moisture content, aloowing the biodried material reutilization. Therefore, metabolic heat release, air convection and oxygen molecular diffusion, and water vapor are the main mechanisms involved in the biodrying process for water removal from waste (Mohammed et al., 2018). Consequently, several parameters can affect the process: These are mainly airflow rate, initial moisture content, waste typology and density, as well as permeability, and retention time (Sutthasil et al., 2022). Moreover, the waste moisture content is a critical parameter for evaluating the biodrying process efficacy. However, high moisture content will limit oxygen transportation and hinder microbial activity. In contrast, low moisture content affects microbial organic components biodegradation and leads to a decrease in biodrying performance (Mohammed et al., 2018; Sutthasil et al., 2022).

6 Industrial Application of Energy from Waste

6.1 Biodrying Advantages and Disadvantages

Biodrying as a waste management technology presents many advantages. Indeed, it can be useful in order to reduce the high solid waste volume and mass reduction, designated for landfilling at relatively short residence time, generating a product more suitable for short-term storage and transport. This waste pre-treatment reduces waste moisture content, which results in its partial bio stabilization, increasing its energy content by generating a high solid fuel in addition to greenhouse emissions reduction (Mohammed et al., 2018; Pecorini et al., 2020; Roy et al., 2022). A recent study reported that around the word, 350 integrated WTE projects reduced the GHG emission equivalent to 51.3 million metric tons of CO_2 located in 52 countries (Roy et al., 2022).

The main biodrying advantages could be resumed in five points:

1. Reduced volume and weight: Biodrying can reduce the volume and weight of organic waste by up to 80%. This volume reduction makes it more cost-effective to transport and dispose of the waste, as fewer trucks or containers are needed.
2. Improved waste stability: The removal of moisture during biodrying helps to stabilize the organic waste, reducing the risk of odors, pathogens, and pests. This makes the waste easier to handle and safer for workers and the environment. In parallel, this moisture content reduction increases the energy content of the waste material.
3. Energy production: Biodrying can generate heat, which could be utilized for various purposes. The generated heat during the process can be used to warm up buildings or to produce electricity through combustion or anaerobic digestion. This can help to offset energy costs and reduce reliance on fossil fuels.

4. Reduced greenhouse gas emissions: By reducing the waste volume and stabilizing it, biodrying can help in minimizing methane emissions, a potent greenhouse gas released during the organic waste decomposition in landfills. By diverting organic waste from landfills, biodrying can contribute to reducing greenhouse gas emissions and combating climate change. Additionally, the high temperatures reached during the process can help inactivating pathogens, reducing the risk of diseases transmission.
5. Versatility: Biodrying can be applied to a wide range of organic waste streams, including MSW, industrial waste, and agricultural residues. This adaptability allows effective management of various types of organic waste and offers flexibility in waste management strategies.

Biodrying presents also some disadvantages that can be resumed in:

1. High capital and operational costs: The installation and operation of biodrying facilities are expensive. The process requires specialized equipment and infrastructure, which can increase the initial investment and ongoing operational costs.
2. Limited moisture content control: Achieving the desired moisture content in the biodried waste can be challenging. The moisture content depends on various factors, such as the initial waste moisture content and the ambient conditions. Ensuring consistent and optimal moisture content may require additional monitoring and adjustments during the process.
3. Potential for odor emissions: While biodrying can help to reduce odors in organic waste, there is still a potential for odor emissions during the process. This can occur if the waste is not properly managed or if there are issues with the biodrying system, such as inadequate ventilation or containment.
4. Nutrient loss: Biodrying can lead to the loss of nutrients, such as nitrogen and phosphorus, from the organic waste. These nutrients are valuable for soil fertility and agricultural applications. If the biodried waste is intended for use as a soil amendment or fertilizer, additional nutrient supplementation may be required.
5. Environmental considerations: Biodrying facilities need to comply with environmental regulations ensuring proper air and water quality management. This includes controlling emissions, such as particulate matter and volatile organic compounds, and implementing proper waste handling and disposal practices for any by-products or residues generated during the process. Additionally, the energy consumption and emissions associated with the operation of biodrying facilities should be carefully considered and minimized to ensure overall environmental sustainability. Proper emission control systems must be in place to mitigate these impacts and meet regulatory standards.
6. Limited waste stream applicability: Biodrying may not be suitable for all types of organic waste. Certain waste streams, such as those with high levels of contaminants or hazardous materials, may not be compatible with biodrying processes. This limitation restricts the range of waste materials that can be effectively treated using this technology.

7. Residual waste management: Biodrying produces a dry, stable material as a byproduct. While this material can be used as a fuel source or feedstock, there may still be some residual waste that requires proper disposal. The management of this residual waste can add complexity and cost to the overall waste management process.
8. Maintenance and operation challenges: Biodrying facilities require regular maintenance and monitoring to ensure optimal performance. Issues such as equipment breakdowns, temperature fluctuations, or improper ventilation can impact the efficiency and effectiveness of the biodrying process.

6.2 Biodrying Product Application in Cement Industry

Biodrying, or the process of drying organic waste through high temperatures and low humidity conditions, produces a dry and stable material known as solid recovered fuel (SRF). The SRF has a range of potential applications, including:

- Energy generation: SRF can be used as a fuel source in various energy conversion processes, such as combustion in power plants or industrial boilers. Its high calorific value makes it an effective alternative to fossil fuels, contributing to renewable energy generation.
- Cement kilns: SRF can be utilized as a substitute for traditional fossil fuels, such as coal or petroleum coke, in cement kilns. This application helps reduce the carbon footprint of cement production and promotes sustainable manufacturing practices.
- Co-firing: SRF can be co-fired with coal or other fuels in existing power plants or industrial boilers. This approach allows for a gradual transition to renewable energy sources by utilizing SRF alongside traditional fuels, reducing greenhouse gas emissions and dependence on fossil fuels.
- Waste-to-energy facilities: SRF can be directly used as a feedstock in waste-to-energy facilities, where it is combusted to generate electricity or heat. This application allows for the efficient and sustainable utilization of organic waste, diverting it from landfills and reducing environmental impacts.
- District heating: SRF can be utilized in district heating systems, where it is burned to produce heat for residential or commercial buildings. This application provides a renewable and cost-effective solution for heating purposes, reducing reliance on fossil fuel-based heating systems.
- Industrial processes: SRF can be used in various industrial processes that require heat, such as drying, melting, or steam generation. By replacing fossil fuels with SRF, industries can reduce their carbon emissions and contribute to a more sustainable production process.

It is important to note that the SRF specific applications can vary depending on local regulations, infrastructure, and market demand. Additionally, the quality and characteristics of the SRF produced through biodrying will also influence its suitability for different applications.

7 Prospective and Challenges of Waste-To-Energy

7.1 SWOT Analysis

The SWOT analysis can vary depending on the specific waste to energy technology being considered, and the local context where it is implemented. However, WTE technologies presents some strengths which are:

1. Renewable energy generation: Waste-to-energy technologies utilize renewable energy sources such as organic waste, to generate electricity or heat, reducing dependence on fossil fuels.
2. Waste reduction and diversion: WTE technologies help divert waste from landfills, reducing the environmental and health impacts associated with waste disposal.
3. Energy independence: WTE technologies provides an opportunity for local communities to generate their own energy and reduce reliance on external energy sources.
4. Reduction of greenhouse gas emissions: By converting organic waste into energy, WTE technologies help in reducing methane emissions from landfills, a potent greenhouse gas.
5. Resource recovery: WTE technologies can recover valuable resources from waste streams, such as metals, plastics, and organic fertilizers, contributing to a circular economy.

Weaknesses:

1. Capital-intensive: Implementing WTE technologies requires significant capital investment for infrastructure, equipment, and operational costs, which can be a barrier for some communities or organizations.
2. Limited waste stream applicability: Not all types of waste can be effectively converted into energy, and certain waste streams may require additional pre-treatment or sorting to be suitable for WTE technologies.
3. Emission concerns: Some WTE technologies, particularly incineration, may generate emissions, including greenhouse gases and air pollutants, if not properly controlled and managed.
4. Public perception and acceptance: WTE technologies, particularly incineration, may face public opposition due to concerns about emissions, potential health impacts, and the perception of waste incineration as an outdated or unsustainable solution.

Opportunities:

1. Policy support: Governments and regulatory bodies can provide incentives, subsidies, or favorable policies to promote the development and implementation of WTE technologies, encouraging their adoption.

2. Technological advancements: Ongoing research and development can lead to improvements in waste to energy technologies, making them more efficient, cost-effective, and environmentally friendly.
3. Integration with other renewable energy sources: Waste to energy technologies can be integrated with other renewable energy sources, such as solar or wind, to create hybrid systems that enhance energy generation and grid stability.
4. Waste management infrastructure development: As waste management practices evolve, waste to energy technologies can be integrated into comprehensive waste management systems to optimize resource recovery and minimize environmental impacts.
5. Job opportunities however waste management need employee for the collection and the recycling which contribute to the development of society.

Threats:

1. Economic viability: The economic feasibility of waste to energy technologies can be influenced by factors such as fluctuating energy prices, waste disposal costs, and competition from other renewable energy sources.
2. Regulatory challenges: Compliance with stringent environmental regulations and emission standards can pose challenges for waste to energy projects, potentially increasing costs and operational complexities.
3. Public perception and opposition: Community opposition or negative public perception can hinder the development and acceptance of waste to energy projects, leading to delays or cancellations.
4. Availability and quality of waste feedstock: The availability and quality of suitable waste feedstock can impact the efficiency and performance of waste to energy technologies. Inconsistent waste streams or high levels of contaminants can pose challenges.

7.2 Comparison Between Waste-To-Energy and Composting

Biodrying and composting are both methods of organic waste treatment, but they differ in terms of processes and end-products (Table 5).

Biodrying is a process that involves the removal of moisture from organic waste, typically through the use of mechanical or biological means. The waste is subjected to high temperatures and low humidity conditions, which accelerate the drying process. Biodrying is often used for high-moisture waste streams, such as sewage sludge or food waste. The end-product of biodrying is a dry partially stabilized material that can be used as a fuel source or as a feedstock for other processes.

Composting, on the other hand, is a natural process that involves the decomposition of organic waste materials, such as food scraps, yard waste, or agricultural residues. Composting relies on the activity of microorganisms, such as bacteria and

Table 5 SWOT analysis of waste to energy management system (Khan & Kabir, 2020; Ouda et al., 2016; Tulebayeva et al., 2020)

Strengths	Weaknesses
-Regular waste collection -Waste purchase centers establishment -Creation of recycling centers -Strong information and training program to promote waste management in the division -Waste tax system -Financial Availability -Power generation -Alleviation of greenhouse gas emissions from landfills -Eenergy and natural resources savings	-Poor dumping management -Inefficient sorting of food waste Shortcomings in the composting process -Lack of processing capability for lunch production -Failures in sludge fertilizer production -Process efficiency -Scale of commercialization -Availability of feedstock -Infrastructure requirements -End-use applications
Opportunities	**Threats**
-Biogas plant installation in a SWM center -Receiving external support from government and industry associations -Five waste separation bins are presented -Application of new technologies for MSW processing -New businesses and subsequent job creation -Alleviation of landfill costs -Electricity generation	-Insufficient attention to promote environmental research -Invalidity of compost standardization -Filtrate poisoning -Economic viability -Regulatory challenges -Availability and quality of waste feedstock

fungi, to break down the organic matter. This process requires a balance of carbon-rich (e.g., leaves, straw) and nitrogen-rich (e.g., food waste, manure) materials, as well as proper moisture and oxygen levels. The end-product of composting is a nutrient-rich soil amendment called compost, which can be used to improve soil fertility and structure.

In summary, biodrying focuses on removing moisture from organic waste using a high ventilation rate to produce a dry, partially stabilized material, with high calorific value, while composting involves the organic waste decomposition producing nutrient-rich stabilized compost (Mohammed et al., 2018).

Also biodrying process is of short duration and consequently, the emission factors occurs rapidly. This bioprocess is based on the auto thermal heat generated by microbial activity. As a result, it is an energy saving process compared to drying because it effectively consumes the biological energy (Tom et al., 2016); the requisite airflow amount is usually higher than that of the composting process. Therefore, forced ventilation is usually used for biodrying owing to its powerful aeration capacity, and ventilation mode is critical for biodrying. Also, it does not aim towards a complete waste mineralization.

8 Conclusion

Municipal solid waste generation is an inevitable result of human activities that can lead to serious consequences for the environment and the quality of life and health of people, if not adequately addressed. The current global generation of MSW is 2.01 billion tons/year, which reach 3.40 billion tons/year by 2050. This chapter is an overview of MSW composition and presenting the common management technologies practiced, then emphasizing on waste-to-energy technology, its advantages and its challenges' as a sustainable and clean bioresources with a comparison between WTE and composting process.

Furthermore, MSW composition is dependent and varies according to the geographic and economic status of a country. However, in developing countries, MSW is rich in organic matter. Waste dumping is the most popular disposal technique worldwide due to financial and logistical raisons. Incineration is a technology that can be adequate for reducing the non-biodegradable waste. In developed countries suffering from low MSW management budget, composting process seems to be an adequate solution reducing the continued increase in organic waste and allowing a new source of fertilizers contributing to develop a sustainable agriculture and indirectly economic benefits. During the last decades, energy recovery from MSW has attended a great attention in several countries due to the continued increase in energy demand, and that waste-to-energy constitutes a renewable energy source. Biodrying as a WTE technology allows the waste volume and moisture content reduction to produce a solid fuel in relatively short time but its installation and operation cost is high. Anaerobic digestion was the most adopted and feasible WTE technology which generate biogas and a digestate that can be used for preparation of organic fertilizer. Regarding to SWOT analysis, WTE technologies for MSW management are a promising alternatives and all the threats and weakness can be solved by a simple cooperation between governments and policymakers who are in charge with the selection of the best way to manage their waste.

References

Abid, W., Mahmoud, I., Masmoudi, S., Triki, M. A., Mounier, S., & Ammar, E. (2020). Physicochemical and spectroscopic quality assessment of compost issued from date palm waste valorization. *Journal of Environmental Management, Elsevier, 264*, 110492.

Abid, W., Magdich, S., Ben Mahmoud, I., Medhioub, K., & Ammar, E. (2018). Date palm wastes co-composted product: An efficient substrate for tomato (*Solanum lycopercicum* L.) seedling production. *Waste Biomass Valorisation, 9*, 45–55.

Abubakar, I. R., Maniruzzaman, K. M., Dano, U. L., AlShihri, F. S., AlShammari, M. S., Ahmed, S. M. S., Al-Gehlani, W. A. G., & Alrawaf, T. I. (2022). Environmental sustainability impacts of solid waste management practices in the global south. *International Journal of Environmental Research Public Health, 19*, 12717.

Abdulrahman, A. (2021). Solid waste management in Tunisia. Recycling, Waste Management.

Abylkhani, B., Guney, M., Aiymbetov, B., Yagofarova, A., Sarbassov, Y., Zorpas, A. A., Venetis, C., & Inglezakis, V. (2020). Detailed municipal solid waste composition analysis for Nur-Sultan City, Kazakhstan with implications for sustainable waste management in Central Asia. *Environmental Science and Pollution Research, 28*, 24406–24418.

Adeboye, B. S., Idris, M. O., Adedeji, W. O., Adefajo, A. A., Oyewusi, T. F., & Adelekun, A. (2022a). Characterization and energy potential of municipal solid waste in Osogbo Metropolis. *Cleaner Waste Systems, 21*, 100020.

Adeboye, B. S., Idris, M. O., Adedeji, W. O., Adefajo, A. A., Oyewusi, T. F., & Adelekun, A. (2022). Characterization and energy potential of municipal solid waste in Osogbo metropolis. *Cleaner Waste Systems 2*(1), 100020.

Ahluwalia, I. J., & Patel, U. (2018). *Solid waste management in india. An assessment of resource recovery and environmental impact*. (Working Paper No. 356). Indian council for research on international economic relation.

Al-Ghouti, M. A., Khan, M., Nasser, M. S., Al-Saad, K., & Heng, O. E. (2021). Recent advances and applications of municipal solid wastes bottom and fly ashes: Insights into sustainable management and conservation of resources. *Environmental Technology & Innovation, 21*, 101267.

Alam, O., & Qiao, X. (2019). An in-depth review on municipal solid waste management, treatment and disposal in Bangladesh. *Sustainable Cities and Society, 52*(1), 101775.

Almendro-Candel, M. B., Navarro-Pedreño, J., Gómez Lucas, I., Zorpas, A. A., Voukkali, I., & Loizia, P. (2019). The use of composted municipal solid waste under the concept of circular economy and as a source of plant nutrients and pollutants. *Municipal Solid Waste Management*. https://doi.org/10.5772/intechopen.83386

Agence Nationale de Gestion des Déchets: ANGED (2018). Evaluation nationale des indicateurs H2020/PAN Tunisie. http://www.anged.nat.tn/.

Arafat, H. A., Jijakli, K., & Ahsan, A. (2015). Environmental performance and energy recovery potential of five processes for municipal solid waste treatment. *Journal of Cleaner Production, 105*, 233–240.

Ashani, P., Shafiei, M., & Karimi, K. (2020). Biobutanol production from municipal solid waste: Technical and economic analysis. *Bioresource Technology, 308*, 123267.

Bello, A. S., Al-Ghouti, M. A., & Abu-Dieyeh, M. H. (2022). Sustainable and long-term management of municipal solid waste: A review. *Bioresource Technology Reports, 18*, 101067.

Cao, X., Williams, P. N., Zhan, Y., Coughlin, S. A., McGrath, J. W., Chin, J. P., & Xu, Y. (2023). Municipal solid waste compost: Global trends and biogeochemical cycling. *Soil and Environmental Health, 1*, 100038.

Chaabane, W. (2020). *Solid waste management in tourism destinations in Tunisia: diagnostic and improvement approaches*. Faculty of Agriculture and Environmental Sciences, 185.

Chiocchio, I., Mandrone, M., Tomasi, P., Marincich, L., & Poli, F. (2021). Plant secondary metabolites: An opportunity for circular economy. *Molecules, 26*(2), 495.

Cucchiella, F., D'Adamo, I., & Gastaldi, M. (2017). Sustainable waste management: Waste to energy plant as an alternative to landfill. *Energy Conversion and Management, 131*, 18–31.

Chen, Y. C. (2018). Effects of urbanization on municipal solid waste composition. *Waste Management, 79*, 828–836. https://doi.org/10.1016/j.wasman.2018.04.017

Das, S., Lee, S. H., Kumar, P., Kim, K. H., Lee, S. S., & Bhattacharya, S. S. (2019). Solid waste management: Scope and the challenge of sustainability. *Journal of Cleaner Production, 228*, 658–678.

Demirarslan, K. O., & Yener, I. (2021). Municipal solid waste in Turkey and its relationship with the population. *Research in Ecology, 3*(01), 31–40.

Deus, R. M., Mele, F. D., Bezerra, B. S., & Battistelle, R. A. G. (2020). A municipal solid waste indicator for environmental impact: Assessment and identification of best management practices. *Journal of Cleaner Production, 242*, 118433.

Dou, X., Ren, F., Nguyen, M., Ahamed, A., Yin, K., Chan, W., & Chang, V. (2017). Review of MSWI bottom ash utilization from perspectives of collective characterization, treatment and existing application. *Renewable and Sustainable Energy Reviews, 79*, 24–38.

EPA United States Environmental Protection Agency. (2018). National overview: Facts and figures on materials, wastes and recycling.

FAO (Food and Agriculture Organization of the United Nations). (2019). The State of Food and Agriculture 2019; FAO.

FAO, Food and Agriculture Organization of the United Nations. (2020). Food wastage: Kkey facts and fgures. http://www.fao.org/news/story/en/item/196402/icode/. Accessed on April 21, 2020.

Farooq, A., Haputta, P., Silalertruksa, T., & Gheewala, S. H. (2021). A framework for the selection of suitable waste to energy technologies for a sustainable municipal solid waste management system. *Frontiers in Sustainability, 2*, 681–690.

Federal Ministry for the Environment, Nature Conservation and Nuclear Safety. (2018). Waste management in Germany 2018 Facts, data, diagrams.

Fedotkina, O., Ghorbashko, E., & Vetolkina, N. (2019). Circular economy in Russia: Drivers and barriers for waste management development. *Sustainability, 11*(20), 5837.

Fernondo, S. J., & Zutshi, A. (2023). Municipal solid waste management in developing economies: A way forward. *Cleaner Waste Systems, 5*, 100103.

Ferronato, N., & Torretta, V. (2019). Waste mismanagement in developing countries: A review of global issues. *International Journal of Environmental Research and Public Health, 16*(6), 1060.

Fouguira, F., El Haji, M., Benhra, J., & Ammar, E. (2023). Optimization of olive oil extraction wastes co composting procedure based on bioprocessing parameters. *Heliyon, 9*(9), e19645.

Funatsu, T. (2019). Chapter 4: Municipal solid waste management in Thai local governments: The state of the problem and prospects for regional waste management. Michikazu Kojima, "Toward Regional Cooperation of Local Governments in ASEAN", ERIA Collaborative/Support Research Report, IDE-JETRO.

Hasib, A., Ouigmane, A., Boudouch, O., Kacmi, R., Bouzaid, M., & Berkani, M. (2020). *Sustainable solid waste management in Morocco: Co-Incineration of RDF as an alternative fuel in cement kilns*. Intechopen, https://doi.org/10.5772/intechopen.93936

Hemidat, S., Achouri, O., El Fels, L., Elagroudy, S., Hafidi, M., Chaouki, B., Ahmed, M., Hodgkinson, I., & Guo, J. (2022). Solid waste management in the context of a circular economy in the MENA region. *Sustainability, 14*(1), 480.

Jayasinghe, P. A., Hettiaratchi, J. P. A., Mehrotra, A. K., & Kumar, S. (2014). Reaction mechanisms and rate constants of waste degradation in landfill bioreactor systems with enzymatic-enhancement. *Bioresource Technology, 162*, 279–282.

Jodar, J. R., Ramos, N., Carreira, J. A., Pacheco, R., & Fernandez-Hernandez, A. (2017). Quality assessment of compost prepared with municipal solid waste. *Open Engineering, 7*(1), 221–227.

Kaur, A., Bharti, R., & Sharma, R. (2023). Municipal solid waste as a source of energy. *Materials Today: Proceedings, 81*, 904–915.

Kechaou, N., & Ammar, E. (2018). Biodrying process: A sustainable technology for treatment of municipal solid wastes organic fraction. In 21st International Drying Symposium, Valencia, Spain.

Khan, I., & Kabir, Z. (2020). Waste-to-energy generation technologies and the developing economies: A multi-criteria analysis for sustainability assessment. *Renewable Energy, 150*, 320–333.

Khatoon, A. (2020). Waste Management—A case study in Nepal. In S. Ghosh (Eds.), *Solid Waste Policies and Strategies: Issues, Challenges and Case Studies* (pp. 185–196). Springer. https://doi.org/10.1007/978-981-15-1543-9_17

Kumar, A., & Agrawal, A. (2020). Recent trends in solid waste management status, challenges, and potential for the future Indian cities—A review. *Current Research in Environmental Sustainability, 2*, 100011.

Kumar, S., Smith, S. R., Fowler, G., Velis, C., Kumar, S. J., Arya, A., Kumar, R., & Cheeseman, C. (2017). Challenges and opportunities associated with waste management in India. *Royal Society Open Science, 4*, 160764.

Lakhdar, A., Rabhi, M., & GHnaya, T., Montemurro, F., Jedidi, N., Abdelly, C. (2009). Effectiveness of compost use in salt-affected soil. *Journal of Hazardous Materials, 17*, 29–37.

Lin, P. T., Zhao, Y., Jiang, Y., Ou, J., Hussain, A., Chen, L., & Lin, C. (2023). Succession of microbial communities with VOC changes during composting process. *Bioresource Technology Reports, 22*, 101390.

Lu, J. S., Chang, Y., Poon, C. S., & Lee, D. J. (2020). Slow pyrolysis of municipal solid waste (MSW): A review. *Bioresource Technology, 312*, 123615.

Maniruzzaman, K. M., Dano, U. L., AlShihri, F. S., AlShammari, M. S., Ahmed, S. M. S., Al-Gehlani, W. A. G., & Alrawaf, T. I. (2022). Environmental sustainability impacts of solid waste management practices in the global South. *International Journal of Environmental Research Public Health, 19*, 12717.

Magdich, S., Ben Rouina, B., Ammar, E. (2022). Combined management of olive mill wastewater and compost in olive grove. Effects on soil chemical properties at differents layers depth. Ecological Engineering, 184, 106769, 13 p.

Mbachu, A. E., Okoli, A. N., Mbachu, N. A., Okoli, F. A., & Ozoh, C. N. (2022). A Review of municipal solid waste composting: Benefits and prospects. *Himalayan Journal of Agriculture, 3*(1), 29–36.

Mirabella, N., Castellani, V., & Sala, S. (2014). Current options for the valorization of food manufacturing waste: A review. *Journal of Cleaner Production, 65*, 28–41.

Mohammed, M., Donkor, A., Ozbay, I. (2018). Biodrying of biodegradable waste for use as solid fuel: a sustainable approach for green waste management. In *Agricultural waste and residues* (pp. 99–104). https://doi.org/10.5772/intechopen.77957.

Moletta, R. (2002). *La biométhanisation* (2ème ed.). Lavoisier.

Moya, D., Aldás, C., Jaramillo, D., Játiva, E., & Kaparaju, P. (2017). Waste to-energy technologies: An opportunity of energy recovery from municipal solid waste, using quito-ecuador as case study. *Energy Proceeding, 134*, 327–336.

Mukherjee, C., Denney, J., Mbonimpa, E. G., Slagley, J., & Bhowmik, R. (2020). A review on municipal solid waste-to-energy trends in the USA. *Renewable and Sustainable Energy Reviews, 119*, 109512.

Nanda, S., Berruti, F., (2020). Municipal solid waste management and landflling technologies: A review. *Environmental Chemistry Letters, 19*, 1433–1456.

Nanda, S., & Berruti, F. (2021). Thermochemical conversion of plastic waste to fuels: A review. *Environmental Chemistry Letters, 19*, 123–148.

Omari, A. M., Kichonge, B. N., John, G. R., Njau, K. N., & Mtui, P. L. (2014). Potential of municipal solid waste, as renewable energy source—A case study of Arusha, Tanzania. *International Journal of Renewable Energy Technology Research, 3*(6), 1–9.

Ouda, O. K. M., Raza, S. A., Nizami, A. S., Rehan, M., Al-Waked, R., & Korres, N. E. (2016). Waste to energy potential: A case study of Saudi Arabia. *Renewable and Sustainable Energy Reviews, 61*, 328–340.

Pecorini, I., Bacchi, D., & Iannelli, R. (2020). Biodrying of the light fraction from anaerobic digestion pretreatment in order to increase the total recovery rate. *Processes, 8*, 276.

Przydatek, G. (2020). Assessment of changes in the municipal waste accumulation in Poland. *Environmental Science and Pollution Research, 27*, 25766–25773.

Qazi, W. A., Abushammala, M. F. M., & Azam, M. H. (2018). Multi criteria decision analysis of waste-to-energy technologies for municipal solid waste management in sultanate of Oman. *Waste Management Research, 36*, 594–605.

Qin, Y., Higashimori, A., Wu, L.-J., Hojo, T., Kubota, K., & Li, Y.-Y. (2017). Phase separation and microbial distribution in the hyperthermophilic-mesophilic-type temperature-phased anaerobic digestion (TPAD) of waste activated sludge (WAS). *Bioresource Technology, 245*, 401–410.

Raharjo, S., Matsumoto, T., Ihsan, T., Rachman, I., & Gustin, L. (2017). Community-based solid waste bank program for municipal solid waste management improvement in Indonesia: A case study of Padang city. *J Mater Cycles Waste Management, 19*, 201–212.

Richard, E. N., Hilonga, A., Machunda, R. L., & Njau, K. N. (2021). Life cycle analysis of potential municipal solid wastes management scenarios in Tanzania: The case of Arusha City. *Sustainable Environment Research, 31*, 1.

Roy, H., Alam, S. R., Bin-Masud, R., Prantika, T. R., Pervez, M. N., Islam, M. S., & Naddeo, V. A. (2022). Review on characteristics, techniques, and waste-to-energy aspects of municipal solid waste management: Bangladesh perspective. *Sustainability, 14*, 10265.

Sailer, G., Eichermüller, J., Poetsch, J., Paczkowski, S., Pelz, S., Oechsner, H., & Müller, J. (2021). Dataset for a full-year time series characterization of separately collected organic fraction of municipal solid waste from rural and urban regions in Germany. *Data in Brief, 39*, 107543.

Sharma, Er. P., Verma, Er. N., Kumar, Er. M. (2020). A review on solid waste management: methods and practices. *International Journal of Advance Science and Technology, 29*, 5431–5436.

Sikalidis, A., & Emmanouil, C. (2019). Description and Economic Evaluation of a "Zero-waste mortar-producing process" for municipal solid waste management in Greece. *Journal of Open Innovation: Technology, Market, and Complexity, 5*, 46.

Singhal, A., Kumar Gupta, A., Dubey, B., & Ghangrekar, M. M. (2022). Seasonal characterization of municipal solid waste for selecting feasible waste treatment technology for Guwahati city, India. *Journal of the Air & Waste Management Association, 72*(2), 147–160.

Soni, A., Das, P. K., & Kumar, P. (2022). A review on the municipal solid waste management status, challenges and potential for the future Indian cities. *Environment, Development and Sustainability, 4*(3), 160764.

Souza Lima Cano, N. S., Iacovidou, E., Rutkowski, E. W. (2022). Typology of municipal solid waste recycling value chains: A global perspective. *Journal of Cleaner Production, 336*, 130386.

Stunžėnas, E., & Kliopova, I. (2021). Industrial ecology for optimal food waste management in a region. *Journal of Environmental Research, Engineering and Management, 77*(1), 7–24.

Suthar, S., Rayal, P., & Ahada, C. P. S. (2016). Role of different stakeholders in trading of reusable/recyclable urban solid waste materials: A case study. *Economics Sustainable Cities and Society, 22*, 104–115.

Sutthasil, N., Ishigaki, T., Ochiai, S., Yamada, M., & Chiemchaisri, C. (2022). Carbon conversion during biodrying of municipal solid waste generated under tropical Asian conditions. *Biomass Conversion and Biorefinery*, 15. https://doi.org/10.1007/s13399-021-02284-1

Tetra Tech. (2021). Clean Cities, Blue Ocean Initial Solid Waste Management Assessment (Indonesia).

Tiseo, I. (2022). Global waste generation—Statistics & facts. Statista. Retrieved October 5, 2022, from https://www.statista.com/topics/4983/waste-generation-worldwide/.

Tom, A. P., Pawels, R., & Haridas, A. (2016). Biodrying process: A sustainable technology for treatment of municipal solid waste with high moisture content. *Waste Management, 49*, 64–72.

Traven, L. (2023). The potential for a combined heat and power (CHP) generation from municipal solid waste: A case study of Croatia. *Case Studies in Chemical and Environmental Engineering, 8*, 100450.

Tulebayeva, N., Yergobek, D., Pestunova, G., Mottaeva, A., & Sapakova, Z. (2020). Green economy: Waste management and recycling methods. E3S Web of Conferences 159, 01012

United Nations. (2023). Department of Economic and Social affairs—Population division: World Population Prospects 2022, https://population.un.org/wpp/DataSources/ .Site visited on September 4, 2023.

Wang, N., Chen, D., Arena, U., & He, P. (2017a). Hot char-catalytic reforming of volatiles from MSW pyrolysis. *Applied Energy, 191*, 111–124.

Wang, D., Zhou, X., Meng, Y., & Chen, Z. (2017b). Durability of concrete containing fly ash and silica fume against combined freezing-thawing and sulfate attack. *Construction and Building Materials, 147*, 398–406.

World Bank. (2019). Solid waste management. The World Bank Group, https://www.worldbank.org/en/topic/urbandevelopment/brief/solid-wastemanagement

Xu, X. -J., Wang, W. -Q., Chen, C., Xie, P., Liu, W. -Z., Zhou, X., … Ren, N. -Q. (2020). Bioelectrochemical system for the enhancement of methane production by anaerobic digestion of alkaline pretreated sludge. *Bioresource Technology, 304*, 123000.

Zainu, Z. A., & Songip, A. R. (2017). Policies, challenges and strategies for municipal waste management in Malaysia. *Jostip, 3*(1).

Zaman, B., Oktiawan, W., Hadiwidodo, M., Sutrisno, E., & Purwono. (2018). Bio-drying technology of solid waste to reduce greenhouse gas. *E3S Web of Conferences, 73*, 05019.

Zhao, Y., & Zhang, J. (2023). Literature mapping of waste sorting and recycling behavior research: A visual analysis using Cite Space. *Environmental Science and Pollution Research, 30*, 67953–67973.

Zhu, Y., Zhang, Y., Luo, D., Chong, Z., Li, E., & Kong, X. (2021). A review of municipal solid waste in China: Characteristics, compositions, influential factors and treatment technologies. *Environment, Development and Sustainability, 23*, 6603–6622.

Zouaoui, L. (2021). Gestion des déchets, innovation et territoires Retours d'expériences et recherche contextuelle. *CITET, DSI n 854 du 8au* novembre 18, 2021.

Sustainable Energy Generation from Municipal Wastewater Using High-Rate Activated Sludge Process

Elham Abdollahzadeh Sharghi and Ghazale Faridizad

Abstract In recent years, growing concerns have arisen within communities, driven by factors such as population expansion, dwindling oil reserves, and the substantial production of municipal wastewater. The increasing scarcity and limitations associated with fossil fuels emphasize the critical significance of renewable energy sources. Notably, municipal wastewater has emerged as a noteworthy energy resource due to its continuous generation and global-scale treatment. Traditionally, the conventional activated sludge process has been employed for over a century in municipal wastewater treatment. While this method offers advantages such as high-quality effluent and reliability, it falls short of being a sustainable wastewater treatment solution due to issues such as the need for aeration, organic matter oxidation in wastewater, and the generation of waste products. To efficiently recover the energy present in wastewater, it is necessary to modify the process and increase the concentration of organic matter within it. Various physical, chemical, and biological processes are deployed to enhance the up-concentration of organic substances in wastewater and capture them in the sludge. These improvements contribute to enhanced energy recovery through anaerobic digestion. Given the drawbacks associated with physical and chemical methods, attention has turned towards biological processes. This paper aims to provide a comprehensive evaluation of the energy potential inherent in municipal wastewater and the techniques employed to increase organic matter concentration. Furthermore, it will assess the limitations of physical and chemical approaches and delve into the biological process of high-rate activated sludge, examining its mechanisms, operational parameters, and recent advancements in this field.

E. A. Sharghi (✉)
Department of Energy, Materials and Energy Research Center, Karaj, Iran
e-mail: E.abdollahzadeh@merc.ir

G. Faridizad
University of Tehran, Tehran, Iran

A. Anouzla and S. Souabi (eds.), *Generation of Energy from Municipal Solid Waste*,
https://doi.org/10.1007/978-3-031-74334-4_6

1 Introduction

Water, nourishment, and energy constitute three critical global concerns (Daelman et al., 2015). Achieving these essential objectives, alongside the challenges posed by population growth and economic development, represents a significant worldwide endeavor that numerous nations confront today (Bhaduri et al., 2015). These challenges are further exacerbated by concerns regarding their future accessibility and long-term viability. To guide society towards a sustainable path, the European Union has formulated a bio-economic plan designed to produce bio-renewable resources and transform these resources and waste streams into value-added products, including food, livestock feed, bio-based commodities, and bioenergy. Among the resources available in this context, municipal wastewater emerges as a prominent candidate (Alloul et al., 2018). The rapid proliferation of populations and urbanization has led to an increase in municipal wastewater, underscoring the recent importance of sustainable wastewater treatment methods (Sarker et al., 2021). Municipal wastewater is regarded as a potential energy source because of its organic content. The chemical energy inherent in municipal wastewater is quantified using the chemical oxygen demand (COD), which measures the concentration of carbonaceous materials and is approximately 12.6 kJ/g COD. In essence, COD serves as a suitable indicator for assessing the energy reclamation potential of wastewater. For more than a hundred years, the conventional activated sludge process (CAS) has been the preferred choice for municipal wastewater treatment due to its capacity to produce high-quality treated wastewater. Nonetheless, this method commonly removes COD from wastewater by employing intensive aeration. Consequently, it is not deemed a viable wastewater treatment method for energy recovery (Guven et al., 2019a, 2019b, 2019c; Yagci et al., 2016). The primary energy consumption in municipal wastewater treatment plants (WWTP) employing the CAS process is primarily associated with its aeration phase, accounting for approximately from 50% to 60% of the overall energy usage. Conversely, within the CAS process, the efficiency of sludge growth typically ranges from 0.5 to 0.3 gr of biomass produced/gr of COD removed. This indicates that for every kilogram of COD removed, approximately from 0.5 to 0.3 kg of dry biomass is generated. Consequently, the substantial production of sludge amplifies the expenses associated with its treatment and disposal, encompassing a range from 25% to 65% of the total operational costs. Furthermore, the incineration of sludge is considered a costly and environmentally incompatible approach, owing to its greenhouse gas emissions and ash production (Liu et al., 2018).

The energy consumption in the CAS process typically falls within the range from 0.3 to 0.6 Kwh/m^3, with an average from 0.45 Kwh/m^3 (equivalent to 1620 kJ/m^3). Consequently, for municipal wastewater containing an average COD concentration of approximately 500 mg/L, the energy utilized to oxidize one gram of COD equals 3.2 kJ/g of COD. In contrast, the potential energy content within municipal wastewater typically ranges from 14.7 to 17.8 kJ/g COD, with an average of 16.2 kJ/g COD. These instances illustrate that the potential energy within wastewater is approximately five times greater than the electrical energy required in the CAS process for wastewater treatment (Liu et al., 2018). Anaerobic techniques are commonly employed to recuperate the chemical energy stored in wastewater. Through anaerobic wastewater treatment or sludge digestion, biogas is generated. However, due to the relatively modest to low COD levels in municipal wastewater, direct anaerobic treatment of it is less feasible. A more recent approach, known as the two-stage absorption/biological oxidation process, has been proposed as a promising method for concentrating organic substances in municipal wastewater and harvesting energy from it. The initial phase, referred to as Step A, is specifically designed to increase the organic matter concentration in wastewater for direct use in anaerobic digestion before biological oxidation. The subsequent phase, known as Stage B, primarily focuses on nutrient removal from wastewater (Guven et al., 2019a, 2019b, 2019c; Liu et al., 2018). To recover energy from wastewater and up-concentrate organic matter (called carbon redirection) in Stage A, various methods are employed, including dissolved air flotation (DAF), filtration, chemical coagulation pretreatment (CEPT), and high-rate activated sludge (HRAS). While DAF, filtration, and CEPT processes are associated with issues like the requirement for extensive aeration, membrane fouling, and the use of chemicals, along with their subsequent treatment, the HRAS process emerges as a more economically sustainable alternative (Guven et al., 2019a, 2019b, 2019c). In recent research, the HRAS process has been integrated into systems such as contact-stabilization systems (CS) (Meerburg et al., 2015), sequential batch reactors (SBR) (Ge et al., 2013, 2017), membrane bioreactors (MBR) (Akanyeti et al., 2010; Faridizad et al., 2022; Faust et al., 2014b), and membrane sequential batch reactors (MSBR) (Shafaei et al., 2022).

In this chapter, we will explore the subject of urban wastewater, including its characteristics and energy content, as well as the conventional method employed for treating municipal wastewater and its associated drawbacks. Additionally, we will conduct a comprehensive evaluation of the processes utilized to concentrate municipal wastewater, discussing their respective advantages and disadvantages, with a particular emphasis on the HRAS process. Furthermore, to gain deeper insights into the HRAS process, we will compare it to the CAS process, delve into the mechanisms responsible for removing and absorbing organic matters, highlight important operational parameters, and review relevant case studies. Ultimately, we will address the challenges and future prospects pertaining to the enhancement of energy recovery from municipal wastewater.

2 Scarcity of Water and the Emergence of Fresh Challenges

As mentioned previously, the growth of the population, increased economic activity, and rapid urbanization in recent decades have led to significant increases in the volume of municipal wastewater (Wastewater, 2017). A recent study, using regression analyses, estimated global wastewater collection, generation, and treatment, revealing values between 225.6, 359.4, and 188.1 km^3/year, respectively (Jones et al., 2021). In simpler terms, these findings suggest that only 63% of the wastewater generated is collected, and a mere 52% of it undergoes treatment. While one might anticipate that improvements in wastewater treatment and technology would lead to a decrease in untreated wastewater discharge, this isn't always the case in developing countries, where the challenges are more significant. For example, in India, untreated wastewater discharge, which was 2 km^3/year in 1982, grew fivefold over the next 30 years (Valipour & Singh, 2016). In Fig. 1, the proportions of untreated and treated wastewater across various regions are depicted. It is evident that Sub-Saharan Africa and South Asia exhibit the highest rates of untreated wastewater (84%), whereas Western Europe is at the forefront of wastewater treatment (86%). Considering that undeveloped and developing countries are expected to experience increased urbanization in the future (Liddle, 2017), it is clear that the volume of untreated wastewater will rise in these regions. Given this information, it's clear that the current operational model of the global wastewater sector cannot meet the challenges posed by rapid urbanization and population growth, resulting in a continued increase in untreated wastewater. The discharge of untreated wastewater unquestionably has a direct adverse impact on the water quality of the receiving bodies of water. Additionally, the growing trend of inadequate wastewater treatment will further degrade overall water quality, posing a significant risk, especially to undeveloped countries in arid regions (Wastewater, 2017).

Ample evidence suggests that global warming is already playing a significant role in exacerbating water scarcity on a global scale (Hoegh-Guldberg et al., 2019). However, while greenhouse gas emissions and global warming are considered the primary causes of water stress (Shen et al., 2014), factors influencing water availability extend beyond climate conditions. Factors such as population size, available technology, socio-economic considerations, and cultural practices also play pivotal roles in determining available water (Şen, 2009). Furthermore, unexpected crises like the COVID-19 pandemic have shown that household water demand can surge due to changes in circumstances or behavior (e.g., distance education, working from

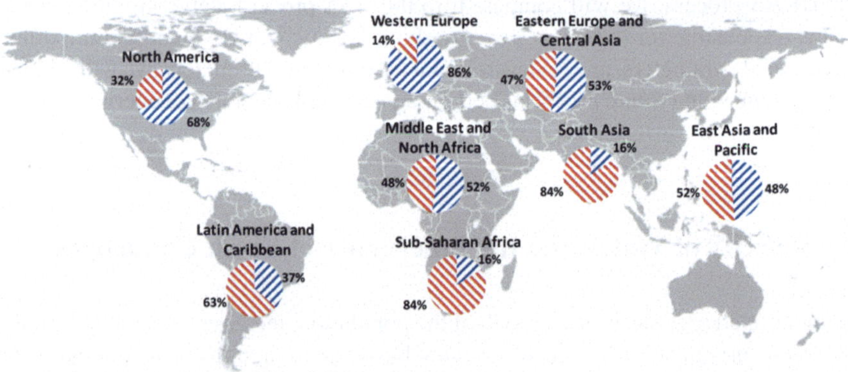

Fig. 1 The global distribution of treated and untreated wastewater (Jones et al., 2021). In the visual representation, untreated wastewater is represented by red diagonal stripes, while treated wastewater is indicated by blue diagonal stripes

home, hygiene practices). Consequently, the data used to forecast water availability provides valuable insights into how water scarcity will spread globally (Lüdtke et al., 2021).

Based on current estimations and analyses, it is projected that India, Africa, China, and various regions in Asia will encounter the most pronounced water scarcity issues in the twenty-first century. Indeed, the findings from hydrological and climate models, coupled with the increasing trend in untreated wastewater generation, underscore the significant vulnerability of water resources in the coming decades, particularly in developing countries (Hanasaki et al., 2013).

3 Scarcity of Energy and the Emergence of Fresh Challenges

The energy crisis is spiraling out of control, and simultaneously, the world is confronting a significant development crisis. It is predicted that the global population will reach 9.6 billion by 2050, with the majority of these individuals living in poverty and lacking access to energy resources if we do not change our current development trajectory (Pathak, 2014). Currently, 1.4 billion people lack access to electricity, with 85% of them residing in rural areas. Consequently, the number of rural communities relying on traditional biomass for their energy needs is projected to increase from 2.7 billion today to 2.8 billion by 2030 (Kaygusuz, 2012). The economic growth, industrial civilization, and lifestyle of the developed world heavily rely on the energy derived from oil and gas sources. Inexpensive natural gas and oil supplies are essential for various aspects of industrial agriculture, including fertilizers, pesticides, machinery for planting, cultivating, harvesting, processing, packaging, transportation, and marketing, all of which are necessary for affordable food production (Pathak, 2014) The dominance of fossil fuel-based power generation (coal, oil, and gas) and rapid population growth in recent decades have created a substantial demand for energy, resulting in global challenges related to a significant rise in carbon dioxide (CO_2) emissions (Asumadu-Sarkodie & Owusu, 2016). As a result, scientists have been actively searching for sustainable energy sources to address the issues mentioned earlier. In the meantime, municipal wastewater has attracted a lot of attention (Derakhshesh et al., 2022).

4 Energy in Municipal Wastewater

Municipal wastewater primarily contains two forms of energy: Chemical and thermal energy. The chemical energy within wastewater originates from its organic content and resides within the chemical bonds of organic molecules. The largest share of total wastewater energy, approximately 1.66 kWh/m3, is attributed to the carbon compounds present in the wastewater. Additionally, the energy associated with nitrogen compounds amounts to 0.3 kWh/m3. When the chemical bonds of organic molecules are broken, a certain amount of energy is released. This chemical energy content of municipal wastewater can be quantified using the COD, which is approximately 12.6 kJ/g COD. Indeed, the COD parameter serves as a suitable indicator for assessing the potential energy recovery from wastewater (Guven et al., 2019a, 2019b, 2019c; Lazarova et al., 2012). Moreover, municipal wastewater is also regarded as a source of thermal energy. The temperature of municipal wastewater remains relatively constant throughout the year. As the wastewater is conveyed through sewer lines, variations in ambient temperature do not exert a substantial impact on the temperature of the wastewater. The thermal energy content of municipal wastewater is approximately 7 kWh/m3, which can be recovered by heat pumps for heating and cooling purposes. The annual energy production from municipal wastewater totals around 140-70 GW, equivalent to 104-52 million tons of oil (Guven et al., 2019a, 2019b, 2019c; Heidrich et al., 2011).

5 Energy Consumption in WWTP

Various approaches for wastewater treatment and sludge disposal, along with their electricity consumption, also entail the use of natural gas or other fuel sources. The energy demands of a wastewater treatment facility rely on factors such as its location, size, the treatment method employed, and the aeration system in use (Gu et al., 2017).

5.1 Energy Consumption in WWTP with Different Technologies

The WWTP typically encompass primary, secondary, and advanced treatment phases. The initial treatment stage, depending on its design and operation, generally demands less energy compared to the other stages. Energy consumption in the secondary stage varies based on the technology employed. Notably, in the CAS process, aeration in the secondary treatment stage constitutes the most energy-intensive aspect of the entire process. In most WWTP, approximately from 50 to 60% of the energy usage is allocated to the aeration sector, while sludge treatment accounts from 15% to 25%, and secondary sedimentation utilizes 15% of the total electrical energy (Gu et al., 2017).

Contrasting with the CAS process, the oxidation ditch system also exhibits higher energy requirements due to the longer hydraulic retention time (HRT) of the wastewater and increased oxygen demand (DO). Fig. 1 illustrates the energy demands of processes within the secondary treatment unit, while Fig. 2 presents the proportion of energy use associated with different components of a CAS treatment system. As

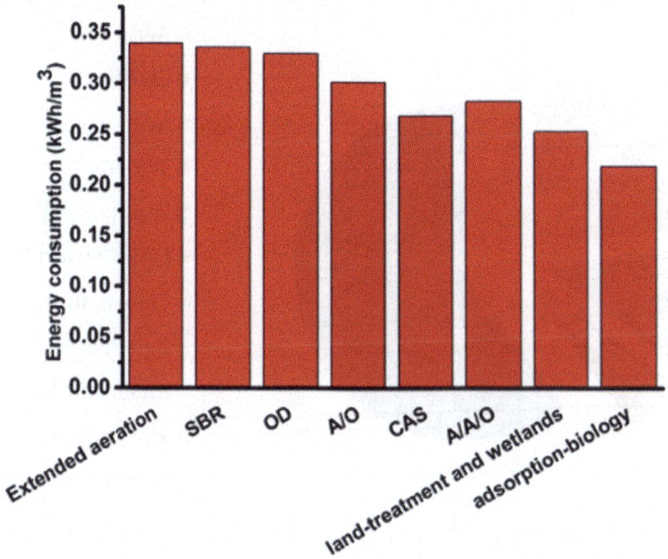

Fig. 2 The energy usage in secondary treatment plants employing various treatment methods in China (Gu et al., 2017)

depicted in Fig. 2, among the mentioned processes, the CAS process boasts the lowest energy consumption, and Fig. 3 further underscores that approximately 60% of this energy is attributed to the aeration section. Advanced wastewater treatment methods, such as reverse osmosis, also register elevated energy consumption due to nutrient removal (Gu et al., 2017).

5.2 *Energy Consumption in WWTP with Different Sizes*

The size of a WWTP has a substantial impact on energy consumption. As the plant's size and incoming wastewater flow increase, the treatment plant's energy consumption decreases. Increased wastewater volume enables equipment to operate with greater efficiency. Moreover, smaller wastewater treatment plants may lack trained specialists, unlike their larger counterparts, which can result in notable disparities in energy management (Gu et al., 2017).

5.3 *Energy Consumption in WWTP at Different Locations*

The energy usage at the WWTP is contingent upon its geographical location. Factors like the quantity of WWTP in a given area, the desired effluent quality, and the

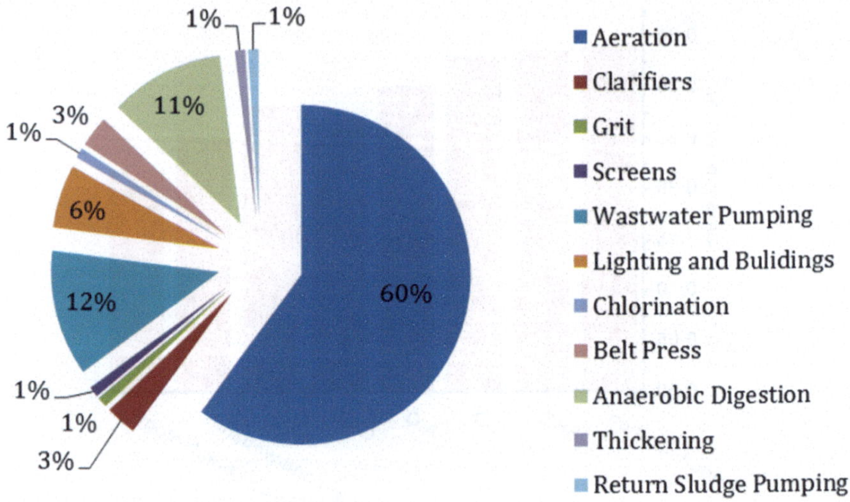

Fig. 3 The distribution of energy consumption associated with various units within a CAS treatment system (Gu et al., 2017)

prevailing treatment approach collectively influence the energy consumption of the WWTP (Gu et al., 2017).

6 Conventional Process of Municipal Wastewater Treatment

The treatment of municipal wastewater has attracted considerable attention due to its composition, including organic materials, mineral substances, and pathogenic biomass. The common method for treating municipal wastewater involves the use of the CAS process for aerobic wastewater treatment, followed by anaerobic digestion to decompose the resulting aerobic sludge (also known as waste sludge) (Liu et al., 2018).

6.1 Conventional Activated Sludge Process

In the CAS process, both COD and nitrogen (nutrients) are concurrently eliminated. Essentially, this involves the oxidation of organic substances and the nitrification process, which converts ammonia into nitrite and nitrate, taking place within an aerobic tank. Subsequently, the treated effluent is directed to a sedimentation tank to facilitate the separation of activated sludge from the effluent. To eliminate nitrogen-containing waste materials, the resulting effluent is conveyed to an anoxic tank, where nitrate is transformed into nitrogen gas through denitrification (Liu et al., 2018).

6.1.1 Challenges of CAS Process

- High energy consumption

As previously noted, the energy consumption of the CAS process is high due to the aeration unit, which accounts for roughly 60% of the total energy consumption. This reliance on electricity generation through the combustion of fossil fuels has raised apprehension regarding diminishing oil reserves, the emission of greenhouse gases, and the issue of global warming (Liu et al., 2018).

- Oxidation of organic matter

The primary objective of the CAS process is to eliminate organic compounds through oxidation. Consequently, it demands a substantial amount of energy for this oxidation and organic substance removal, with no provision for energy recovery (Guven et al., 2019a, 2019b, 2019c). Fig. 4 illustrates the distribution of COD (Chemical Oxygen Demand) across different stages of the activated sludge process. As depicted in Fig. 3, approximately 30% of the COD undergoes oxidation within the CAS process and exits the wastewater as CO2, while only about 60% is separated from the wastewater as sludge. Subsequently, only around 33% of the COD can be reclaimed as energy by converting it into methane gas through an anaerobic digestion process.

- Production of large waste sludge

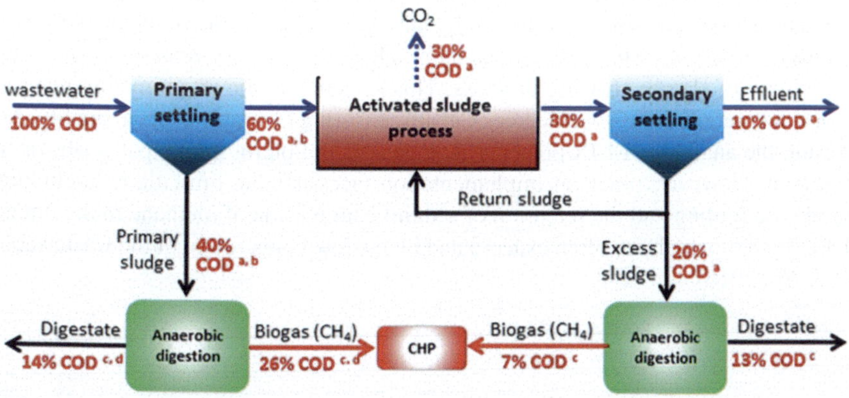

Fig. 4 COD mass flow through a conventional WWTP (Ren et al., 2016)

Typically, the sludge growth yield is about 0.3–0.5 g of dry biomass/g of COD removed, resulting in a significant volume of waste sludge. Disposing of this waste is costly, covering roughly from 25% to 65% of total operational expenses. Additionally, as population and urbanization rates rise, soil disposal options for waste sludge become limited, and incineration contributes to greenhouse gas emissions and ash formation (Liu et al., 2018).

6.2 *Anaerobic Digestion Process*

Anaerobic digestion is a complex sequence of steps during which organic substances are transformed into methane and carbon dioxide. In the initial stages, bacteria carry out the hydrolysis and fermentation of organic materials, followed by the anaerobic conversion of organic acids and alcohol into acetate in subsequent phases (Derakhshesh et al., 2022).

6.2.1 Challenges of Anaerobic Digestion Process

The utilization of anaerobic processes has been hindered by several challenges, including the low quality of effluent, the presence of residual dissolved methane in the effluent, limited nutrient removal, and decreased performance at lower temperatures due to reduced microbial activity for solid material hydrolysis and methane production. However, the central issue with anaerobic digestion is the necessity for a feed with a COD concentration exceeding 1000 mg/L to maintain system stability and proper process efficiency. Given the low organic substance concentration in municipal wastewater, it poses significant challenges for direct use in this process and subsequent biogas production. Consequently, innovative technologies are required to enhance biogas production efficiency. One such solution is the implementation of an anaerobic MBR system, which maintains a high biomass concentration, enabling the activity of slowly growing biomass. This system has shown promising results, even at temperatures below 6 degrees Celsius, ensuring favorable effluent quality. As a result, the anaerobic MBR presents an appealing option for municipal wastewater treatment. However, practical implementation faces notable limitations, including membrane fouling and the retention of a significant portion of methane in the liquid phase, both of which are often exacerbated by the low temperature of the wastewater (Li & Yu, 2016).

7 Methods of Up-Concentration of Organic Matters in Municipal Wastewater for Energy Recovery

In recent years, there has been a growing interest in the two-stage biological adsorption/oxidation (A/B) process for concentrating municipal wastewater and increasing the organic substance concentration. This process, tailored specifically for energy recovery from municipal wastewater, consists of two stages, between A and B, each with different organic loading rates. In stage A, municipal wastewater is concentrated, and COD is absorbed by the sludge. Stage B focuses on nutrient removal (Fig. 5–a). Fig. 5–b illustrates that the energy produced in stage A exceeds the energy consumed in stage B, making it a highly promising avenue for energy recovery. Various methods, including DAF, filtration, CEPT, and the HRAS process, can be employed in stage A to concentrate organic substances (Ren et al., 2016).

7.1 Carbon Redirection by DAF Process

The DAF process essentially combines two processes: Liquid separation and sludge thickening. This method proves highly effective in eliminating suspended particles. The core concept behind this approach is to float particles with a density greater

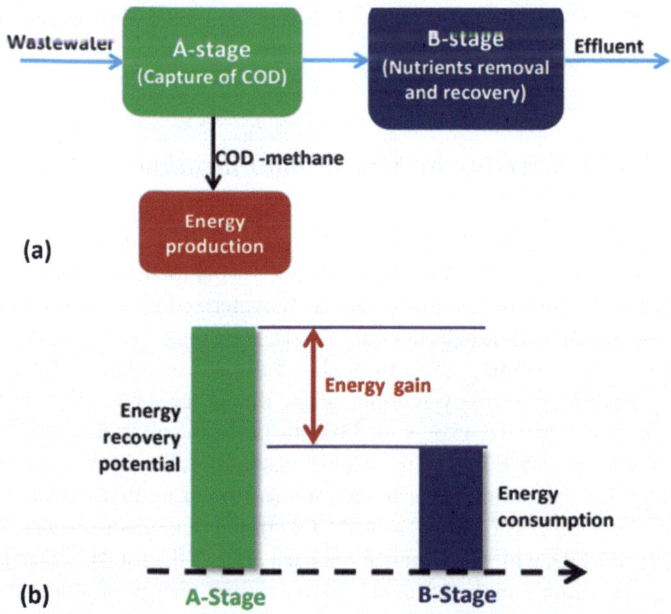

Fig. 5 Absorption/biological oxidation process; (a) A and B processes and (b) energy produced in A and B processes (Ren et al., 2016)

than that of water. Here, air bubbles adhere to the particles, decreasing their density, causing them to rise to the surface of the tank, thereby increasing sludge concentration. To enhance the efficiency of this process, commercial polymers can be utilized. However, the high cost of these materials and their associated environmental concerns pose challenges for this method (Sancho et al., 2019).

7.2 Carbon Redirection by Membrane Filtration Process

The membrane filtration process can be used to concentrate wastewater by dividing it into two parts: A concentrated portion with a higher organic content and a filtered portion. However, a drawback of this method is the tendency for membrane fouling, primarily due to the absence of aeration and fouling control (Jin et al., 2016). In an effort to address this issue, Kimura et al. (2017) employed membrane stimulation and vibration techniques, in conjunction with the use of granular materials, to mitigate membrane fouling during wastewater treatment via membrane filtration, with the aim of concentrating wastewater and recovering energy. In this experiment, aeration was intentionally avoided to prevent microbial oxidation. While individual methods did not yield satisfactory results, the combination of membrane stimulation and vibration proved effective. Through efficient physical cleaning and reverse chemical washing using citric acid, membrane fouling was entirely eliminated, obviating the need for membrane cleaning for up to one month. The research findings indicated that with a hydraulic retention time (HRT) of 1.3 h, a remarkable 75% recovery of organic matter was achieved.

7.3 Carbon Redirection by Membrane Filtration Process

The CEPT process involves the addition of coagulants and chemicals to transform suspended and colloidal particles into larger flocs. This method consumes minimal energy, has a high removal efficiency, and is characterized by straightforward operation and maintenance (Murugesan et al., 2014). It enables the recovery of approximately 60% of the incoming COD through the use of coagulants. However, when compared to traditional primary sedimentation, this process results in an increased sludge production of around 45%, with 33% attributed to solids absorption. Furthermore, the resulting sludge is rich in organic matter suitable for biogas production through anaerobic digestion. Nevertheless, it's worth noting that the CEPT process removes only a small portion of dissolved COD in municipal wastewater. Nearly all of the dissolved COD, which accounts for at least 30% of the total COD in municipal wastewater, may remain in the effluent. The maximum energy production from this method equals 2.09 kJ/g of COD. It's also reported that the addition of chemical coagulants can adversely affect the digestion efficiency of waste sludge (Ren et al., 2016). Importantly, adding coagulants to CEPT results in an overall increase in operational costs (Zeng et al., 2020). Fig. 5 illustrates the COD mass balance in this process. The prudent selection of coagulant type and quantity is crucial to minimize adverse effects on the biological process, reduce COD, and control operational expenses. The jar test, involving various coagulants and doses, is used for this purpose (Dong et al., 2019). Effective coagulants in this context include aluminum sulfate (alum) (Patel et al., 2021), iron chloride (Bezirgiannidis et al., 2019; Patel et al., 2021), polyaluminum chloride (Bezirgiannidis et al., 2019) and lime (Patel et al., 2021). However, it's worth noting that dewatering the sludge produced by alum can be challenging. In a comparison between aluminum sulfate and iron chloride, Sarparastzadeh et al. (2007) found that aluminum sulfate had lower carbon removal than iron chloride, but iron chloride resulted in the production of a substantial amount of difficult-to-dispose-of waste sludge, thus increasing costs. Employing an electrochemical pre-treatment process could potentially address the issue of managing the sludge produced by this method (Zeng et al., 2020).

7.4 Carbon Redirection by Anaerobic Process

In contrast to the CAS process, anaerobic methods have consistently garnered attention as an economical choice due to several advantages, including the absence of aeration requirements, reduced waste sludge production, and the generation of methane biogas. Today, various anaerobic techniques such as the upflow anaerobic fixed bed reactor (UASB) (Derakhshesh et al., 2022), fixed film anaerobic bioreactor, and anaerobic baffled reactors are employed for municipal wastewater treatment (Ren et al., 2016). However, the slow hydrolysis of suspended solids can lead to the accumulation of solids, particularly in the UASB reactor, adversely affecting the activity of methane-producing anaerobic biomass (Guven et al., 2019a, 2019b, 2019c). Despite its numerous applications at industrial and laboratory scales, this system consistently presents challenges, including extended start-up periods, sensitivity to seasonal temperature fluctuations, slow growth rates due to the low organic substance concentration in municipal wastewater, and residual dissolved methane in the effluent (Ren et al., 2016).

7.5 Carbon Redirection by HRAS Process

The HRAS process proves to be a favorable option for wastewater treatment when compared to the CAS method, particularly in terms of energy recovery and organic material absorption. Approximately 55–65% (with an average of 66%) of organic substances are absorbed in this process (Ren et al., 2016). The primary distinction between HRAS and CAS is in their operating parameters, including organic loading rate (OLR), HRT, and sludge retention time (SRT). Notably, the SRT in HRAS process is less than 2 days, and the HRT falls within the range of 2–4 h (Jimenez et al., 2015). A low SRT minimizes the mineralization of wastewater organic matter, while a short HRT increases the concentration of organic matter in the waste sludge stream. In such conditions, only a small fraction of the dissolved organic matter in wastewater is utilized by biomass for extracellular polymeric substances (EPS) production (Emaminejad et al., 2019). Another significant feature of this system is its minimal need for aeration to effectively remove a substantial portion of organic compounds from wastewater (Modin et al., 2015). Low energy consumption for aeration, efficient biological absorption of organic substances, and the generation of organic-rich sludge are among the notable advantages of HRAS (Ge et al., 2013). Smith et al. (2014) conducted a comparison between the anaerobic MBR process and the HRAS process. They found that the anaerobic MBR recovers 49% more energy than HRAS. However, the amount of energy required and the emission of greenhouse gases were higher than the HRAS process due to the presence of dissolved methane in the effluent. Moreover, if disposal methods like incineration are employed for the HRAS process's sludge waste, its economic viability surpasses that of the anaerobic MBR. In another study, Taboada-Santos et al. (2020) compared CEPT processes with an HRT of 30 min to HRAS with an HRT of 2 h, both with a sludge age of 1 h. Iron chloride was used as a coagulant in the CEPT process. The results demonstrated that while the CEPT process had lower energy requirements than HRAS, its operating costs were higher due to the high coagulant concentration (at least 150 mg/L). Additionally, HRAS exhibited much higher COD removal efficiency from (80%–90%) compared to CEPT (55%–4%). Consequently, HRAS stands out as an economical and effective choice for municipal wastewater treatment.

7.5.1 Mechanism of COD Removal and Biosorption in HRAS Process

Ulrich and Smith were the first to report the term "bio-adsorption" in 1951 to describe the adsorption process, which effectively removes a substantial amount of organic matter. Surface absorption involves the attachment of suspended and colloidal organic substances to the surface of sludge flocs through physico-chemical interactions, a phenomenon known as bioflocculation. This mechanism is considered the primary driver of COD removal in HRAS processes (Huang et al., 2020).

Intracellular storage is another facet of microbial metabolism and functions as a COD removal mechanism, similar to absorption. This mechanism transforms soluble organic matter into insoluble polymers, including Polyhydroxyalkanoate (PHA), glycogen, Triacylglycerol, glyceride, and fat. Such storage enhances the sedimentation of sludge flocs and is regarded as a beneficial mechanism (Kinyua, Miller et al., 2017a, 2017b). Yet another crucial mechanism in carbon removal during this process involves the oxidation of organic materials. This mechanism entails the consumption and oxidation of organic substances for biomass growth, resulting in the production of water and carbon dioxide. It's important to note that this mechanism leads to the oxidation of wastewater energy. Therefore, in HRAS processes aimed at energy recovery, efforts are made to minimize the impact of this mechanism (Huang et al., 2020). In the bioflocculation process, colloidal and suspended organic compounds

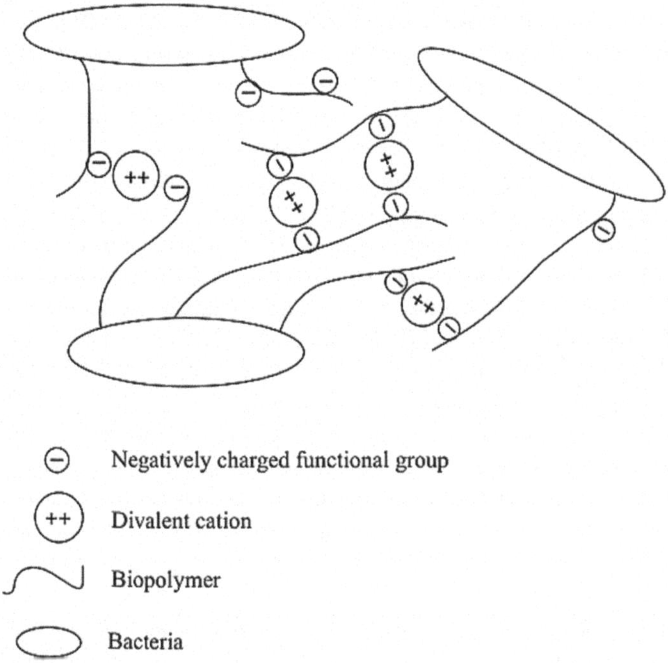

Fig. 6 Schematic diagram of the bioflocculation process (Leong et al., 2020)

in wastewater are combined to form larger compounds with the assistance of EPS, which are generated by the biomass. This process creates a concentrated flow of organic materials. Notably, only a small portion of the COD is utilized for microbial growth and EPS production. EPS, due to their adhesive and cohesive properties, serve as a binding agent between biomass or wastewater particles. At a neutral or near-neutral pH, EPS carry a net negative charge owing to the presence of acidic functional groups. Then attracts positively charged binary cations such as calcium and magnesium, forming bridges with EPS, which have a negative charge. This interaction causes cells and particles to adhere to one another. As a result, EPS fill the spaces between cells and particles, forming biofloc and, in essence, ensuring the structural and functional integrity of activated sludge masses (Fig. 6) (Leong et al., 2020).

7.5.2 Factors Affecting Bioflocculation

Bioflocculation is a complex process influenced by various factors. Among these, the most crucial factor affecting bioflocculation is the presence of bacteria within the floc ecosystem, as they play a vital role in forming and maintaining the floc structure. A specific category of bacteria, known as filamentous bacteria, significantly contributes to the strength and stability of these flocs. These bacteria grow in the form of long strings and cause the lumps in the sludge to connect (Fig. 6). It's worth noting, however, that an excessive presence of these bacteria can lead to operational issues like sludge bulking, increased sludge discharge, and foam production. Hence, controlling their concentration becomes essential by managing the food-to-biomass (F/M) ratio and dissolved oxygen (DO) concentration (Suresh et al., 2018).

Another significant factor that plays a pivotal role in bioflocculation is the composition and structure of EPS, which includes both Tightly Bound EPS (TB-EPS) and Loosely Bound EPS (LB-EPS) components (Fig. 7). The composition of EPS, such as proteins and polysaccharides, can vary considerably. EPS forms a complex matrix encompassing carbohydrates, proteins, humic acid, nucleic acids, and phospholipids (Kinyua, Elliott et al., 2017a, 2017b; Suresh et al., 2018). Increasing the hydrophobicity of sludge results in enhanced bioflocculation. Proteins, for instance, boost the hydrophobicity of EPS and foster connections with cations and heavy metals, thereby promoting better agglomeration and the formation of more robust flocs (Leong et al., 2020). Furthermore, whether the EPS bonds are Tightly or Loosely also influences bioflocculation. Various studies have shown that TB-EPS has no impact on bioflocculation, but an increase in LB-EPS concentration can worsen the biological bioflocculation process due to LB-EPS's high hydrophilic nature (Hong et al., 2018; Li et al., 2019).

Among the various factors influencing the bioflocculation process, we can consider the source of carbon and nitrogen, SRT (Sludge Retention Time), temperature, alkalinity, pH, and salinity.

The carbon-to-nitrogen (C/N) ratio significantly influences sludge characteristics. A lower C/N ratio results in higher protein content, while a higher ratio leads to a greater concentration of carbohydrates. The choice of carbon source in the feed

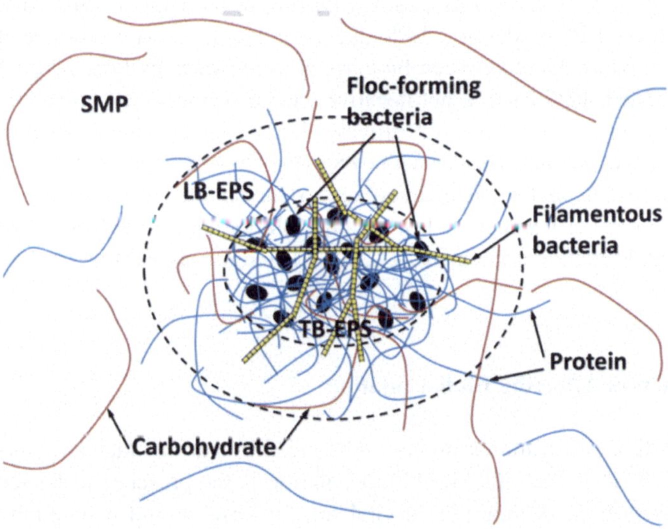

Fig. 7 Schematic of bioflocculation by EPS and bacteria (Suresh et al., 2018)

also affects EPS properties and, subsequently, bioflocculation. For example, a feed containing acetate, glucose, and peptone tends to result in higher protein content compared to one composed solely of glucose.

Increasing SRT tends to increase particle size and enhance bioflocculation. With rising SRT, protein concentration typically decreases, while carbohydrate concentration rises. Elevated carbohydrate levels contribute to larger particle sizes, ultimately improving bioflocculation (Faust et al., 2014b; Silva et al., 2017).

Temperature also plays a role in influencing EPS characteristics and, consequently, bioflocculation. Higher temperatures encourage the production of polysaccharides and filamentous bacteria, which may lead to sludge bulking and reduced bioflocculation (Suresh et al., 2018).

At low pH levels, hydroxyl and carboxylic functional groups with negative charges and hydrophilic properties in EPS decrease, resulting in increased aggregation. Conversely, at high pH levels, flocs break apart, releasing colloidal and suspended materials. High or moderate salt concentrations reduce enzyme activity, microbial growth, and can cause plasmolysis. Therefore, increased salt concentration and salinity levels typically lead to reduced bioflocculation (Leong et al., 2020).

7.5.3 Operating Parameters

Operational parameters that impact the HRAS process encompass SRT, HRT, and DO concentration. Among these, a short SRT stands out as a critical design parameter in the HRAS process, primarily because it reduces the energy needed for increased energy recovery (Ge et al., 2017). Research indicates that younger sludge is more effective at converting biogas into thermal and electrical energy. Additionally, sludge age can significantly influence COD absorption efficiency. Changes in SRT can have a substantial impact on bioflocculation and the physical and chemical properties of sludge, including surface charge and hydrophobicity, which govern EPS production (Sancho et al., 2019).

DO and HRT represent other crucial operational parameters affecting organic material absorption and energy recovery. They play a role in biomass formation and energy consumption. An increase in HRT and a decrease in DO levels lead to reduced microbial activity, lower EPS production, and, consequently, a slower bioflocculation process and de-bioflocculation (Faust et al., 2014a; Guven et al., 2017).

7.5.4 Recommended Design Parameters in HRAS Process

Table 1 provides recommended operating parameter ranges for an HRAS system. It's important to note that these suggested parameters may require adjustments depending on factors such as influent conditions, reactor setups, and site-specific circumstances. However, they offer a starting point based on widely accepted values in existing literature. The optimal operating parameters for the A-Stage ultimately hinge on its specific objectives, which typically involve maximizing COD removal.

Table 1 Suggested operating and design parameters in the HRAS process (Guthi et al., 2022)

Parameter	Value
HRT	30 min
SRT	0.3–0.5 days
DO	0.5 mg/L
MLSS	2–3 g/L
Recirculation rate	0.5–1
Waste strategy/Sludge extraction	Target MLSS based control

To maintain a DO level of 0.5 mg/L, a common approach involves using compressed air delivered through diffusers located at the bottom of the reactor. The efficiency of oxygen transfer under actual process conditions, compared to that in fresh water, is denoted as the α-factor, which is a critical design parameter. In contrast, nitrifying CAS systems typically exhibit α-factors ranging from 0.60 to 0.75. Notably, a higher α-factor of 0.8 has been observed for B-Stage, attributed to the removal of surfactants and other inhibitory substances in the A-Stage. In HRAS reactors, constituents that impede oxygen transfer, particularly surfactants, degrade less efficiently than in CAS systems, thus reducing oxygen transfer efficiency.

An effective approach to mitigate diurnal fluctuations in influent COD concentrations is to implement sludge extraction control, which maintains a constant concentration of mixed liquor suspended solids (MLSS) in the reactor. However, when it comes to the design and sizing of separation units, the available literature offers limited information. The ratio of clarifier to reactor volume varies significantly across the reviewed studies, ranging from 0.5 to 7 (Guthi et al., 2022).

7.5.5 Comparing Waste Sludge from HRAS Process to CAS Process

As previously mentioned, the primary distinction between the HRAS process and the CAS process is in the system's operational parameters. Consequently, the characteristics of the waste sludge produced by these two processes also differ. One of the primary objectives of the HRAS process is to generate more sludge waste to enhance energy production from wastewater. The content of suspended solids and COD in these two types of sludge varies significantly. Moreover, sludge from the HRAS process is younger, and its volatile organic matter content surpasses that of sludge from the CAS process. Operating conditions also influence the size and structure of flocs. Research indicates that decreasing SRT, increasing DO, and HRT result in the formation of smaller flocs with a less robust structure (Guven et al., 2019a, 2019b, 2019c). Another parameter related to particle size distribution (PSD) is the Sludge Volumetric Index (SVI), which reflects the compressibility and sedimentation behavior of activated sludge. SVI indicates the compressibility and sedimentation of activated sludge. Higher sludge settling and compressibility (lower SVI) indicates better bioflocculation. According to the research done, with the reduction of SVI number, PSD and sludge settling properties increase (Guven et al., 2019a, 2019b, 2019c; Miller et al., 2016; Seuntjens et al., 2016).

7.5.6 Systems Used in HRAS Process

In recent years, numerous studies have delved into the HRAS process within various system configurations, such as CAS, CS, SBR, MBR, and MSBR. These investigations aim to pinpoint optimal parameters and elucidate the microbial communities functioning under high OLR. Table 2 provides a concise overview of these studies, briefly described.

As previously noted, the operational parameters of SRT, HRT, and DO exert a profound influence on energy recovery and bioflocculation. For instance, Jimenez et al. (2015) delved into these parameters' effects in real municipal wastewater treatment within the HRAS-CAS system. Their study revealed that reducing SRT from 2 to 0.1 days led to a remarkable increase in COD absorption, ranging from 23 to 48%. This heightened organic matter absorption corresponded to a decrease in mineralization. Additionally, they observed a direct correlation between DO levels and EPS content. While reducing SRT from 2 to 1-day enhanced sludge yield, further reduction below from 1 day to 0.1 day resulted in a drastic decline in sludge yield and bioflocculation. This substantial drop could be attributed to the diminished production of EPS under very short SRT conditions. Guven et al. (2019a, 2019b, 2019c) conducted a study examining the impact of HRT on the HRAS-CAS system's performance and bioflocculation during real municipal wastewater treatment. Their findings revealed that as HRT decreased, there was an increase in COD removal efficiency and bioflocculation. They attributed the primary mechanism for removing colloidal substances is sedimentation and bioflocculation. Rahman et al. (2016) conducted research to assess the performance and energy recovery potential of two HRAS-CAS systems (continuous stirred reactor and pipe flow reactor) and a high-rate CS system during real municipal wastewater treatment with low COD concentration. Their findings indicated that energy recovery and organic matter absorption were most effective at low SRT. When SRT was less than 1.1 days, the CS system exhibited high COD removal efficiency between (52–59%) and energy recovery (0.33–0.34 g of COD/g of added COD), outperforming the stirred reactor and pipe flow systems (which achieved 28–37% COD removal with no organic material absorption). A key factor contributing to improved bioflocculation, sedimentation, and biosorption was the low concentration of DO. Rahman et al. (2016) also noted that up to a certain threshold SRT, biosorption efficiency increased as SRT decreased. They determined biological absorption efficiency by measuring the rate of oxygen consumption and emphasized the role of reduced mineralization in promoting organic substance absorption. Preventing the mineralization of organic matter in wastewater leads to its absorption into the sludge and enhances bioflocculation. Reducing SRT from 2.2 days to 0.8 days resulted in a rise in adsorption efficiency, from 50 to 224mg O_2/VSS, while further reducing SRT from 0.8 to 0.2days significantly decreased sludge yield. In their study, an SRT of 0.8days was identified as the optimal SRT. Meerburg et al.

Table 2 A summary of the operational conditions of the studies conducted in the HRAS process

Ref	COD removal (%)	Effluent COD (mg/L)	Influent COD (mg/L)	DO (mg/L)	OLR(kg/m^3.day)	HRT (min)	SRT(day)	Wastewater	System
Meerburg et al. (2015)	40–55	360–271	600	2	7.2	120	0.41–1.3	Synthetic municipal wastewater	CS
Ge et al. (2013)	70–80	870–580	2900	1.2–5	2.5–9.8	720–1440	2–3	Slaughterhouse wastewater	SBR
Ge et al. (2017)	54–62	181–150	393	3–3.5	18.9	30	0.3–5	Municipal wastewater	SBR
Faust et al. (2014b)	88–92	43–71	614	–	21.92	40	0.125–5	Municipal wastewater	MBR
Akanyeti et al. (2010)	88–77	64–71	276–529	6–8	10.5–5.5	72	0.25–1	Municipal wastewater	MBR
Jimenez et al. (2015)	–	–	480	2	11.42–24	30–60	0.1–2	Municipal wastewater	CAS
Emaminejad et al. (2019)	87–89	44–52	387	8–6	6.2	90	0.5–2	Gray water	MBR
Kinyua, Elliott et al. (2017a, 2017b)	80	120–132	600	0.5–1.5	14.28–28.5	30–60	0.56–0.28	Municipal wastewater	CAS
Faust et al. (2014a)	88–92	41–62	515	1–4	12.3	60	0.5	Municipal wastewater	MBR
Guven, Ozgun et al. (2019a, 2019b, 2019c)	37–49	230–290	463	0.3–0.4	5.12–11.11	60–130	0.35	Municipal wastewater	CAS
Rahman et al. (2016)	62–66	144–161	425	0.3–0.5	2–3.67	30–90	0.2–2.2	Municipal wastewater	CAS

(continued)

Table 2 (continued)

Ref	COD removal (%)	Effluent COD (mg/L)	Influent COD (mg/L)	DO (mg/L)	OLR (kg/m^3.day)	HRT (min)	SRT(day)	Wastewater	System
Leal et al. (2010)	90	58–101	603–1014	–	12.9–11.3	78–114	0.2–1	Gray water	MBR
Başaran et al. (2012)	66–68	11–25	35–78	–	–	30–120	0.5–2	Synthetic municipal wastewater	MBR
Dai et al. (2018)	78	96.6	441	1–1.5	8.8	72	0.5–1.8	Municipal wastewater	CS-MBR
Faridizad et al. (2022)	93–94	42–36	600	1	5.76	150	0.5–2	Synthetic municipal wastewater	MBR
Shafaei et al. (2022)	96	72	1987	1	4–16	12–3	2	Synthetic wastewater	MSBR
Taboada-Santos et al. (2020)	87	48–97	375–750	3–3.5	4.5–9	120	2.5–3	Municipal wastewater	CSRT coupled to a settler

(2016) conducted a study to assess the impact of SRT and HRT on energy recovery and COD removal from synthetic municipal wastewater using a high-rate SBR. They achieved COD removal efficiency in the range of 80% to 93% under various operating conditions. Their results demonstrated that as the SRT decreased, sludge yield increased, with values ranging from 2.8 to 0.24 days. In another study, Meerburg et al. (2015) noted that the high-rate CS system exhibited significantly higher sludge yield compared to the CAS process due to much lower SRT. Ge et al. conducted research on energy recovery from slaughterhouse wastewater with a high input OLR. They explored SRTs ranging from 0.5 to 3 days (Ge et al., 2017) and 2 to 3 days (Ge et al., 2013) in an SBR system. These investigations revealed that SRT significantly influenced sludge degradability and anaerobic digestion. For instance, reducing SRT from 3 days to 0.5 days increased sludge anaerobic degradability from 66 to 80% (Ge et al., 2013). The highest organic matter removal rate, reaching 80%, was observed at an SRT of 2 days, primarily attributed to absorption and accumulation on flocs (Ge et al., 2017). Additionally, the highest nitrogen, ammonia, and phosphorus removal occurred at an SRT of 3 days, with values between 50%, 36%, and 34%, respectively (Ge et al., 2013).

MBR is another widely used system in the HRAS process. Unlike other systems such as SBR, which may exhibit unfavorable and low-quality performance (Belli et al., 2017), or the CAS process, which demands a complex structure and significant space, MBR offers advantages such as high-effluent quality, a simple design, and a minimal footprint (Hamedi et al., 2019). However, membrane fouling remains a significant challenge in these systems. Consequently, numerous studies have investigated MBR systems under HRAS conditions.

Faust et al. conducted an investigation into the impact of SRT (Faust et al., 2014b) and DO concentration (Faust et al., 2014a) on bioflocculation in high loaded MBR (HL-MBR) systems during real municipal wastewater treatment. They made several noteworthy observations, including that at the highest SRT tested (5 days), both bioflocculation and mineralization reached their maximum levels, standing between 98% and 32%, respectively. Their findings indicated that within an SRT range from 0.5 to 1 day, minimal mineralization of organic matter occurred, but bioflocculation was notably high. Consequently, this SRT range was identified as the optimal SRT. Faust et al. (2014b) also noted that as SRT increased up to 1 day, the total EPS content increased. However, further increases in SRT, up to 5 days, led to a decrease in total EPS content. Their research delved into the impact of liquid mixture properties such as EPS and PSD on membrane fouling. Surprisingly, they did not find a positive correlation between EPS and membrane fouling. Instead, they discovered that the concentration of sub-micron particles (ranging from 450 to 45 microns) served as a reliable indicator of membrane fouling. Notably, membrane fouling was most pronounced at the lowest SRT. Furthermore, Faust et al. (2014a) observed that increasing DO concentration from 1 to 4 mg/L resulted in improved bioflocculation efficiency, rising from 69 to 92%. This increase in efficiency was attributed to the production of EPS and the presence of more polyvalent metal cations in the system. Emaminejad et al. (2019) examined the performance of a HL-MBR system for treating synthetic gray water with SRT ranging from 0.5 to 2 days and HRT at

1.5 h. They achieved a COD removal efficiency of 89–87%. Similar to Faust et al. (2014b), they observed that at the highest SRT, there was the most mineralization and bioflocculation. Furthermore, they found that the highest concentrations of EPS, TB-EPS, and the protein component were obtained with an SRT of 2 days. The study also delved into PSD to explore bioflocculation. The PSD was normal at SRTs between 1 and 2 days, but at an SRT of 0.5 days, it exhibited a bimodal pattern. This variation in PSD with SRT was attributed to the interaction between SRT and the F/M ratio. Increasing the SRT led to higher mixed liquor suspended solid (MLSS) concentrations, resulting in a decrease in the F/M ratio. Therefore, as the influence of SRT or F/M dominated the system, a decrease in SRT, followed by an increase in F/M, either increased or decreased PSD. Akanyeti et al. (2010), in their investigation of real municipal wastewater treatment in an HL-MBR system, managed to recover 35% of the incoming COD as methane. They suggested that SRT, DO, and HRT were crucial parameters for energy recovery and bioflocculation. Hernandez et al. (2010) also treated gray water using an HL-MBR system, examining three SRT/HRT ratios between 3, 8, and 12. They observed that with an increase in this ratio, bioflocculation also increased. Additionally, they achieved a substantial recovery of 73% of the incoming COD as methane and proposed that the bioflocculation process held promise for gray wastewater treatment and energy recovery. Basaran et al. (2012) referred to the HL-MBR as the "ultra-fast MBR" and managed to capture energy equivalent to 54%–77% of the input synthetic municipal wastewater energy. They noted that under different SRTs ranging from 2 to 0.5 days, the microbial population and morphology exhibited variations. Their research primarily focused on investigating and quantifying the soluble COD fraction. Dai et al. (2018) implemented a hybrid system comprising a high-rate CS-MBR system for municipal wastewater treatment. Their findings revealed that this combined system outperformed the high rate MBR alone, particularly at SRTs from 0.5 to 1.8 days, exhibiting higher COD removal efficiency and enhanced bioflocculation. Increased content of EPS contributed to improved bioflocculation. An additional significant advantage of the combined system was the reduction in membrane fouling due to lower concentrations of colloidal and suspended materials inside the reactor. Faridizad et al. (2022) conducted an investigation into the performance and fouling characteristics of a HL-MBR system employed in the treatment of municipal wastewater. They carried out experiments at different SRTs between 0.5, 1, and 2 days. The findings revealed that as the SRT increased, the rate of membrane fouling also increased. The primary cause of membrane fouling was identified as the formation of a cake layer, which Fourier Transform Infrared analysis (FTIR) indicated was primarily composed of protein and carbohydrate. As the SRT increased, there was a more significant increase in soluble microbial products (SMP) and a more pronounced reduction in particle size within the cake layer compared to the mixed liquor. Conversely, EPS remained relatively constant in the cake layer but increased in the mixed liquor. Several factors contributed to the rise in membrane fouling rate and cake resistance with increasing SRT, including SMP and its components, the protein-to-carbohydrate ratio in the EPS, the relative hydrophobicity of SMP, bacterial floc, and submicron particles. The results of multiple regression analysis indicated that SMP was the primary factor responsible for the escalation in membrane fouling rate with increasing SRT. Ultimately, concerning membrane fouling and system performance, the study found that the lowest SRT examined yielded the best performance for the HL-MBR system. He et al. (2023) conducted an investigation involving a pilot-scale chemically enhanced HL-MBR (CE-HL-MBR) to enhance the practical applicability of this process. To address issues like membrane fouling, reliance on intensive aeration, and insufficient concentration factors, they introduced polyaluminum chloride (PAC) and activated carbon (AC). They designed eight different conditions to study the impact of chemicals and aeration on filtration, pollutant removal, and organic recovery. They successfully achieved stable filtration rates of 8–10 L/m^2.h at MLSS concentrations ranging from 15 to 20 g/L. This was attributed to the dynamic formation and shedding of a protective cake layer on the membrane surface, as well as a reduction in EPS. CE-HL-MBR managed to remove over 75% of soluble organics and 80% of soluble phosphorus, with the primary influence coming from the chemicals used. Furthermore, the process led to simultaneous nitrification and denitrification due to uneven oxygen distribution. The influent organics were concentrated by factors ranging from 27 to 80, with a loss rate of less than 15%. However, it was noted that methane production might not be the most efficient method for resource extraction. The practical feasibility of the proposed CE-HL-MBR was confirmed through an additional three months of stable operation.

Shafaie et al. (2022) conducted a pioneering study involving the use of a MSBR in a HRAS process. Their research focused on evaluating the impact of the OLR, a crucial operational parameter in HRAS processes, across four different levels within the range of 4–16 kg COD/m^3.d. They aimed to understand its effects on the performance in terms of COD and energy recovery, as well as the membrane fouling characteristics of a high rate MSBR operating under HRAS conditions. As the OLR increased from 4 to 16 kg COD/m^3.d, the COD recovery reached its peak at 84%, surpassing previously reported values for bioreactors without membranes. Additionally, the study achieved a total energy recovery of 0.25 g COD CH_4/g COD_{inf} for the waste activated sludge (WAS) stream produced under HL-MSBR operating conditions. However, it was noted that the highest OLR also led to the most rapid membrane fouling, with the primary fouling mechanism being the formation of a cake layer, observed across all OLR levels. The research identified soluble microbial products, particularly their protein fraction, as the primary contributor to fouling in the HL-MSBR. In light of these findings, the study proposed the utilization of membrane relaxation and the addition of $FeCl_3$ as strategies to control membrane fouling when employing the HL-MSBR in HRAS processes.

Taboada-Santos et al. (2020) conducted a comparative analysis of CEPT and HRAS methods in terms of their energy requirements, operational expenses, and efficiency in removing viruses. In the study, a pilot-scale CEPT plant operated with an HRT of 30 min, while a laboratory-scale HRAS reactor operated with an HRT of 2 h and SRT of 1 day in continuous mode. The HRAS reactor yielded a slightly lower COD recovery, primarily due to the partial oxidation of incoming COD in the influent (15%). Notably, the CEPT configuration exhibited lower energy consumption (0.07 kWh/m^3 of wastewater) compared to the HRAS-based setup (0.13 kWh/m^3

of wastewater), although it incurred significantly higher operational costs. Finally, when it comes to virus removal, HRAS outperformed CEPT. In conclusion, the combination of HRAS followed by post-treatment with $FeCl_3$ demonstrated itself as a more effective alternative for pre-concentrating COD in modern WWTPs compared to CEPT.

8 Challenges and Future Prospects

Given the growing concerns in communities regarding water scarcity and the depletion of fossil fuel reserves, the adoption of the HRAS process has emerged as a promising solution. However, the primary challenge lies in increasing methane production and enhancing energy recovery within this process. The PHA (Polyhydroxyalkenoate) generated within the initial sludge stage (Stage A) holds significant potential due to its biodegradable nature and mechanical properties similar to synthetic plastics. It can be leveraged for bioplastic production, offering a means to address issues associated with sludge digestion and disposal (Guven et al., 2019a, 2019b, 2019c). Simultaneous digestion of Stage A waste sludge alongside various organic materials, such as kitchen waste, presents an opportunity to boost energy recovery (Guven, Ersahin et al., 2019a, 2019b, 2019c). Despite numerous studies conducted on HRAS systems, a universal relationship between operational parameters and system performance remains elusive, necessitating further research in this domain (Sancho et al., 2019). The aeration employed in the HRAS process can lead to the oxidation of certain organic substances. Consequently, an alternative perspective is to explore HRAS without aeration, a concept that requires extensive research and investigation in terms of process application and energy recovery (Cao et al., 2020). The adoption of MBR technology introduces a novel approach due to the high-quality effluent produced in the HRAS process. Nevertheless, membrane fouling remains a major challenge in these systems. Therefore, there is a pressing need for additional studies focused on effective control strategies to mitigate membrane fouling issues (Emaminejad et al., 2019). Moreover, to gain deeper insights into the performance of the HRAS, it is essential to conduct additional research on the operational factors that influence it, with particular emphasis on temperature and mixing. These factors have not received adequate attention in existing literature. Also, incorporating the biosorption mechanism into numerical models can expedite research in this field. This requires the identification of novel state variables and a more detailed breakdown of the influent fractionation (Guthi et al., 2022).

9 Conclusion

Municipal wastewater treatment has long piqued interest due to the presence of harmful biomass and environmental pollution. Meanwhile, in light of the energy shortage, many nations are actively seeking renewable energy sources. Municipal wastewater, rich in organic matter, holds promise as a potential energy reservoir capable of fulfilling diverse community needs. The CAS process falls short in terms of economic and environmental sustainability due to its substantial energy consumption for aeration and organic material oxidation. On the other hand, anaerobic digestion and energy recovery also encounter challenges in treating municipal wastewater, owing to its low organic substance concentration and wastewater dilution. Consequently, attention has shifted towards altering the municipal wastewater treatment approach. A two-stage process has emerged as a viable alternative. In this process, energy is harnessed within the sludge during the initial stage (Stage A), followed by nutrient recovery and removal in the subsequent stage (Stage B).

Various methods have been employed in Stage A, including DAF, membrane filtration, anaerobic processes, CEPT, and the HRAS process. DAF and membrane filtration methods encounter issues like high coagulation requirements and membrane fouling, which drive up operational costs. CEPT faces challenges such as excessive sludge waste production and increased operational expenses due to chemical coagulant use. The anaerobic process is deemed uneconomical due to the residual soluble methane in the effluent.

In contrast, the HRAS process, operating at low SRT and HRT, has garnered significant attention recently. Its unique operating conditions lead to reduced mineralization and bioflocculation, ultimately enhancing energy recovery. Over the past decade, this approach has found application in various systems like SBR, CS, MBR, and MSBR. Considering the suboptimal effluent quality in SBR and the space requirements of CS systems, MBR emerges as a suitable technology in this field. However, the persistent challenge in these systems, namely membrane fouling, demands further research and investigation.

References

Akanyeti, I., Temmink, H., Remy, M., & Zwijnenburg, A. (2010). Feasibility of bio flocculation in a high-loaded membrane bioreactor for improved energy recovery from sewage. *Water Science and Technology, 61*(6), 1433–1439.

Alloul, A., Ganigué, R., Spiller, M., Meerburg, F., Cagnetta, C., Rabaey, K., & Vlaeminck, S. E. (2018). Capture–ferment–upgrade: A three-step approach for the valorization of sewage organics as commodities. *Environmental Science & Technology, 52*(12), 6729–6742.

Asumadu-Sarkodie, S., & Owusu, P. A. (2016). Feasibility of biomass heating system in middle east technical university. *Northern Cyprus Campus. Cogent Engineering, 3*(1), 1134304.

Başaran, S. T., Aysel, M., Kurt, H., Ergal, I., Kumru, M., Akarsubaşı, A., Sözen, S., & Orhon, D. (2012). Removal of readily biodegradable substrate in super fast membrane bioreactor. *Journal of Membrane Science, 423*, 477–486.

Belli, T. J., Bernardelli, J. K. B., da Costa, R. E., Bassin, J. P., Amaral, M. C. S., & Lapolli, F. R. (2017). Effect of solids retention time on nitrogen and phosphorus removal from municipal wastewater in a sequencing batch membrane bioreactor. *Environmental Technology, 38*(7), 806–815.

Bezirgiannidis, A., Plesia-Efstathopoulou, A., Ntougias, S., & Melidis, P. (2019). Combined chemically enhanced primary sedimentation and biofiltration process for low cost municipal wastewater treatment. *Journal of Environmental Science and Health, Part A, 54*(12), 1227–1232.

Bhaduri, A., Ringler, C., Dombrowski, I., Mohtar, R., & Scheumann, W. (2015). Sustainability in the water–energy–food nexus. In (Vol. 40, pp. 723–732). Taylor & Francis.

Cao, S., Lu, D., Phua, K., Yan, W., Le, C., Tao, G., & Zhou, Y. (2020). Organics transformation and energy production potential in a high rate a-stage system: A demo-scale study. *Bioresource Technology, 295*, 122300.

Daelman, M. R., van Voorthuizen, E. M., van Dongen, U. G., Volcke, E. I., & van Loosdrecht, M. C. (2015). Seasonal and diurnal variability of N2O emissions from a full-scale municipal wastewater treatment plant. *Science of the Total Environment, 536*, 1–11.

Dai, W., Xu, X., & Yang, F. (2018). High-rate contact stabilization process-coupled membrane bioreactor for maximal recovery of organics from municipal wastewater. *Water, 10*(7), 878.

Derakhshesh, S., Sharghi, E. A., Bonakdarpour, B., & Khoshnevisan, B. (2022). Integrating electrocoagulation process with up-flow anaerobic sludge blanket for in-situ biomethanation and performance improvement. *Bioresource Technology, 360*, 127536.

Dong, T., Shewa, W. A., Murray, K., & Dagnew, M. (2019). Optimizing chemically enhanced primary treatment processes for simultaneous carbon redirection and phosphorus removal. *Water, 11*(3), 547.

Emaminejad, S. A., Avval, S. S., & Bonakdarpour, B. (2019). Gaining deeper insights into the bioflocculation process occurring in a high loaded membrane bioreactor used for the treatment of synthetic greywater. *Chemosphere, 230*, 316–326.

Faridizad, G., Sharghi, E. A., & Bonakdarpour, B. (2022). The use of membrane bioreactors in high rate activated sludge processes: How and why sludge retention time affects membrane fouling. *Journal of Water Process Engineering, 47*, 102807.

Faust, L., Temmink, H., Zwijnenburg, A., Kemperman, A. J., & Rijnaarts, H. (2014a). Effect of dissolved oxygen concentration on the bioflocculation process in high loaded MBRs. *Water Research, 66*, 199–207.

Faust, L., Temmink, H., Zwijnenburg, A., Kemperman, A. J., & Rijnaarts, H. (2014b). High loaded MBRs for organic matter recovery from sewage: Effect of solids retention time on bioflocculation and on the role of extracellular polymers. *Water Research, 56*, 258–266.

Ge, H., Batstone, D. J., & Keller, J. (2013). Operating aerobic wastewater treatment at very short sludge ages enables treatment and energy recovery through anaerobic sludge digestion. *Water Research, 47*(17), 6546–6557.

Ge, H., Batstone, D. J., Mouiche, M., Hu, S., & Keller, J. (2017). Nutrient removal and energy recovery from high-rate activated sludge processes–Impact of sludge age. *Bioresource Technology, 245*, 1155–1161.

Gu, Y., Li, Y., Li, X., Luo, P., Wang, H., Robinson, Z. P., Wang, X., Wu, J., & Li, F. (2017). The feasibility and challenges of energy self-sufficient wastewater treatment plants. *Applied Energy, 204*, 1463–1475.

Guthi, R.-S., Tondera, K., Gillot, S., Buffière, P., Boillot, M., & Chazarenc, F. (2022). A-Stage process–challenges and drawbacks from lab to full scale studies: A review. *Water research, 226*, 119044.

Guven, H., Dereli, R. K., Ozgun, H., Ersahin, M. E., & Ozturk, I. (2019a). Towards sustainable and energy efficient municipal wastewater treatment by up-concentration of organics. *Progress in Energy and Combustion Science, 70*, 145–168.

Guven, H., Ersahin, M. E., Dereli, R. K., Ozgun, H., Isik, I., & Ozturk, I. (2019b). Energy recovery potential of anaerobic digestion of excess sludge from high-rate activated sludge systems co-treating municipal wastewater and food waste. *Energy, 172*, 1027–1036.

Guven, H., Ersahin, M. E., Dereli, R. K., Ozgun, H., Sancar, D., & Ozturk, I. (2017). Effect of hydraulic retention time on the performance of high-rate activated sludge system: A pilot-scale study. *Water, Air, & Soil Pollution, 228*, 1–10.

Guven, H., Ozgun, H., Ersahin, M. E., Dereli, R. K., Sinop, I., & Ozturk, I. (2019c). High-rate activated sludge processes for municipal wastewater treatment: The effect of food waste addition and hydraulic limits of the system. *Environmental Science and Pollution Research, 26*, 1770–1780.

Hamedi, H., Ehteshami, M., Mirbagheri, S. A., Rasouli, S. A., & Zendehboudi, S. (2019). Current status and future prospects of membrane bioreactors (MBRs) and fouling phenomena: A systematic review. *The Canadian Journal of Chemical Engineering, 97*(1), 32–58.

Hanasaki, N., Fujimori, S., Yamamoto, T., Yoshikawa, S., Masaki, Y., Hijioka, Y., Kainuma, M., Kanamori, Y., Masui, T., & Takahashi, K. (2013). A global water scarcity assessment under Shared Socio-economic Pathways-Part 2: Water availability and scarcity. *Hydrology and Earth System Sciences, 17*(7), 2393–2413.

He, C., Wang, K., Wang, W., Luo, J., & Fang, K. (2023). Chemically enhanced high-loaded membrane bioreactor (CE-HLMBR) for A-stage municipal wastewater treatment: Pilot-scale experiments and practical feasibility evaluation. *Separation and Purification Technology, 307*, 122853.

Heidrich, E., Curtis, T., & Dolfing, J. (2011). Determination of the internal chemical energy of wastewater. *Environmental Science & Technology, 45*(2), 827–832.

Hoegh-Guldberg, O., Jacob, D., Taylor, M., Guillén Bolaños, T., Bindi, M., Brown, S., Camilloni, I. A., Diedhiou, A., Djalante, R., & Ebi, K. (2019). The human imperative of stabilizing global climate change at 1.5 C. *Science, 365*(6459), 6974.

Hong, P.-N., Taing, C., Phan, P.-T., & Honda, R. (2018). Polarity-molecular weight profile of extracellular polymeric substances in a membrane bioreactor: Comparison between bulk sludge and cake layers. *Journal of Water and Environment Technology, 16*(1), 40–53.

Huang, B.-C., He, C.-S., Fan, N.-S., Jin, R.-C., & Yu, H.-Q. (2020). Envisaging wastewater-to-energy practices for sustainable urban water pollution control: Current achievements and future prospects. *Renewable and Sustainable Energy Reviews, 134*, 110134.

Jimenez, J., Miller, M., Bott, C., Murthy, S., De Clippeleir, H., & Wett, B. (2015). High-rate activated sludge system for carbon management–Evaluation of crucial process mechanisms and design parameters. *Water Research, 87*, 476–482.

Jin, Z., Gong, H., Temmink, H., Nie, H., Wu, J., Zuo, J., & Wang, K. (2016). Efficient sewage pre-concentration with combined coagulation microfiltration for organic matter recovery. *Chemical Engineering Journal, 292*, 130–138.

Jones, E. R., Van Vliet, M. T., Qadir, M., & Bierkens, M. F. (2021). Country-level and gridded estimates of wastewater production, collection, treatment and reuse. *Earth System Science Data, 13*(2), 237–254.

Kaygusuz, K. (2012). Energy for sustainable development: A case of developing countries. *Renewable and Sustainable Energy Reviews, 16*(2), 1116–1126.

Kimura, K., Honoki, D., & Sato, T. (2017). Effective physical cleaning and adequate membrane flux for direct membrane filtration (DMF) of municipal wastewater: Up-concentration of organic matter for efficient energy recovery. *Separation and Purification Technology, 181*, 37–43.

Kinyua, M. N., Elliott, M., Wett, B., Murthy, S., Chandran, K., & Bott, C. B. (2017a). The role of extracellular polymeric substances on carbon capture in a high rate activated sludge A-stage system. *Chemical Engineering Journal, 322*, 428–434.

Kinyua, M. N., Miller, M. W., Wett, B., Murthy, S., Chandran, K., & Bott, C. B. (2017b). Polyhydroxyalkanoates, triacylglycerides and glycogen in a high rate activated sludge A-stage system. *Chemical Engineering Journal, 316*, 350–360.

Lazarova, V., Choo, K.-H., & Cornel, P. (2012). *Water-energy interactions in water reuse.* IWA publishing.

Leal, L. H., Temmink, H., Zeeman, G., & Buisman, C. (2010). Bioflocculation of grey water for improved energy recovery within decentralized sanitation concepts. *Bioresource Technology, 101*(23), 9065–9070.

Leong, K.-Y., Pooi, C.-K., Yeap, T.-S., & Ng, H. Y. (2020). Influence of bio (de) flocculation on activated sludge processes in membrane bioreactors. In *Current Developments in Biotechnology and Bioengineering* (pp. 375–396). Elsevier.

Li, W.-W., & Yu, H.-Q. (2016). Advances in energy-producing anaerobic biotechnologies for municipal wastewater treatment. *Engineering, 2*(4), 438–446.

Li, Y., Chen, W., Liu, Q., Xiang, W., Zheng, X.-C., & Qu, J.-X. (2019). Effect on composition and quantity of extracellular polymeric substances in an innovative hybrid membrane bioreactor under different sludge retention times. *Environmental Pollutants and Bioavailability, 31*(1), 103–111.

Liddle, B. (2017). Urbanization and inequality/poverty. *Urban Science, 1*(4), 35.

Liu, Y.-J., Gu, J., & Liu, Y. (2018). Energy self-sufficient biological municipal wastewater reclamation: Present status, challenges and solutions forward. *Bioresource Technology, 269*, 513–519.

Lüdtke, D., Luetkemeier, R., Schneemann, M., & Liehr, S. (2021). Increase in Daily Household Water Demand during the First Wave of the COVID-19 Pandemic in Germany. *Water, 13*, 260.

Meerburg, F. A., Boon, N., Van Winckel, T., Vercamer, J. A., Nopens, I., & Vlaeminck, S. E. (2015). Toward energy-neutral wastewater treatment: A high-rate contact stabilization process to maximally recover sewage organics. *Bioresource Technology, 179*, 373–381.

Meerburg, F. A., Vlaeminck, S. E., Roume, H., Seuntjens, D., Pieper, D. H., Jauregui, R., Vilchez-Vargas, R., & Boon, N. (2016). High-rate activated sludge communities have a distinctly different structure compared to low-rate sludge communities, and are less sensitive towards environmental and operational variables. *Water Research, 100*, 137–145.

Miller, M., De Armond, J., Elliott, M., Kinyua, M., Kinnear, D., Wett, B., & Bott, C. (2016). Settling and dewatering characteristics of an A-stage activated sludge process proceeded by shortcut biological nitrogen removal. *The International Journal of Water Wastewater Treat, 2*(5), 1–8.

Modin, O., Saheb Alam, S., Persson, F., & Wilén, B.-M. (2015). Sorption and release of organics by primary, anaerobic, and aerobic activated sludge mixed with raw municipal wastewater. *PLoS ONE, 10*(3), e0119371.

Murugesan, K., Selvam, A., & Wong, J. W. (2014). Flocculation and dewaterability of chemically enhanced primary treatment sludge by bioaugmentation with filamentous fungi. *Bioresource Technology, 168*, 198–203.

Patel, H. V., Brazil, B., Lou, H. H., Jha, M. K., Luster-Teasley, S., & Zhao, R. (2021). Evaluation of the effects of chemically enhanced primary treatment on landfill leachate and sewage co-treatment in publicly owned treatment works. *Journal of Water Process Engineering, 42*, 102116.

Pathak, S. (2014). Energy crisis: A review. *International Journal of Engineering Research and Applications, 4*, 845–851.

Rahman, A., Meerburg, F. A., Ravadagundhi, S., Wett, B., Jimenez, J., Bott, C., Al-Omari, A., Riffat, R., Murthy, S., & De Clippeleir, H. (2016). Bioflocculation management through high-rate contact-stabilization: A promising technology to recover organic carbon from low-strength wastewater. *Water Research, 104*, 485–496.

Ren, C., Liu, X., Zhang, Z., Wang, Y., Duan, W., Li, S., & Liang, Z. (2016). CRISPR/Cas9-mediated efficient targeted mutagenesis in Chardonnay (Vitis vinifera L.). *Scientific Reports, 6*(1), 32289.

Sancho, I., Lopez-Palau, S., Arespacochaga, N., & Cortina, J. (2019). New concepts on carbon redirection in wastewater treatment plants: A review. *Science of the Total Environment, 647*, 1373–1384.

Sarker, B., Keya, K. N., Mahir, F. I., Nahiun, K. M., Shahida, S., & Khan, R. A. (2021). Surface and ground water pollution: Causes and effects of urbanization and industrialization in South Asia. *Scientific Review, 7*(3), 32–41.

Sarparastzadeh, H., Saeidi, M., Naeimpour, F., & Aminzadeh, B. (2007). Pretreatment of municipal wastewater by enhanced chemical coagulation.

Şen, Z. (2009). Global warming threat on water resources and environment: A review. *Environmental Geology, 57*, 321–329.

Seuntjens, D., Bundervoet, B. L., Mollen, H., De Mulder, C., Wypkema, E., Verliefde, A., Nopens, I., Colsen, J. G., & Vlaeminck, S. (2016). Energy efficient treatment of A-stage effluent: Pilot-scale experiences with shortcut nitrogen removal. *Water Science and Technology, 73*(9), 2150–2158.

Shafaei, N., Bonakdarpour, B., & Faridizad, G. (2022). The application of membrane sequencing batch reactors in high rate activated sludge processes: The effect of the organic loading rate on the bioflocculation and fouling phenomenon. *Environmental Science: Water Research & Technology, 8*(7), 1561–1578.

Shen, Y., Oki, T., Kanae, S., Hanasaki, N., Utsumi, N., & Kiguchi, M. (2014). Projection of future world water resources under SRES scenarios: An integrated assessment. *Hydrological Sciences Journal, 59*(10), 1775–1793.

Silva, A. F., Antunes, S., Freitas, F., Carvalho, G., Reis, M. A., & Barreto Crespo, M. T. (2017). Impact of sludge retention time on MBR fouling: Role of extracellular polymeric substances determined through membrane autopsy. *Biofouling, 33*(7), 556–566.

Smith, A. L., Stadler, L. B., Cao, L., Love, N. G., Raskin, L., & Skerlos, S. J. (2014). Navigating wastewater energy recovery strategies: A life cycle comparison of anaerobic membrane bioreactor and conventional treatment systems with anaerobic digestion. *Environmental Science & Technology, 48*(10), 5972–5981.

Suresh, A., Grygolowicz-Pawlak, E., Pathak, S., Poh, L. S., bin Abdul Majid, M., Dominiak, D., Bugge, T. V., Gao, X., & Ng, W. J. (2018). Understanding and optimization of the flocculation process in biological wastewater treatment processes: A review. *Chemosphere, 210*, 401–-416

Taboada-Santos, A., Rivadulla, E., Paredes, L., Carballa, M., Romalde, J., & Lema, J. M. (2020). Comprehensive comparison of chemically enhanced primary treatment and high-rate activated sludge in novel wastewater treatment plant configurations. *Water Research, 169*, 115258.

Valipour, M., & Singh, V. P. (2016). Global experiences on wastewater irrigation: Challenges and prospects. *Balanced urban development: Options and strategies for liveable cities* (pp. 289–327).

Wastewater, W. (2017). The Untapped Resource, The United Nations World Water Development Report. In UNESCO World Water Assessment Programme.

Yagci, N., Konuk, M., Sozen, S., Meriç, S., & Orhon, D. (2016). Chemically enhanced membrane process–towards a novel sewage treatment concept to potentially replace biological processes. *Desalination and Water Treatment, 57*(35), 16238–16249.

Zeng, Q., Hao, T., Yuan, Z., & Chen, G. (2020). Dewaterability enhancement and sulfide mitigation of CEPT sludge by electrochemical pretreatment. *Water Research, 176*, 115727.

A Proposed Model for Energy Generation from MSW of Kolkata Through Gasification Route

S. K. Mahasin and Pradip Mondal

Abstract Kolkata, famously known as the City of Joy, generates around 4000MT of municipal solid waste (MSW) on daily basis. Recent research trends impart that energy can be generated from such huge amount of generated waste and thus a thermal model of a combined cycle thermal power plant using gasification of MSW in a downdraft gasifier has been developed in this study. The plant consists of three modules: Gasifier Module, Toping GT Module and Bottoming ST Module. Plant's performance has been analysed by varying the pressure ratio of the compressor and equivalence ratio of combustion chamber of the topping gas turbine module. Based on the analysis and in consideration of metallurgical limit of GT material, most suitable pressure ratio (of 6–8) and equivalence ratio (of 2) have been suggested. Plant's energetic and exergetic efficiency have been noted as 37.14% and 32.87%, respectively at the pressure ratio of 8 and oxygen to equivalence ratio of 2 while gasifier bed temperature is 800 C and equivalence ratio is 0.3 with atmospheric pressure.

Nomenclatures

C	Carbon
H	Hydrogen
O	Oxygen
N	Nitrogen
w	No. of mole of moisture
$\dot{m}$	Mass flow rate of feedstock
C_v	Calorific value of feedstock

S. K. Mahasin
Solid Waste Management Department, Kolkata Municipal Corporation, Kolkata, West Bengal 700013, India

P. Mondal (✉)
Indian Institute of Science Engineering and Technology Shibpur, Howrah, West Bengal, India
e-mail: mondal.pradip87@gmail.com

A. Anouzla and S. Souabi (eds.), *Generation of Energy from Municipal Solid Waste*,
https://doi.org/10.1007/978-3-031-74334-4_7

m	No. of mole of oxygen for gasification
$m\prime$	No. of mole of oxygen for combustion
X_1	No. of mole of hydrogen
X_2	No. of mole of CO
X_3	No. of mole of CO_2
X_4	No. of mole of H_2O
X_5	No. of mole of CH_4
X_6	No. of mole of N_2
T	Temperature
ΔG	Change of Gibbs Function
ΔH	Change of enthalpy
ΔS	Change of entropy
H_f^0	Standard state enthalpy
S_f^0	Standard state entropy
C_p	Specific heat at constant pressure
W	Work Output
M	Mass
HHV	Higher heating value
MC	Moisture content
M_w	Molecular weight of the feedstock
C_1	Compressor 1
C_2	Compressor 2

Greek Letters

Φ	Equivalence ratio
λ	Equivalence ratio in combustor
η	Efficiency
r_p	Pressure ratio

Sub-scripts

x	No. of hydrogen atom per mole of the feedstock
y	No. of oxygen atom per mole of the feedstock
z	No. of nitrogen atom per mole of the feedstock
shift	Water gas shift reaction
meth	Methanation reaction
fds	Feedstock
eq	Equivalent
o	Reference state

B	Bed
G	Gasifier
Comp	Compressor
Out	Outlet
In	Inlet
GT	Gas Turbine
ST	Steam Turbi
air	Standard air
Stm	Steam
Flue	Flue gas
FP	Feed pump
Gen	Generator
Stk	Stack

1 Introduction

Solid waste management is an important obligatory function for all the Urban Local Bodies (ULB) and need to perform scientifically for avoiding environmental and health hazards. It comprises of generation of waste, collection, storage, transportation, processing and finally it's disposal. About 95% of the world's municipal solid waste (MSW) is disposed of by landfilling or dumping openly on land or on riverbank or thrown into the sea (2016). But it can be used for the benefit of the society if energy contained in it is harnessed by employing waste to energy technologies (WtE) thereby reducing environmental damages.

Chakraborty et al. (2013) assessed energy generation potential of MSW in Delhi taking five WtE technologies such as bio-methanation, incineration, gasification/pyrolysis, RDF and plasma arc gasification. Zainal et al. (2001) developed equilibrium model of a downdraft gasifier to predict producer gas composition and its calorific value. Effect of moisture content and gasification temperature on the calorific value of the producer gas have been presented. Melgar et al. (2007) developed a mathematical model for gasification of biomass in a downdraft gasifier. Effect of fuel air ratio and moisture content on the producer gas composition has been presented. Barman et al. (2012) presented a thermodynamic model of a fixed bed downdraft gasifier considering tar as a gasification product. Jarungthammachote et al. (2007) developed a thermodynamic equilibrium model using corrected equilibrium constant of water gas shift reaction and methanation reaction to predict producer gas composition with good accuracy during gasification of biomass in a downdraft gasifier. Bhavanam et al. (2013) presented mathematical model of solid waste gasification to predict syngas composition in a downdraft fixed bed reactor. Syed et al. (2012) developed a mathematical model to predict cold gasification efficiency (CGE) of different feedstock in an entrained flow gasifier. Mondal et al. (2020) carried out 4E analysis of GT-ST combined cycle plant of about 300 MWe for Kolkata Metropolitan city, India, integrated with MSW gasification. Dutta et al. (2010) has presented thermal performance and sizing of a biomass based decentralized power generation unit integrated with externally fired gas turbine.

Thus in consultation with the contemporary literature, a novel MSW gasification based combined cycle plant has been modelled in this study which can tackle the waste generated from Kokata. Energetic and exetgetic performance of the plant is accessed via determining the new electrical output, energy efficiency as well as exergy efficiency at different pressure ratio (4–16) of the compressor and at different equivalence ratio ($\lambda = 1$–6) of the combustion chamber.

2 Development of the Proposed Model

2.1 Description

Layout of the proposed model is given in Fig. 1. Segregated MSW (Feedstock) is fed to the gasifier along with controlled amount of standard air. Gasification is carried out in an oxygen deficient environment and generated producer gas in the gasifier is fed to the combustion chamber through the cleaning tank. Combustion of the producer gas is taken place in the combustion chamber in presence of compressed excess air thereby producing high pressure high temperature flue gas which is fed to the gas turbine fitted with generator where it expands up to the atmospheric pressure and generates power. High temperature flue gas after expanding in the gas turbine passes through HRSG where it transfer excess heat to water producing steam in the bottom cycle. Steam expands in the steam turbine and power is generated.

2.2 Feedstock and Assumptions

1. Chemical formula of the feedstock ($C_1H_{1.8121}O_{0.4640}N_{0.0123}$) considered in this model has been derived from the results of the ultimate analysis of the MSW samples of Kolkata metropolitan city and assumptions made as per Mondal et al. (2020). No pressure loss and heat loss is considered for this model.

2.3 Gasifier Module

Air gasification of the feedstock is done in a downdraft gasifier between 800°C and 1.01325 bar with equivalence ratio 0.3. In the gasification zone it is considered that product species attains equilibrium condition and behave like a perfect gas.

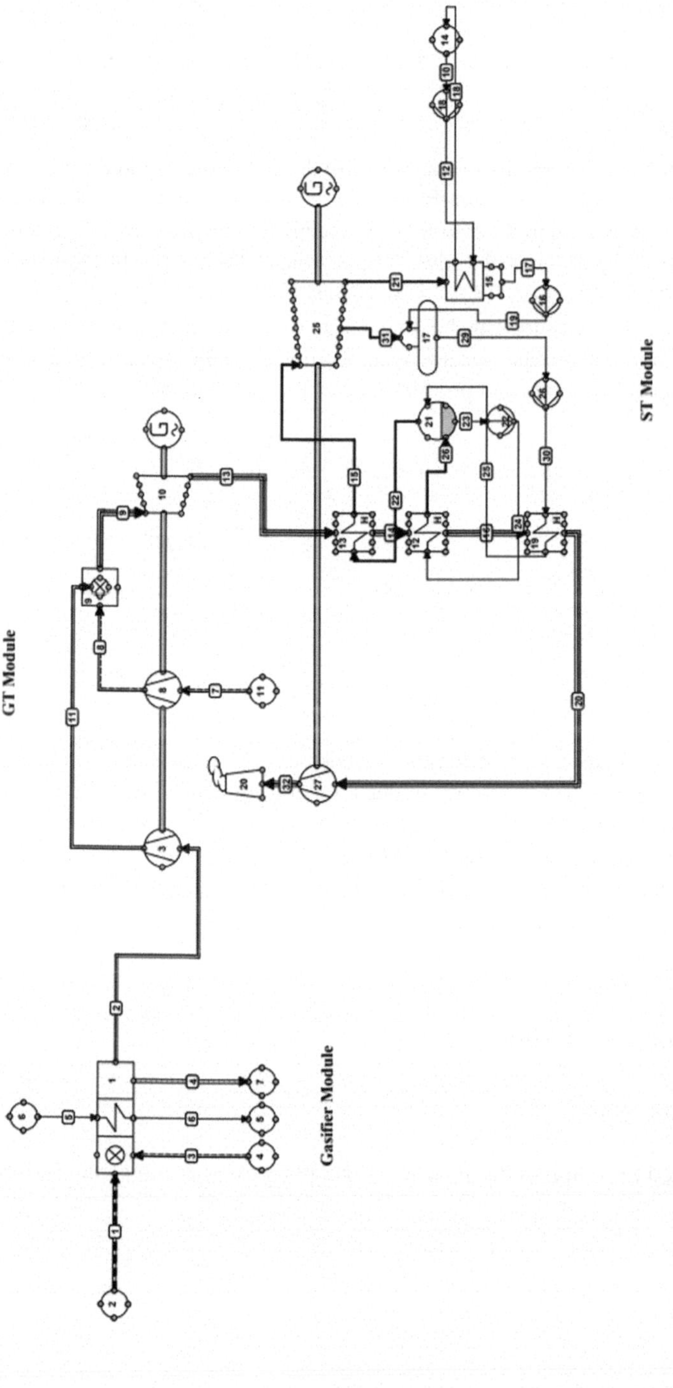

Fig. 1 Layout of the proposed model

Global gasification reaction for the feedstock ($CH_xO_yN_z$) using air can be written as

$$CH_xO_yN_z + wH_2O + m(O_2 + 3.76N_2) = X_1H_2 + X_2CO + X_3CO_2 + X_4H_2O + X_5CH_4 + X_6N_2 \quad (1)$$

where, x, y and z are the number of atoms of hydrogen, oxygen and nitrogen respectively based on a single atom of carbon in the feedstock. w and m are the amount of moisture and oxygen in kmol per kmol of the feedstock respectively. X_i, where i = 1 to 6 in Eq. (1), is the number of moles of the species 'i' of the product gases in kmol per kmol of the feedstock and $X_6 = \frac{z}{2} + 3.76m$.

Gasification temperature can be found out from the energy balance equation between reactants and product species. Considering adiabatic process, energy balance equation between reactants and product species can be written as

$$H^0_{f,fds} + wH^0_{f,H_2O} + m\left(H^0_{f,O_2} + \int_{T_o}^{T_B} C_{P,O_2}dT\right) + 3.76m\left(H^0_{f,N_2} + \int_{T_o}^{T_B} C_{P,N_2}dT\right) = \sum_{i=Product} x_i\left(H^0_{f,i} + \int_{T_o}^{T_G} C_{p,i}dT\right) \quad (2)$$

where,

$$C_{p,i}(T) = a_i + b_iT + c_iT^2 + d_iT^3 \quad (3)$$

Here, $H^0_{f,fds}$ is the enthalpy of formation of feedstock, $H^0_{f,i}$ is the enthalpy of formation of the product species i, T_o is the reference temperature which is 298.15 K, T_B is the bed temperatureand T_G is the gasification temperature. $C_{p,i}$ is the average specific heat of species 'i' which depends on the absolute temperature T and coefficient constant. Enthalpy of formation of feedstock can be calculated from HHV as per Arafat et al. (2013).

2.4 Toping GT Module

2.4.1 Compressor

Power required to compress the air which is supplied to the combustion chamber can be written as

$$W_C = C_{p,air}\left(T_{Comp,Out} - T_{Comp,In}\right) \quad (4)$$

2.4.2 Combuster

Combustion equation in combustion chamber can be written as

$$X_1H_2 + X_2CO + X_3CO_2 + X_4H_2O + X_5CH_4 + X_6N_2 + m'O_2 = X_8CO_2 + X_9H_2O + X_{10}O_2 + X_{11}N_2 + X_{12}SO_2 \quad (5)$$

where $X_{i,i=1to6}$ in Eq. (4), is the number of moles of the species '*i*' of the producer gas obtained during gasification and $X_{j,j=8to12}$ is the number of mole of combustion product, flue gas, which greatly depends on the equivalence ration in combustion chamber.

Combustion temperature which is the inlet temperature of flue gas in gas turbine ($T_{GT,In}$) can be found out from the energy balance equation between producer gas species and flue gas as

$$\sum_{i=1}^{i=6} x_i\left(H_{f,i}^0 + \int_{T_o}^{T_G} C_{p,i}dT\right) + m'\left(H_{f,O_2}^0 + \int_{T_o}^{T_{Comp.\ Out}} C_{P,O_2}dT\right) + 3.76m'\left(H_{f,N_2}^0 + \int_{T_o}^{T_{Comp.\ Out}} C_{P,N_2}dT\right) = \sum_{j=8}^{j=12} x_j\left(H_{f,j}^0 + \int_{T_o}^{T_{GT,In}} C_{p,j}dT\right) \quad (6)$$

2.4.3 Gas Turbine

Power generated in the gas turbine can be written as

$$W_{GT} = C_{p,flue}\left(T_{GT,In} - T_{GT,Out}\right)\eta_{Gen} \quad (7)$$

$$W_{net\_GT} = W_{GT} - (W_{C_1} + W_{C_2}) \quad (8)$$

2.5 *Bottoming ST Module*

It consists of Steam Turbine, a power generation component and Feed Pump, a power consuming component.

Net power outputfrom this module can be written as

$$W_{ST} = M_{Stm}\left(h_{ST,In} - h_{ST,Out}\right)\eta_{Gen} \quad (9)$$

$$W_{net\_ST} = W_{ST} - W_{FP} \tag{10}$$

Equation for finding out mass of steam generated in HRSG which is expanded in the steam turbine can be written as

$$M_{flue}.C_{p,flue}\left(T_{GT,Out} - T_{Stk}\right) = M_{Stm}\left(h_{ST,In} - h_{FP,Out}\right) \tag{11}$$

2.6 *Plant's Output Parameter*

Performance of the plant is judged energetically as well as exergitically. Plant's net output is determined as:

$$W_{net} = W_{net\_GT} + W_{net\_ST} \tag{12}$$

And plant's energy efficiency is calculated as:

$$\frac{W_{net}}{\dot{m}.C_v} \tag{13}$$

whereas plant's exergetic efficiency is evaluated as:

$$\frac{W_{net}}{\dot{m} \cdot C_v \cdot \text{ß}} \tag{14}$$

where β is the exergy multiplication factor of the fuel and calculated as per the work of Mondal et al. (2020).

3 Results and Discussion

Table 1 shows the base case performance of the plant (both input and output parameters). Plant's net output obtained as 152.9MWe and energy efficiency and exergy efficiency between 31.832 and 28.218 respectively. Details are given in the table below. Product gas composition of this gasifier model matches well with the available works in the literature with minimum deviation.

From the Fig. 2. It is seen that output of power of gas turbine is increased with increase in pressure ratio, whereas output of power of steam turbine is decreased with increase in pressure ratio. Combined effect of these two is seen as resultant power

Table 1 Base case performance of the plant

Input Parameter	
Gasifier	
i. Flow of MSW (Tone per hour)	108
ii. Gasification Pressure (bar)	1.01325
iii. Gasification Temperature (°C)	800
iv. Equivalence Ratio	0.3
Combustor	
i. Pressure Ratio	4
ii. Equivalence ratio	1
Steam Turbine	
i. Pressure (bar)	10
ii. Temperature (°C)	250
Condenser	
i. Pressure (bar)	0.0424
ii. Pinch Point (°C)	10
Boiler	
i. Circulation Ratio	4
Stack	
i. Outlet Temperature of Flue Gas (°C)	120
Output Parameter	
Producer Gas Composition (% Mole)	
i. CO_2	0.0674
ii. CO	0.2101
iii. H_2O	0.0772
iv. H_2	0.2625
v. CH_4	0.0003
vi. N_2	0.3777
Combustion Temperature (°C)	2158
Inlet Temperature of Gas Turbine (°C)	2158
Net Work Output from GT (MWe)	60.1
Net Work Output from ST (MWe)	92.8
Total Output of the Plant (MWe)	153
Plant's Energy Efficiency (%)	31.83
Plant's Exergy Efficiency (%)	28.21

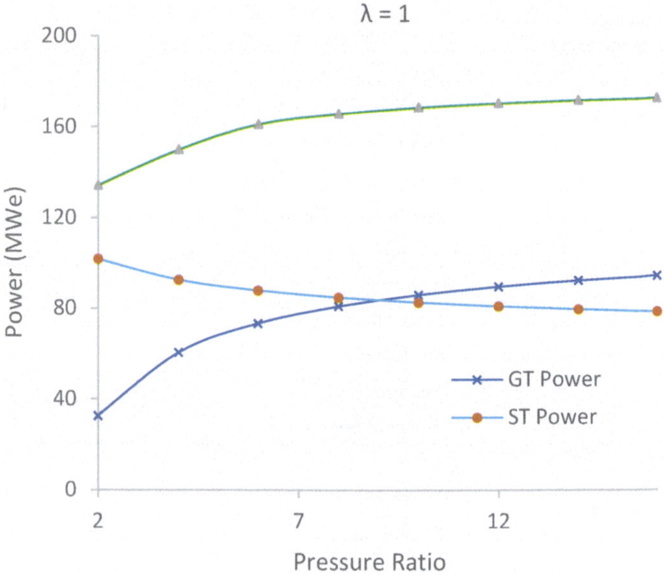

Fig. 2 Output of power with r_p

output of the plant is increased with increase in pressure ratio up to 16, thereafter effect of it is marginal.

From the Fig. 3. It is seen that output of power is greatly depends on the equivalence ratio. It increases with increase in pressure ratio up to 2 and decreases thereafter because of dilution of the flame by excess air due to increased pressure ratio.

From the Fig.4. It is seen that energy efficiency of the plant is increased with increase n pressure ratio. But effect of it becomes negligible beyond pressure ratio16. From the Fig.5. It is seen that exergy efficiency of the plant is also increased with increase in pressure ratio. But effect of it becomes negligible beyond pressure ratio16.

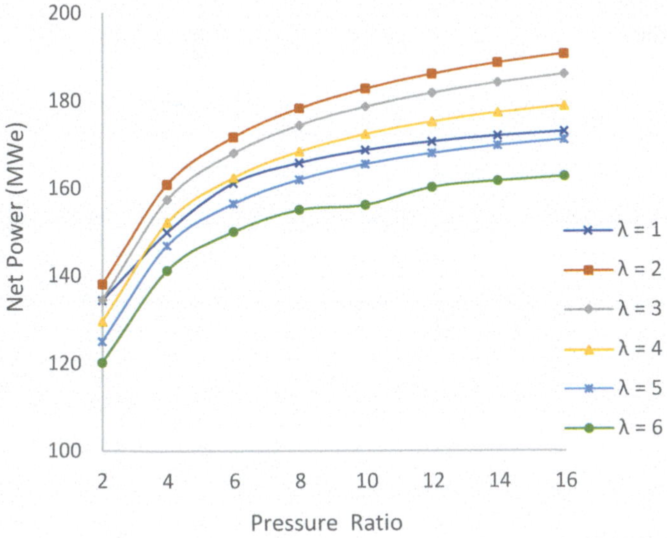

Fig. 3 Net power output with variable λ

Fig. 4 Plant's energy efficiency with r_p

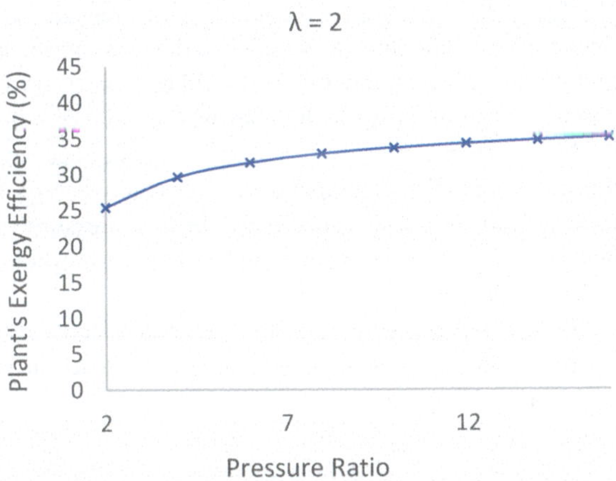

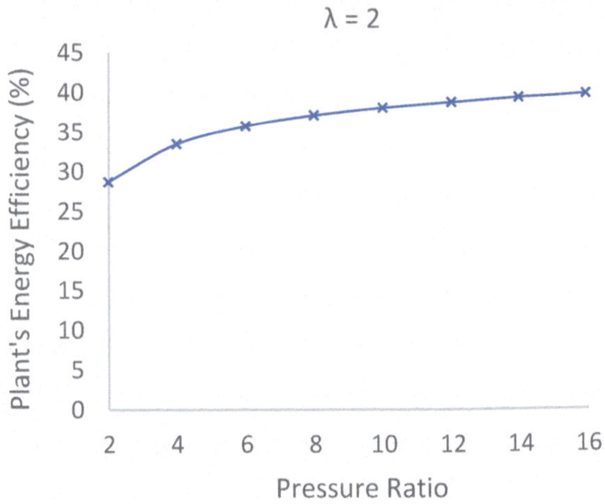

Fig. 5 Plant's exergy efficiency with r_p

4 Conclusion

From the above discussion it is seen that plant's net power output, energy and exergy efficiency are increased with increase in pressure ratio but at the same time inlet temperature of the flue gas which is fed to gas turbine is also increased. Owing to the metallurgical limit temperature should be from 1300°C to 1400 C. This temperature attains in the combustion chamber when pressure ratio remains from 6 to 8.

At gasifier bed temperature 800 C, equivalence ratio 0.3 and feedstock flow 108 tone per hour GT inlet temperature is seen between 1330 C and 1390 C when pressure ratio is from 6 and 8 respectively. In that pressure ratio energy efficiency is seen between 35% & 36.3% respectively and exergy efficiency is seen between 30.9% & 32.1% respectively.

So, it is a promising option to generate electrical power from MSW of Kolkata metropolitan city by using gasification process to curb environmental damage owing to the open dumping.

References

Arafat, H. A., & Jijakli, K. (2013). Modelling and comparative assessment of municipal solid wastegasification for energy production. *Waste Management, 33*, 1704–1713.

Barman, N. S., Ghosh, S., & De, S. (2012). Gasification of biomass in a fixed bed downdraft gasifier—A realistic model including tar. *Bioresource Technology, 107*, 505–511.

Bhavanam, A., & Sastry, R. C. (2013). Modelling of solid waste gasification process for synthesis gas production. *Journal of Scientific & Industrial Research, 72*, 611–616.

Chakraborty, M., Sharma, C., Pandey, J., & Gupta, P. (2013). Assessment of energy generation potentials of MSW in Delhi under different technological options. *Energy Conversion and Management., 75*, 249–255.

Chattopadhyay, S., Dutta, A., & Ray, S. (2016). Municipal solid waste management in Kolkata. *India-A Review. Waste Management, 29*, 1449–1458.

Datta, A., Ganguly, R., & Sarkar, L. (2010). Energy and exergy analyses of an externally fired gas turbine (EFGT) cycle integrated with biomass gasifier for distributed power generation. *Energy, 35*, 341–350.

Jarungthammachote, S., & Dutta, A. (2007). Thermodynamic equilibrium model and second law analysis of a downdraft waste gasifier. *Energy, 32*, 1660–1669.

Melgar, A., Perez, J. F., Laget, H., & Horillo, A. (2007). Thermochemical equilibrium modelling of a gasifying process. *Energy Conversion and Management, 48*, 59–67.

Mondal, P., Ghosh, S., & Das, S. (2020). MSW to energy: A novel and sustainable solution to waste management for Kolkata, India. *International Journal of Global Warming, 22*, 295–313.

Syed, S., Janajreh, I., & Ghenai, C. (2012). Thermodynamics equilibrium analysis within the entrained flow gasifier environment. *The International Journal of Thermal & Environmental Engineering, 4*, 47–54.

Zainal, Z. A., Ali, R., Lean, C. H., & Seetharamu, K. N. (2001). Prediction of performance of a downdraft gasifier using equilibrium modeling for different biomass materials. *Energy Conversion and Management, 42*, 1499–1515.

Gasification of Municipal Solid Waste

Cut Keumala Banaget

Abstract Gasification has been developed as an alternative method for producing cleaner energy. However, over 1,400 unit of operating gasifiers are using coal as energy source, while only 200 unit of gasifiers are using biomass and waste as feedstock (Higman in *State of the Gasification Industry: Worldwide Gasification and Syngas Databases 2016 Update*, 2016.). Feedstock used for gasification process became a main concern to ensure the process produce clean energy from renewable source. Gasification for treating municipal solid waste (MSW) considered as a promising method to solve environmental and energy production problems as it eliminates MSW and generates clean energy at the same time. The gasification process converts MSW into syngas mixture consist of carbon monoxide (CO), hydrogen (H_2) and carbon dioxide (CO_2), methane, and water vapour (Mondal in *International Journal of Hydrogen Energy* 47 20064–20075, 2022; Wiyono et al., in *Case Studies in Thermal Engineering* 20 1–11, 2020), that can be used as fuel for a gas turbine (Mazzoni et al., in *Energy* 196, 2020).

Gasification process needs medium or gasifiying agent to sustain the process, such as oxygen or air and steam. MSW gasification in the presence of air can generate syngas contained 12% of H_2 with lower heating value (LHV) ranged between 2 and 6 MJ/Nm^3 (Couto et al., 2015) and syngas contains about 19% of H_2 (Ozturk & Dincer in *International Journal of Hydrogen Energy* 46 6251–6261, 2021). While steam gasification can increase H_2 gas in syngas mixture significantly up to 55% with higher calorific value of about 13 MJ/Nm^3 (Wang et al., in *International Journal of Hydrogen Energy* 37 6503–6510, 2012). However, there are some challenges in treating MSW using gasification process which are regarding undesirable CO_2 production and tar formation. Both problems related to improvement of syngas quality and can be solved by catalyst use.

C. K. Banaget (✉)
Institut Teknologi Kalimantan, Balikpapan, Indonesia
e-mail: keumala@lecturer.itk.ac.id

A. Anouzla and S. Souabi (eds.), *Generation of Energy from Municipal Solid Waste*,
https://doi.org/10.1007/978-3-031-74334-4_8

1 Definition

Gasification is a thermochemical process that operates at high temperatures (between 700 and 1400°C) to convert organic material into syngas consist of carbon monoxide, hydrogen, methane, carbon dioxide, small amounts of light hydrocarbons and heavier hydrocarbons (tar) (Yang et al., 2021). Gasification has been considered as one of the cleanest energy production methods as it prevents dioxins and furans formation that emits from combustion process (Wang et al., 2022). However, only 200 unit of gasifiers are using biomass and waste as feedstock, while over 1,400 unit of operating gasifiers are still using coal as energy source (Higman, 2016). Therefore, fuel type or feedstock used for gasification process became a main concern to ensure the process produce clean energy from renewable source and gasification for treating municipal solid waste (MSW) considered as a promising method to solve environmental and energy production problems as it eliminates MSW and generates clean energy at the same time.

Different from pyrolysis process, gasification needs medium for its process. It can be carried out in the presence of gases such as air, pure oxygen, water steam, carbon dioxide, nitrogen, or a mixture of these to produce a syngas with a specific composition. Each gasifying agents has varied effects on gasification process. Gasification can be considered as combination of combustion and pyrolysis when oxygen used as a medium as it has beneficial effect on syngas production as it provides heat for the largely endothermic reduction reactions and the initial devolatilization of the biomass within the gasifier (Champion et al., 2014). In addition, oxygen and air are used as medium due to their high oxidising power (high carbon conversion) and low cost, while steam is used to increase the hydrogen content of the syngas (Diboma et al., 2023).

Gasification produces syngas which refers to the gas mixture mainly comprising between H_2, CO and CO_2, which is an intermediate used to produce hydrogen, ammonia, methanol, ethanol and synthetic hydrocarbons (Martínez et al., 2020). The composition of syngas is different depend on the process temperature. Gasification at moderate temperature of 800–900°C produces syngas consist of 10% hydrocarbons, mainly CH_4, C_2H_4, and C_6H_6, while process at higher temperature or about 1100–1300°C, produces a mixture of H_2 and CO with N_2, but containing almost no hydrocarbons like methane (Thomson et al., 2020). The gas product from gasification can be used as fuel for internal combustion engines, turbines and boilers, feedstock for producing chemicals and as fuel source for power generation and the transportation fuel (Fatema et al., 2022). Currently, the use of gasifiers is not only restricted to use in the chemical and petrochemical industries but also has a wide range of applications in many other fields such as other industrial sector and domestic use (Oliveira et al., 2021).

2 Medium

The most common gasifying agent or medium are air, oxygen, steam, or the combination of those medium (Wang et al., 2022). The gas composition derived from gasification process using air and steam can be seen in (Table 1) and Table 2, respectively.

2.1 *Air or Oxygen*

When economic feasibility is main consideration in designing the gasification process, the suitable medium for the process is air or oxygen-enriched air. However, N_2 contained in the air results in syngas with high N_2 content and lower H_2 content and calorific value of syngas (Fatema et al., 2022; Wang et al., 2022). The gas composition resulted from gasification using air with different air flow rate can be seen in Table 1. Gasification process using air with flow rate of 40 m^3/h produced higher CH_4 and H_2 concentration, compared to the process using air with flow rate of 70 and 100 m^3/h.

Table 1 Gas composition based on air flow rate

Air flow rate (m^3/h)	CO (%)	CO_2 (%)	CH_4 (%)	H_2 (%)
40	5	19	2	2
70	4	20	1.8	1.8
100	3	20.5	1.7	1.7

Source (Couto et al., 2015)

2.2 *Steam*

Steam is a gasifying medium that suitable to increase hydrogen production up to 30–60% of total syngas due to N_2-free environment during the process. Steam gasification lower amounts of nitrogen, oxides of sulphur and nitrogen and tar production in syngas mixture (Mondal, 2022). Steam as gasifying agent can also slightly increase LHV of syngas compared to air gasification. Steam mixed with air to improve lower heating value from 4.4 MJ/Nm^3 to 4.5 MJ/Nm^3 during air gasification (H. Wang et al., 2022). Combination of air and steam can be useful as air is an economic gasification agent that can lower the operational costs of gasification and air can supply heat during steam gasification to maintain the endothermic reactions, which can preserve auto-thermal features. Introducing more steam or H_2O to gasifier leads to the increase of CO consumption that promote CO_2 and H_2 production (Lv et al., 2019). The gas composition produced from gasification process at temperature of

Table 2 Gas composition based on steam/MSW ratio

S/MSW ratio	CO (%)	CO_2 (%)	CH_4 (%)	H_2 (%)
0	45	14	16	29
1.2	18	21	5	56
2.1	17	24	4	54
3.1	15	26	5	42

Source (Wang et al., 2012)

800°C using steam as gasifying agent can be seen in (Table 2). Steam as gasifying medium increase the production of H_2 from 29% up to 56%.

3 Mechanism

Gasification process consist of several stages which are drying at temperature of 150 C, pyrolysis at temperature from 150 to 700°C, oxidation or combustion at temperature from 700–1500 C, and reduction or char gasification at temperature from 800–1100 C (Hashem Samadi et al., 2023; Oliveira et al., 2021; Safarian et al., 2019).

- Drying stage is endothermic process of evaporation of compounds with low boiling occurs at temperature around 100°C. At this stage, the moisture is reduced and the feedstock is dried (Tezer et al., 2023). For biomass feed, the moisture can be reduced from about 35% to below 5% (Safarian et al., 2019).
- Pyrolysis stage starts at temperature of 200°C until 700 C, during this stage volatile is vaporised and the feedstock breakdown due to high temperature to synthetic gas which are hydrogen, carbon monoxide, carbon dioxide, methane, tar (heavier hydrocarbon) gases, and water vapor (Tezer et al., 2023). If oxygen used as gasifying agent inserted to the reactor, it reacts with the combustible substances to produce between CO_2 and H_2O (Oliveira et al., 2021; Safarian et al., 2019).
- Combustion or oxidation stage is an exothermic process that occurred when gasification agent injected into the system to reacted with the products obtained after the pyrolysis process and the heat generation required in endothermic reactions takes place (Champion et al., 2014; Tezer et al., 2023). The oxidation is defined as the portion of the gasifier where molecular oxygen has not been completely consumed yet.
- Gasification or reduction stage influences the quality of gas product. Several prominent reactions convert products from previous stages to produce combustible components, such as water–gas shift reaction that reduce H_2O to produce H_2, the Boudouard reaction forms CO from solid char and CO_2, as well as the steam reforming reaction, methanation reaction, and tar reduction reactions

that produces a mixture of non-condensable gasses, light hydrocarbons, and unconverted tar (Safarian et al., 2019).

In general, gasification process reactions are listed in equations below (Cheng et al., 2023).

Devolatilization	
MSW $\rightarrow \alpha_1$ Volatiles + α_2 Tar + α_3 Char	
Volatiles $\rightarrow \alpha_4 CO + \alpha_5 CO2 + \alpha_6 CH4 + \alpha_7 H2$	
Tar $\rightarrow$ Volatiles + Secondary Tar	
Heterogeneous solid–gas reaction	$\Delta H = -131$ kJ/mol
Char-C + $H_2O \rightarrow CO + H_2$ *(Primary water–gas reaction)*	$\Delta H = -90$ kJ/mol
Char-C + $2H_2O \rightarrow CO_2 + H_2$ *(Secondary water–gas reaction)*	$\Delta H = -172$ kJ/mol
Char-C + $CO_2 \rightarrow 2CO$ *(Boudouard reaction)*	$\Delta H = -75$ kJ/mol
$C + 2H_2 \rightarrow CH_4$ *(Methanation)*	
Combustion/partial oxidation	$\Delta H = -111$ kJ/mol
Char-C + $0.5O_2 \rightarrow CO$ *(Oxidation of Char, partial combustion)*	$\Delta H = -394$ kJ/mol
Char-C + $O_2 \rightarrow CO_2$ *(Oxidation of Char, complete combustion)*	$\Delta H = -283$ kJ/mol
$CO + 0.5O_2 \rightarrow CO_2$ *(Oxidation of CO)*	$\Delta H = -242$ kJ/mol
$H_2 + 0.5O_2 \rightarrow H_2O$ *(Oxidation of H_2)*	
Homogeneous gas-phase reaction	$\Delta H = +247$ kJ/mol
$CH_4 + CO_2 \rightarrow 2H_2 + 2CO$ *(Dry reforming reaction)*	$\Delta H = +206$ kJ/mol
$CH_4 + H_2O \rightarrow CO + 3H_2$ *(Steam reforming reaction)*	$\Delta H = -41$ kJ/mol
$CO_2 + H_2 \rightarrow CO + 2H_2O$ *(Water–gas shift reaction)*	

4 Operating Condition

The distribution of products which are gas or volatiles, tars and solid products, as well as composition of syngas is influenced by several factors which are temperature, equivalence ratio (ER), feedstock properties, and catalyst (Wang et al., 2023).

4.1 Temperature

Temperature is the most important parameter for gasification process as the main gasification reactions are endothermic. The gasification temperature also has significant influence on syngas yield and composition. The increase of the gasifier temperature diminishes tar and char yield, and increase carbon conversion efficiency (Thomson et al., 2020). Gasification at temperature of 800 to 900 C promote tar cracking reaction that led to an increase of gas production (Martínez et al., 2020; Thomson et al., 2020). The gas and H2 production from gasification with different temperature process can be seen in (Fig.1).

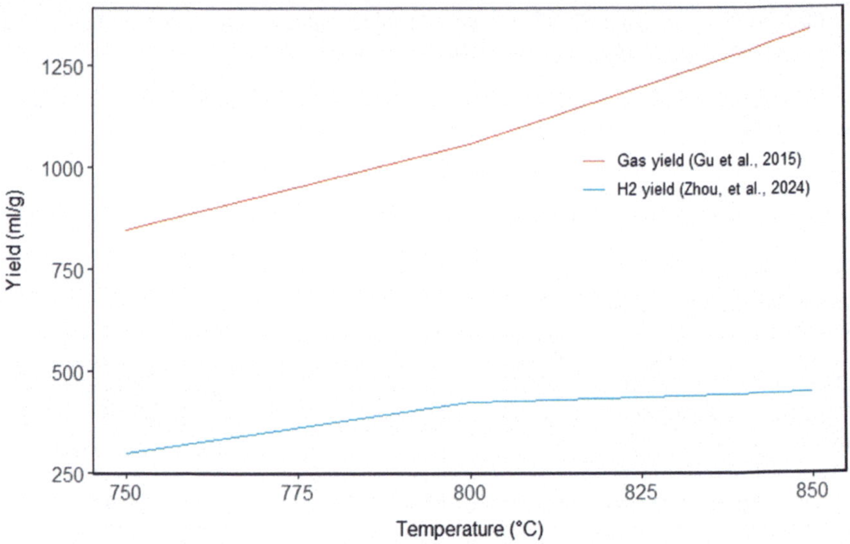

Fig. 1 Gas product from different temperature process

Table 3 Gas composition produced from different temperature process

Temperature	CO (%)	CH_4 (%)	H_2 (%)
500 < T < 700	14.5	19	36.5
700 < T < 900	30	18.5	46

Source (Tezer et al., 2023)

At higher temperature, light PAHs compounds in tar convert into the gas products with the increase of temperature (Lin et al., 2020). The H_2 yield gradually increased in the temperature range, which is mainly attributed to the higher reaction rates of the primary water gas reaction, steam reforming reaction, and tar cracking reaction at higher temperatures (Hu et al., 2015). At higher temperature, CO content also increase due to endothermic reactions and has higher growth rate than H_2 due to water–gas shift reaction where H_2 is converted to CO (Couto et al., 2015). On the other hand, CH_4 decrease due to the strong reaction of the endothermic steam reforming reaction (Martínez et al., 2020). The percentage of gas composition derived from gasification with different temperature can be seen in (Table 3).

4.2 *Air Equivalence Ratio (ER)*

Equivalence Ratio (ER) defined as the quantity of air supplied to gasifier per theoretical amount needed for complete combustion of feedstock (Champion et al., 2014) can be calculated using an equation below.

Table 4 Gas composition produced from different ER

ER	CO (%)	CO_2 (%)	CH_4 (%)	H_2 (%)
0.42	20.74	10.2	1.77	10.53
0.49	26.07	10.9	1.99	16.55
0.58	31.37	19.44	2.07	16.15

Source (Pandey et al., 2022)

$$ER = \frac{\text{Air/Fuel mass}}{\text{Stoichiometric Air/Fuel mass}}$$

Generally, higher ERs in favour to higher gas yield production with higher CO_2 levels, while lower in H_2, CO, tar yield, and the heating value of syngas (Thomson et al., 2020). The ER increase as the increase of air flow and the more air or oxygen introduces to the system promote the oxidation reaction or combustion that led to higher between CO_2 and H_2O content, while reducing CO and H_2 (Couto et al., 2015). Therefore, syngas produce from gasification with lower ER has higher quality due to higher values of the combustible gases (Couto et al., 2015). However, if the ER is too low, the gasification agent cannot supply enough oxygen to convert char into CO or CO_2 as ER promotes partial combustion between char, CO and H_2. The amount of gas composition produced from gasification with different Air Equivalence Ratio can be seen in (Table 4).

4.3 *Feedstock Properties*

Physical properties such as size and moisture content, as well as composition for biomass feedstock can influence gasification process. The smaller particle sizes produced higher product gas volumes and higher proportions of hydrogen. While larger particles in favour to produce to higher char yields and higher tar content in the gas as larger particle size restricts the rate of heat transfer within the particle (Thomson et al., 2020).

The higher moisture content increases steam concentration that promotes the water–gas shift reaction, methane steam reforming reaction, and secondary tar cracking reaction to produce more CO_2 and H_2. The higher moisture content of feedstock increases hydrogen production from 10 to 40%, while decrease CH_4 fraction from about 19.61 vol% to 12.28 vol% with feedstock moisture from 10 wt.% to 60 wt.% (Hu et al., 2015).

For biomass feedstock, the composition of biomass which are xylene, cellulose, and lignin have significant influence on gasification product as lignocellulose component influence incondensable gas production. The yield of incondensable gas products is also influenced by the amount of cellulose and hemicellulose, while the

Table 5 Gas composition produced from gasification with Fe_2O_3 catalyst

Catalyst/Dose	CO (mg/l)	CO_2 (mg/l)	CH_4 (mg/l)	H_2 (mg/l)
Fe_2O_3–0%	320	125	175	290
Fe_2O_3–10%	260	220	265	380
Fe_2O_3–20%	240	230	300	460
Fe_2O_3–30%	235	260	325	475
Fe_2O_3–40%	230	300	335	515

Source (Zhou et al., 2024)

quantity of oil product is favoured by the proportions of lignin (Thomson et al., 2020).

4.4 *Catalysts*

Catalysts use in gasification process has significant impact on enhancing char gasification or tar conversion (Thomson et al., 2020). Either catalyst type or concentration have certain effect to gas production. The gas production from gasification process using Fe_2O_3 catalyst with different dose can be seen in (Table 5).

(Zhou et al., 2024) used three different catalyses which are Fe_2O_3, Al_2O_3, and $CaCO_3$. Fe_2O_3 have significant effect on H_2 production and reduced CO significantly. While $CaCO_3$ did not have significant effect on CO reduction but increase CO_2 production significantly from 120 ml/g to 320 ml/g when 40% added to reactor and the dose matters. Al_2O_3 have impact on increasing CH_4 production significantly from 125 ml/g to 275 ml/g when 10% added to reactor and the dose does not have significant impact.

5 Gasifier Types

Different types of gasification technologies such as fixed bed, entrained-flow, fluidized bed, plasma, and rotary kiln have been developed and upgraded to convert biomass raw materials into syngas (Hashem Samadi et al., 2023). The commonly used types of gasifiers are fixed and fluidized beds (Rodrigues et al., 2022).

5.1 *Fixed Bed Gasifier*

Fixed beds use gravity to move the fuel or feedstock, the equipment operates at relatively low temperatures, between 400°C and 500°C, and works with high density

Table 6 Gas composition produced from different reactor type

Reactor type	CO (%)	CH_4 (%)	H_2 (%)
Updraft	24.5	18	42
Downdraft	30	18.5	46

Source (Tezer et al., 2023)

and grade fuels (Wiyono et al., 2020). The most common type of fixed-bed gasifiers are updraft (counter-current) or down-draft (co-current) (Gómez-Barea & Leckner, 2010). The gas composition derived from gasification process with different reactor type can be seen in (Table 6).

- **Downdraft reactor**

Downdraft gasifier has relatively high efficiency, low-tar product gas, and easy maintenance and operation. However, it is more suitable for small-scale applications with capacity of 10 kW–1 MW (Hashem Samadi et al., 2023). The maximum size of these units is limited to a few MW fuel power because of the problem of maintaining a regular conversion front in a wide fixed bed (Gómez-Barea & Leckner, 2010).

Both gasification agent and fuel are fed from the top part of the system and then both move downward or in the same direction (cocurrent) (Tezer et al., 2023). The reaction of drying, pyrolysis, oxidation, and reduction zones take place sequentially from top to bottom. While the moisture in the biomass at the drying zone is removed by the heat from the oxidation process, this heat can be also utilized for pyrolysis and reduction purpose. The gas leaves the reactor near the hottest zone, which makes the tar concentration much lower than in updraft gasifiers (Gómez-Barea & Leckner, 2010).

- **Updraft reactor**

Solid fuel is feed from the upper part of the reactor while gasification agent injected from bottom part that makes a counter-current and the gas produced from the process leaves the system from upper part (Tezer et al., 2023). Updraft gasifiers can be used for wet fuels and are relatively insensitive to the fuel size. The hot gas with a low oxygen content transmits the heat while it moves upwards, and the gasification process which are drying, devolatilization, gasification and combustion reactions occurs.

The gas produces from updraft gasifier has a high organic content or tar as the gas leaves the reactor near the pyrolysis zone and the solid carbon in the fuel is almost completely converted into gas and tar (Gómez-Barea & Leckner, 2010).

5.2 *Fluidized Beds*

In fluidized bed reactor, the fuel particles are suspended in a bed with inert sand or aluminium particles and then fluidised by the air flow to improve the heat transfer conditions in the combustion chamber, enabling more efficient reactions (Wiyono et al., 2020). This operation of this type of gasifier is more flexible, being more suitable for the density of MSW and tolerating greater variations in water content.

Compared to fixed bed, fluidized bed gasifiers have numerous advantages in terms of mixing, reaction rates, and the possibility of being built in sizes (Gómez-Barea & Leckner, 2010). There are two types of fluidized beds based on the fuel recirculation in gasifier, which are circulating fluidized bed gasifier (Lin et al., 2020) and bubbling fluidized bed (Jiang et al., 2021).

6 Catalyst

Catalytic use in gasification process can enhance syngas quality by removing undesirable CO_2 and preventing tar formation. CaO is introduced together with biomass and steam into the gasifier when temperature inside is reached between 650 and 750°C (Martínez et al., 2020). CaO is preferrable catalyst due to its low cost and abundance, is an excellent CO_2 sorbent to produce hydrogen-rich gas in biomass gasification for hydrogen production and play a vital role as tar reforming catalyst (Hu et al., 2015).

6.1 Syngas Quality

Tar reforming reaction is influenced by CaO as it can accelerate the reaction rate of the steam with tar and char, thus enhancing the total gas and hydrogen yields. During gasification process, CaO absorb CO_2 through water–gas shift reaction and secondary tar cracking reaction that result in lower CO_2 concentration and higher hydrogen production in syngas. A study found the effects of CaO addition on H_2 yield and gas composition by gasification of wet MSW with moisture content of 40 wt.% and gasification temperature of 750°C (Hu et al., 2015). The increase of CaO as catalyst to the feed from 0 to 0.7, increase the H_2 yield from 64.68 ml/g to 277.67 ml/g MSW. However, an excess of CaO in the reactor will adsorb large amounts of heat and hinder heat transfer between reactants which leads to the decrease of hydrogen production (Hu et al., 2015).

6.2 CO_2 and Tar Reduction

CaO can be used as catalyst by mixing it with the fuel (Hu et al., 2015) or put it as the bed (Martínez et al., 2020). Both methods shows a good capacity toward CO_2 adsorption that resulting in hydrogen-rich syngas. CaO can selectively adsorb CO_2 through exothermic gas–solid carbonation and reversibly release the captured CO_2

by endothermic calcination (Hu et al., 2015). CaO produce significantly higher H_2 and lower CO and CO_2 contents in the syngas due to carbonation reaction of the CaO due to the enhancement of the WGS reaction as CO_2 is removed from the gas phase (Martínez et al., 2020).

The CaO as catalyst can decrease CO_2 level in the reactor as it adsorbs CO_2 to form $CaCO_3$. At temperature below 850 C, CO_2 was absorbed by CaO and generate $CaCO_3$ while at temperature below 850 C, $CaCO_3$ decomposed into CaO and CO_2. The CO_2 concentrations decreased from 25.79 vol% to 7.29 vol %, with the increase of Ca ratio to feedstock from 0 to 1.3 (Hu et al., 2015). However, CaO carbonation reaction which have a role in removing CO_2 level from the product gas is limited by the minimum value set either by the equilibrium or the availability of CaO (Hu et al., 2015).

Beside CO_2 production, tar formation is also a problem in gasification process. Tar is a thick and dark viscous liquid which condenses on cold surfaces and clogs downstream equipment (Safarian et al., 2019), which is predominantly composed of heavy organic and inorganic molecules. Tar is considered as undesirable product as it decreases gasification efficiency by causing blockage that leads to equipment fouling and operational interruption (Emiola-Sadiq et al., 2023). CaO promote steam reforming of tar to produce more hydrogen and the presence of CaO may catalyse the tar reduction reactions and meanwhile increase the hydrogen production (Hu et al., 2015).

References

Champion, W. M., Cooper, C. D., Mackie, K. R., & Cairney, P. (2014). Development of a chemical kinetic model for a biosolids fluidized-bed gasifier and the effects of operating parameters on syngas quality. *Journal of the Air and Waste Management Association, 64*(2), 160–174. https://doi.org/10.1080/10962247.2013.845619

Cheng, S., Ding, X., Dong, X., Zhang, M., Tian, X., Liu, Y., Huang, Y., & Jin, B. (2023). Immigration, transformation, and emission control of sulfur and nitrogen during gasification of MSW: Fundamental and engineering review. *Carbon Resources Conversion, 6*(3), 184–204. https://doi.org/10.1016/j.crcon.2023.03.003

Couto, N. D., Silva, V. B., Monteiro, E., & Rouboa, A. (2015). Assessment of municipal solid wastes gasification in a semi-industrial gasifier using syngas quality indices. *Energy, 93*(PA1), 864–873. https://doi.org/10.1016/j.energy.2015.09.064

Diboma, B. S., Atiotsia, V. H., Che, L. C., Essomba, P. B., Bot, B. V., & Tamba, J. G. (2023). Gasification of charcoal derived from tropical wood residues in an updraft fixed bed reactor. *Bioresource Technology Reports, 21*, 101308. https://doi.org/10.1016/j.biteb.2022.101308

Emiola-Sadiq, T., Zhang, L., & Dalai, A. (2023). Catalytic steam gasification of soy hull pellets in a fluidized bed gasifier. *Journal of the Energy Institute, 111*, 1. https://doi.org/10.1016/j.joei.2023.101385

Fatema, J., Ahmed, T., Islam, M. M., Sakib, M. N., Chowdhury, A. M. S., & Haque, P. (2022). Gasification of kitchen wastes in an updraft fluidized bed gasifier and simulation of the process with Aspen Plus. *Journal of Cleaner Production, 371*, 133670. https://doi.org/10.1016/j.jclepro.2022.133670

Gómez-Barea, A., & Leckner, B. (2010). Modeling of biomass gasification in fluidized bed. In *Progress in energy and combustion science* (Vol. 36, pp. 444–509). Elsevier Ltd. https://doi.org/10.1016/j.pecs.2009.12.002

Hashem Samadi, S., Ghobadian, B., Nosrati, M., & Rezaei, M. (2023). Investigation of factors affecting performance of a downdraft fixed bed gasifier using optimized MLP neural networks approach. *Fuel, 333*, 126249. https://doi.org/10.1016/j.fuel.2022.126249

Higman, C. (2016). *State of the gasification industry: Worldwide gasification and syngas databases 2016 update.*

Hu, M., Guo, D., Ma, C., Hu, Z., Zhang, B., Xiao, B., Luo, S., & Wang, J. (2015). Hydrogen-rich gas production by the gasification of wet MSW (municipal solid waste) coupled with carbon dioxide capture. *Energy, 90*, 857–863. https://doi.org/10.1016/j.energy.2015.07.122

Jiang, P., Mahmud Parvez, A., Meng, Y., Dong, X., Xu, M., Luo, X., Shi, K., & Wu, T. (2021). Novel two-stage fluidized bed-plasma gasification integrated with SOFC and chemical looping combustion for the high efficiency power generation from MSW: A thermodynamic investigation. *Energy Conversion and Management, 236*, 114066. https://doi.org/10.1016/j.enconman.2021.114066

Lin, C., Zhang, J., Zhao, P., Wang, Z., Yang, M., Cui, X., Tian, H., & Guo, Q. (2020). Gasification of real MSW-derived hydrochar under various atmosphere and temperature. *Thermochimica Acta, 683*, 178470. https://doi.org/10.1016/j.tca.2019.178470

Lv, L., Zhang, Z., & Li, H. (2019). SNG-electricity cogeneration through MSW gasification integrated with a dual chemical looping process. *Chemical Engineering and Processing—Process Intensification, 145*, 107665. https://doi.org/10.1016/j.cep.2019.107665

Martínez, I., Grasa, G., Callén, M. S., López, J. M., & Murillo, R. (2020). Optimised production of tailored syngas from municipal solid waste (MSW) by sorption-enhanced gasification. *Chemical Engineering Journal, 401*, 126067. https://doi.org/10.1016/j.cej.2020.126067

Mazzoni, L., Janajreh, I., Elagroudy, S., & Ghenai, C. (2020). Modeling of plasma and entrained flow co-gasification of MSW and petroleum sludge. *Energy, 196*, 117001. https://doi.org/10.1016/j.energy.2020.117001

Mondal, P. (2022). From municipal solid waste (MSW) to hydrogen: Performance optimization of a fixed bed gasifier using Box-Benkhen method. *International Journal of Hydrogen Energy, 47*(46), 20064–20075. https://doi.org/10.1016/j.ijhydene.2022.04.150

Oliveira, M., Ramos, A., Monteiro, E., & Rouboa, A. (2021). Modeling and simulation of a fixed bed gasification process for thermal treatment of municipal solid waste and agricultural residues. *Energy Reports, 7*, 256–269. https://doi.org/10.1016/j.egyr.2021.07.124

Ozturk, M., & Dincer, I. (2021). An integrated system for clean hydrogen production from municipal solid wastes. *International Journal of Hydrogen Energy, 46*(9), 6251–6261. https://doi.org/10.1016/j.ijhydene.2020.11.145

Pandey, B., Sheth, P. N., & Prajapati, Y. K. (2022). Air-CO2 and oxygen-enriched air-CO2 biomass gasification in an autothermal downdraft gasifier: Experimental studies. *Energy Conversion and Management, 270*, 116216. https://doi.org/10.1016/j.enconman.2022.116216

Rodrigues, L. F., dos Santos, I. F. S., dos Santos, T. I. S., Barros, R. M., & Tiago Filho, G. L. (2022). Energy and economic evaluation of MSW incineration and gasification in Brazil. *Renewable Energy, 188*, 933–944. https://doi.org/10.1016/j.renene.2022.02.083

Safarian, S., Unnþórsson, R., & Richter, C. (2019). A review of biomass gasification modelling. In *Renewable and sustainable energy reviews* (Vol. 110, pp. 378–391). Elsevier Ltd. https://doi.org/10.1016/j.rser.2019.05.003

Tezer, Ö., Karabağ, N., Öngen, A., & Ayol, A. (2023). Syngas production from municipal sewage sludge by gasification Process: Effects of fixed bed reactor types and gasification agents on syngas quality. *Sustainable Energy Technologies and Assessments, 56*, 103042. https://doi.org/10.1016/j.seta.2023.103042

Thomson, R., Kwong, P., Ahmad, E., & Nigam, K. D. P. (2020). Clean syngas from small commercial biomass gasifiers; a review of gasifier development, recent advances and performance evaluation. In *International journal of hydrogen energy* (Vol. 45, pp. 21087–21111). Elsevier Ltd. https://doi.org/10.1016/j.ijhydene.2020.05.160

Wang, B., Gupta, R., Bei, L., Wan, Q., & Sun, L. (2023). A review on gasification of municipal solid waste (MSW): Syngas production, tar formation, mineral transformation and industrial challenges. In *International journal of hydrogen energy* (Vol. 48, pp. 26676–26706). Elsevier Ltd. https://doi.org/10.1016/j.ijhydene.2023.03.086

Wang, H., Ren, R., Liu, B., & You, C. (2022). Hydrogen production with an auto-thermal MSW steam gasification and direct melting system: A process modeling. *International Journal of Hydrogen Energy, 47*(10), 6508–6518. https://doi.org/10.1016/j.ijhydene.2021.12.009

Wang, J., Cheng, G., You, Y., Xiao, B., Liu, S., He, P., Guo, D., Guo, X., & Zhang, G. (2012). Hydrogen-rich gas production by steam gasification of municipal solid waste (MSW) using NiO supported on modified dolomite. *International Journal of Hydrogen Energy, 37*(8), 6503–6510. https://doi.org/10.1016/j.ijhydene.2012.01.070

Wiyono, A., Gandidi, I. M., Berman, E. T., & Mutaufiq, & Pambudi, N. A. (2020). Design, development and testing of integrated downdraft gasifier and multi IGCS system of MSW for remote areas. *Case Studies in Thermal Engineering, 20*, 1–11. https://doi.org/10.1016/j.csite.2020.100612

Yang, Y., Liew, R. K., Tamothran, A. M., Foong, S. Y., Yek, P. N. Y., Chia, P. W., Van Tran, T., Peng, W., & Lam, S. S. (2021). Gasification of refuse-derived fuel from municipal solid waste for energy production: A review. In *Environmental chemistry letters* (Vol. 19, pp. 2127–2140). Springer Science and Business Media Deutschland GmbH. https://doi.org/10.1007/s10311-020-01177-5

Zhou, X., Cheng, S., Li, Y., Yang, C., Zhang, L., Xiao, X., & Wang, T. (2024). Characterization and mechanism of red-mud-catalyzed steam gasification of corn stover. *Fuel, 356*, 129611. https://doi.org/10.1016/j.fuel.2023.129611

Thermal Modeling and Analysis of a Novel CHP Plant Employing Solid Oxide Fuel Cell and MSW Gasification

S. Chowdhury, S. Ghosh, and P. Mondal

Abstract The paper deals with the thermal modeling and performance assessment of a municipal solid waste (MSW) gasification fuelled combined heat and power (CHP) plant employing solid oxide fuel cell (SOFC). Both air and steam are considered as gasifying agents for the MSW gasification. The SOFC has been used for the generation of power while the residual heat in the burnt gases is used to produce steam from using a heat recovery steam generator (HRSG) where a certain amount of steam is fed to the gasifier and the rest is utilized for process heating. Thermal and exergetic efficiencies, fuel energy savings ratio (FESR), SOFC output voltage and sustainability index (SI) are the key performance parameters for the plant. The plant offers best performance at an equivalence ratio of 0.25, fuel utilization factor of 0.9 and current density of 2000 A/m^2. The maximum values of the thermal and exergetic efficiencies are 64.49% and 57.77%, respectively while maximum FESR is 64% and environmental sustainability index is 2.368, observed at the favorable operating conditions. However, the maximum net power obtainable from the proposed CHP plant is found to be 1821 kW and the maximum heat available for process heating is 360 kW when the current density is raised to as high as 6000 A/m^2.

1 Introduction

Municipal solid waste (MSW) is a very serious issue due to the problem of land-filling as the latter causes the contamination of land, air, water and soil for a long-term basis. With the ever increasing urban population, there is more generation of MSW from every household. So, the researchers were bound to think of some alternatives for MSW disposal and to curb the problem of land-filling by MSW. So gasification of MSW is proved to be one of the solutions addressing to the said problems. Some of the research works attributed to the gasification of MSW has been reported in the literature. Baggio et al. (2008) have assessed MSW-based pyrolysis and combined cycle plant energetically and environmentally. Murphy and Mckegh (2004) have

S. Chowdhury · S. Ghosh · P. Mondal (✉)
Department of Mechanical Engineering, IIEST Shibpur, Howrah, West Bengal 711103, India
e-mail: mondal.pradip87@gmail.com

A. Anouzla and S. Souabi (eds.), *Generation of Energy from Municipal Solid Waste*,
https://doi.org/10.1007/978-3-031-74334-4_9

Table 1 Ultimate analysis of MSW (wet basis) (Mondal et al., 2021)

Constituents	C	H	N	O	Cl	Ash	Moisture
Percentage by mass	40	5.5	0.6	30	0.9	10.1	12.9

performed the 3E analyses of the MSW based combined heat and power (CHP) plant. Awasthi et al. (2022) presented a brief overview on MSW treatment and various waste-to-energy conversion technologies for disposal and management of MSW in various places of China. Different MSW segregation and management techniques like incineration, pyrolysis, gasification, anaerobic digestion technology and composting were addressed. The method that is reported by the authors is MSW gasification considering both air and steam as gasifying agents. Mondal et al. (2021) modelled a novel MSW fired combined power plant integrated with an organic Rankine cycle (ORC). Roy et al. (2018) thermodynamically assessed the performance of a natural gas-fuelled SOFC module collaborated with an ORC.

As noted from contemporary literature sources that quite a good number of works are attributed to MSW gasification and CHP plants derived from the gasification of MSW. However, very few works have been attributed to generation of a combined heating and power plant derived from gasification of MSW by employing SOFC. So, in this paper, MSW gasification based CHP by employing SOFC plant is proposed with both air and steam as gasifying agents. The composition of MSW has been presented in Table 1 following the work of Mondal et al. (2021).

2 System Modeling and Assumptions

In MSW gasification process, the solid waste when fed to the gasifier undergoes thermo-chemical conversion to produce combustible gas mixture rich in hydrogen, carbon monoxide, carbon dioxide, water vapor and some traceable amount of methane. For assessing the thermodynamic performance of the CHP plant, a model has been established in software Cycle Tempo 5.0. The schematic diagram of the model is shown in Fig. 1. There are some assumptions involved and input parameters considered in Table 2 with the modeling in order to make the analysis more feasible and compatible.

a. The plant is functioning in steady-state condition.
b. Every gas considered in this study is assumed to be a perfect gas.
c. The mechanical and isentropic efficiency of the air compressor and feed pump of the HRSG are assumed to be 85% each.
d. The stack is totally insulated.

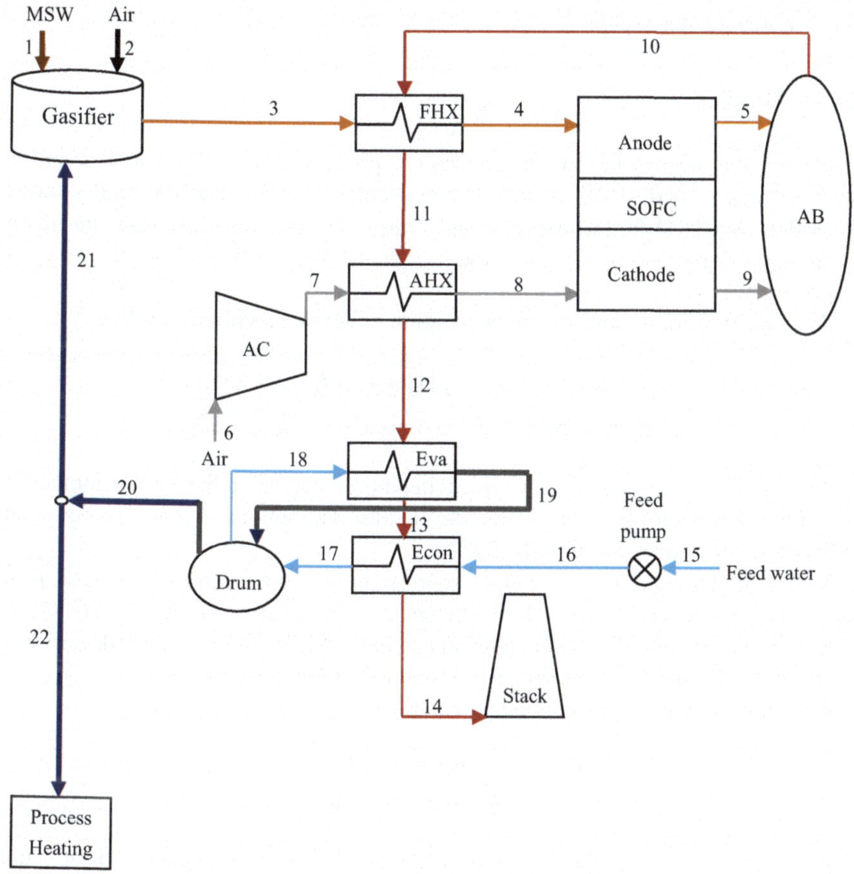

Fig. 1 Schematic of CHP plant

Table 2 Input parameters at the base case level

Input Parameters	Value(s)	Units
Steam inlet condition	120	°C
LHV of MSW	17,120	kJ/kg
Gasifier temperature	800	°C
Air–fuel equivalence ratio (λ)	0.25	–
Steam-to-fuel (*S/F*) ratio	1.5	–
Fuel utilization factor range	0.8	–
Current density value range	2000	A/m^2
SOFC inlet temperature	700	°C
Temperature difference between inlet and exit of SOFC	80	°C
Feed pump pressure	2	bar
Circulation ratio	8	–

2.1 *Energy Analysis*

2.1.1 Gasifier

A gasifier has been selected for gasification process with both air and steam as gasifying agents and MSW as fuel. The molecular formula of MSW (in dry form) is written as $CH_xO_yN_z$, where x, y and z are constants obtained from the main gasification equation by equating the weight of hydrogen, carbon and oxygen respectively.

The equation for air and steam gasification of MSW is written as follows:

$$\begin{aligned} & CH_xO_yN_z + wH_2O + \gamma H_2O + m(O_2 + 3.76N_2) \\ & \rightarrow aH_2 + bCO + cCO_2 + dCH_4 + eH_2O + fN_2 \end{aligned} \tag{1}$$

As the gasification process is very typical in nature, there are quite a good number of complex reactions taking place inside the gasifier. The set of reactions can be found in the paper of Zaman and Ghosh (2022):

Where the letters *a, b, c, d, e* and *f* represent the number of moles of hydrogen, carbon monoxide, carbon dioxide, methane, steam and nitrogen respectively. Here, w denotes the amount of moisture content present in MSW and γ denotes the amount of steam that is supplied for every unit mole of MSW fed.

The amount of moisture content, w can be evaluated as:

$$w = \frac{\dot{m}_{\mathrm{MSW}} \times \mathrm{MC}}{M_{\mathrm{H_2O}} \times (1 - \mathrm{MC})} \tag{2}$$

Here, $\dot{m}_{\mathrm{MSW}}$ is the molecular weight of MSW and MC is the percentage of moisture present in the MSW. The steam to fuel (S/F) ratio in terms of γ can be written as:

$$^{\mathrm{S}}/_{\mathrm{F}} = \frac{\mathrm{M_{H_2O}} \times \gamma}{\dot{m}_{\mathrm{MSW}} + \mathrm{M_{H_2O}} \times w} \tag{3}$$

The lower heating value (LHV) and higher heating value (HHV) value of the MSW are computed with the help of following correlations as obtained from the work of Zaman and Ghosh (2022):

$$\mathrm{LHV_{MSW}} = \mathrm{HHV_{MSW}} - h_{fg}(9 \times \mathrm{H}_B + \mathrm{MC}) \tag{4}$$

$$\mathrm{HHV_{MSW}} = 349.1 \times \mathrm{C}_B + 1178.3 \times \mathrm{H}_B - 103.4 \times \mathrm{O}_B - 15.1 \times \mathrm{N}_B - 21.1 \times \mathrm{Ash} \tag{5}$$

where h_{fg} represents latent heat of vaporization of water, H_B represents mass fraction of hydrogen, C_B represents mass fraction of carbon, O_B represents mass fraction of oxygen, N_B represents mass fraction of nitrogen, *Ash* represents the amount of ash.

The expression for cold gas efficiency is as follows:

$$\eta_{cg} = \frac{\dot{m}_{gas} \times LHV_{gas}}{\dot{m}_{MSW} \times LHV_{MSW} + \dot{m}_{agent} \times H_{agent} + \dot{Q}_{in}} \quad (6)$$

where, $\dot{m}_{MSW}$ and $\dot{m}_{agent}$ are the mass flow rates of MSW and gasifying agents (air and steam) respectively, $\dot{m}_{gas}$ is the mass flow rate of product gas obtained.

2.1.2 SOFC

An internal reforming SOFC is considered for the present investigation. Three chemical reactions, viz, steam reforming, shifting and electro-chemical reaction are presented in chronological order.

The internal reforming taking place inside the fuel cell is as follows:

$$CH_4 + H_2O \rightarrow CO + 3H_2 \quad (7)$$

The water gas shift reaction proceeds as follows:

$$CO + H_2O \rightarrow CO_2 + H_2 \quad (8)$$

The electro-chemical reaction taking place inside the fuel cell is as follows:

$$H_2 + \frac{1}{2}O_2 \rightarrow H_2O \quad (9)$$

The fuel utilization factor in the SOFC stack is written as follows:

$$U_f = \frac{n_{H_2,utilised}}{n_{H_2,in} + n_{CH_4,in} + n_{CO,in}} \quad (10)$$

The Nernst equation for the reversible cell potential for the electrochemical reaction is written as:

$$E_{nernst} = -\frac{\Delta G_0}{2F} - \frac{RT_{cell}}{2F}\left(\frac{p_{H_2O}}{p_{H_2} \cdot p_{O_2}^{0.5}}\right) \quad (11)$$

where, F is the Faraday's number, R and T_{cell} are universal gas constant and operating cell temperature respectively.

However this equation doesn't care of the losses occurring in the SOFC. The effective cell after considering all the losses in the main governing equation is given by:

$$E_{net} = E_{nernst} - E_{loss} \quad (12)$$

Table 3 Energy balance equations across every component of the CHP plant

Component	Mass balance	Energy balance
Gasifier	$\dot{m}_1 + \dot{m}_2 + \dot{m}_{21} = \dot{m}_3$	$\dot{m}_1 h_1 + \dot{m}_2 h_2 + \dot{m}_{21} h_{21} = \dot{m}_3 h_3$
Fuel heat exchanger (FHX)	$\dot{m}_3 + \dot{m}_{10} = \dot{m}_4 + \dot{m}_{11}$	$\dot{m}_3 h_3 + \dot{m}_{10} h_{10} = \dot{m}_4 h_4 + \dot{m}_{11} h_{11}$
Air heat exchanger (AHX)	$\dot{m}_7 + \dot{m}_{11} = \dot{m}_8 + \dot{m}_{12}$	$\dot{m}_7 h_7 + \dot{m}_{11} h_{11} = \dot{m}_8 h_8 + \dot{m}_{12} h_{12}$
Air compressor (AC)	$\dot{m}_6 = \dot{m}_7$	$\dot{m}_6 h_6 + W_{AC} = \dot{m}_7 h_7$
SOFC	$\dot{m}_4 + \dot{m}_8 = \dot{m}_5 + \dot{m}_9$	$\dot{m}_4 h_4 + \dot{m}_8 h_8 = \dot{m}_5 h_5 + \dot{m}_9 h_9 + W_{SOFC}$
Afterburner (AB)	$\dot{m}_5 + \dot{m}_9 = \dot{m}_{10}$	$\dot{m}_5 h_5 + \dot{m}_9 h_9 = \dot{m}_{10} h_{10}$
Evaporator	$\dot{m}_{12} + \dot{m}_{18} = \dot{m}_{13} + \dot{m}_{19}$	$\dot{m}_{12} h_{12} + \dot{m}_{18} h_{18} = \dot{m}_{13} h_{13} + \dot{m}_{19} h_{19}$
Economizer	$\dot{m}_{13} + \dot{m}_{16} = \dot{m}_{14} + \dot{m}_{17}$	$\dot{m}_{13} h_{13} + \dot{m}_{16} h_{16} = \dot{m}_{14} h_{14} + \dot{m}_{17} h_{17}$
Feed pump (FP)	$\dot{m}_{15} = \dot{m}_{16}$	$\dot{m}_{15} h_{15} + W_{FP} = \dot{m}_{16} h_{16}$

The total voltage loss is equal to the sum of three voltage losses; they are, activation loss (E_{act}), ohmic loss (E_{ohmic}) and concentration loss (E_{conc}),

$$E_{loss} = E_{act} + E_{conc} + E_{ohmic} \tag{13}$$

Power developed in every individual stack is given by:

$$W_{SOFC,stack} = N_{cell} E_{net} j A_{cell} \tag{14}$$

where, j is the current density and A_{cell} is the area of each cell

Where, N_{cell} represent the number of cells

$$W_{SOFC} = W_{SOFC,stack} . N_{SOFC} \tag{15}$$

where, N_{SOFC} represent the number of SOFC stacks

2.1.3 Other Components

Apart from the gasifier and SOFC, there are certain other components related to the CHP plant. The mass balance and the energy balance equations across every component are depicted in Table 3.

2.2 *Exergy Analysis*

Exergy analysis is done for exploring best possible thermodynamic performance of a system and mathematically it is equal to the sum total of physical exergy and chemical exergy. The physical exergy at any state point is written as follows:

$$\mathrm{Ex_{phy}} = \dot{m} \times (h - h_0 - \mathrm{T_0} \times (s - s_0)) \tag{16}$$

where the parameters with subscript '*0*' represents the value of the thermodynamic properties at dead state conditions and those without subscript represents the thermodynamic properties at any state point.

The chemical exergy at any state point is evaluated from the given equation:

$$\mathrm{Ex_{che}} = \sum y_i ex_i^{\mathrm{ch}} + \mathrm{RT_0}\left(\sum y_i \ln(y_i)\right) \tag{17}$$

where ex_i^{ch} represents standard chemical exergy of any particular gas and represents y_i the mole fraction of that particular gas.

2.3 *Performance Parameters*

Thermal efficiency, exergetic efficiency, FESR and sustainability index are the important performance parameters of the proposed CHP plant. The thermal efficiency of the plant is written as follows:

$$\eta_{\mathrm{th}} = \frac{\mathrm{W_{SOFC}} - \mathrm{W_{AC}} - \mathrm{W_{FP}} + \mathrm{Q_{process}}}{\dot{m}_{\mathrm{MSW}} \times \mathrm{LHV_{MSW}}} \tag{18}$$

where $Q_{process}$ is the amount of heat available for process heating, W_{SOFC} is the output power obtained from the SOFC, W_{FP} is the input power supplied to the feed pump of the HRSG and W_{AC} is work input to the air compressor (AC).

The amount of fuel saving for the plant is expressed considering MSW gasification based separate heating and power plants is given by the following expression (Mondal & Ghosh, 2016):

$$\Delta\mathrm{F} = \left(\frac{\mathrm{W_{net}}}{\eta_{\mathrm{P,ref}}} + \frac{\mathrm{Q_{process}}}{\eta_{h,\mathrm{ref}}}\right) - \dot{m}_{\mathrm{MSW}} \times \mathrm{LHV_{MSW}} \tag{19}$$

where $\eta_{\mathrm{P,ref}}$ and $\eta_{h,\mathrm{ref}}$ reference efficiency values are set to be 25% and 80% for the individual power and heat plants respectively. The FESR is given by:

$$\mathrm{FESR} = \frac{\Delta F}{\frac{W_{\mathrm{net}}}{\eta_{\mathrm{P,ref}}} + \frac{Q_{\mathrm{process}}}{\eta_{b,\mathrm{ref}}}} \tag{20}$$

The exergetic efficiency of the plant is defined as the ratio of the total output exergy to the amount of input exergy (Horlock, 1987). The expression for exergetic efficiency of the entire CHP plant is mentioned as follows:

$$\eta_{ex} = \frac{W_{SOFC} - W_{FP} - W_{AC} + Q_{process} \times \left(1 - \frac{T_0}{T_{process}}\right)}{Ex_{MSW}} \tag{21}$$

where Ex_{MSW} is the input exergy of MSW (fuel)

The sustainability index (SI), as an environmental performance parameter of the plant can be computed from the following expression:

$$SI = \frac{1}{1 - \eta_{ex}} \tag{22}$$

3 Results and Discussions

In this study, the thermodynamic performance of an MSW gasification based CHP plant is assessed with steam and air as gasifying agents. Effects of several input parameters like fuel utilization factor, current density and equivalence ratio on the output parameters like heat rate available & net power output, thermal & exergetic efficiencies, FESR and sustainability index are shown in the following subsections. At first the CHP plant is allowed to operate at base case scenario, i.e. when U_f is 0.8 and the current density value equal to 2000 A/m^2 for computing all the relevant output parameters of the plant. Then the favorable output conditions are determined from the given input parameters. The data composition of the producer gas has been validated with the reported experimental work of Wang et al. (2012) as shown in Table 4 suggesting that the code developed for mathematical simulation held a good agreement with that of the experimental one. The output performance parameters have at the base-case scenario has been mentioned in Table 5.

Table 4 Validation of the modeled gas data with Wang et al. (2012)

S/F ratio	H_2		CO		CO_2		H_2O	
	Exp	Simul	Exp	Simul	Exp	Simul	Exp	Simul
0.5	40.7	35.52	37.77	39.68	14.8	16.43	6.73	7.56
1	49.84	43.12	22.68	24.45	19.88	18.43	7.6	9.66
1.5	55	48	19.9	22.17	10.22	10.8	14.88	15.78
2	54.13	43.6	17.5	18.16	9.57	9.26	18.8	20.57
2.5	52.25	39	15.66	15.38	9.31	9.11	22.78	24.39
3	51.33	38.96	15	14.33	8.8	8.73	24.87	27.48

Table 5 Output parameters obtained at the base case level ($U_f = 0.8; j = 2000$ A/m^2)

Parameter	Value	Unit
Cold gas efficiency	84.5	%
LHV of producer gas	4342	kJ/kg
SOFC voltage	0.734	V
SOFC power output	607	kW
Heat rate for process heating	100	kW
FESR	61.78	%
Thermal efficiency	61.2	%
Exergetic efficiency	54.82	%
Sustainablity Index (*SI*)	2.577	–

3.1 Effect of Equivalence Ratio on Thermodynamic Performance of the Plant

Figure 2 shows the effect of equivalence ratio on the thermodynamic performance, i.e. over the FESR and exergetic performance of the plant. Trends reveal that both the FESR and exergetic efficiency of the plant deteriorates with the increment of equivalence ratio. This reason behind this observation is that when the equivalence ratio is enhanced the air–fuel mixture is made leaner which implies more amount of air for the same amount of fuel.

Figure 3 shows the influence of equivalence ratio over the net power produced and the available process heat from the plant. It is observed that with equivalence ratio, the net power decreases but the available process heat enhances. This is so because when the equivalence ratio is enhanced, the air–fuel mixture is made leaner which gives rise to lesser yield of gases like hydrogen, carbon dioxide from the producer gas that can be used as fuel to the SOFC. So there is lesser generation of power. On the other hand, there is more amount of flue gases produced due to less power output obtained from the SOFC. So the residual heat obtained from the burnt gases increase which in turn increases the heat load available for process heating.

3.2 Effect of Fuel Utilization Factor on Thermodynamic Performance of the Plant

The effect of fuel utilization factor on FESR and the exergetic efficiency of the plant have been depicted in Fig. 4. It is noticed that the FESR as well as exergetic efficiency of the plant increases with the fuel utilization factor. This is so because when more amount of fuel is utilized, the losses associated with the cell voltage gets reduced which results in the generation of more power output for a given input. This raises the

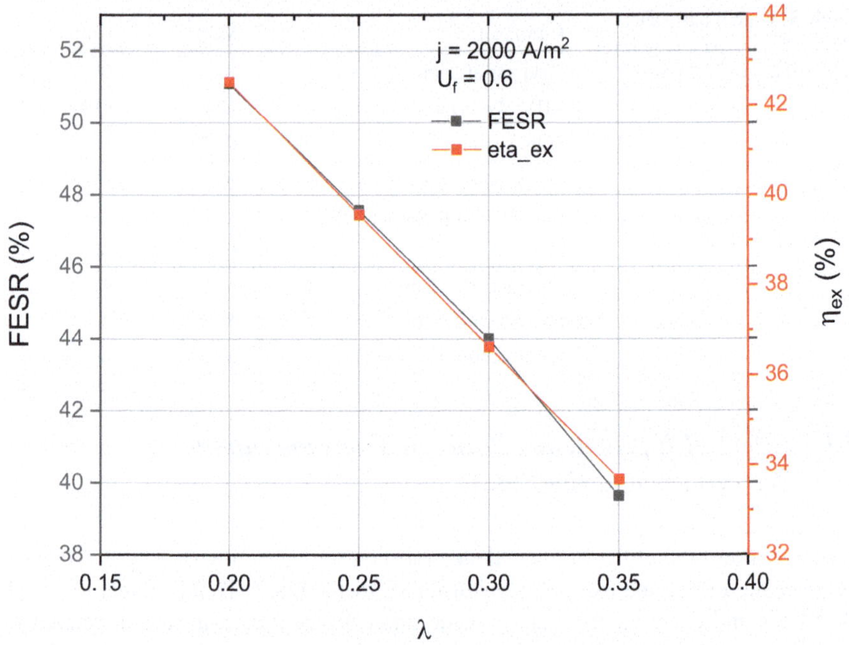

Fig. 2 Effect of equivalence ratio on FESR and exergetic efficiency

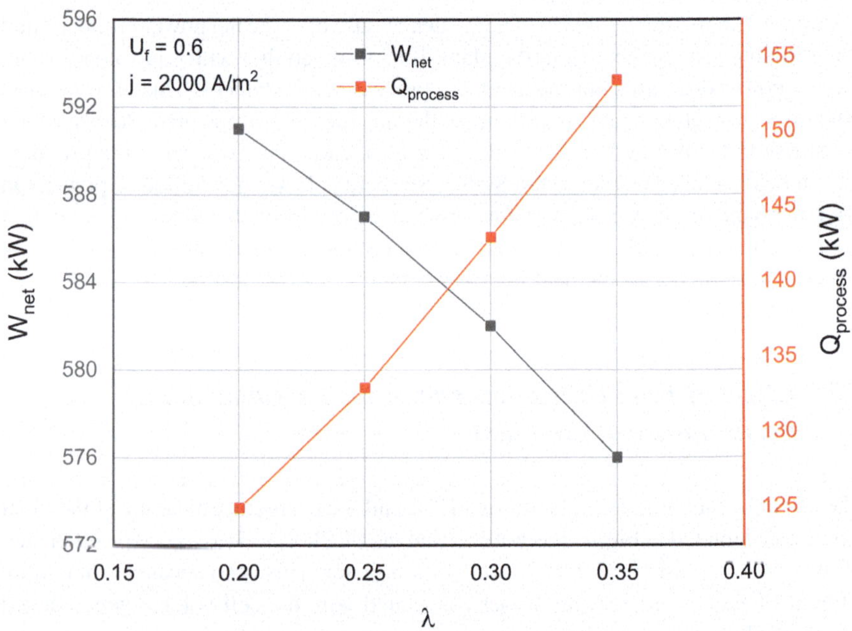

Fig. 3 Effect of equivalence ratio on net power and process heat obtained

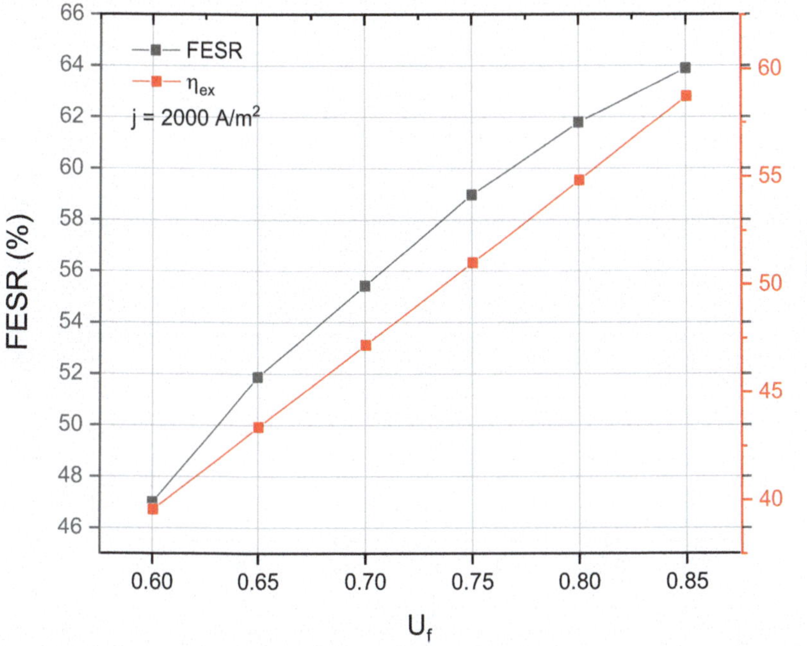

Fig. 4 Effect of U_f on FESR and exergetic efficiency of the plant

power generated from the SOFC which in turn raises the FESR as well as exergetic efficiency of the plant.

Figure 5 shows the variation of fuel utilization factor with the SOFC power and the heat load available for process heating applications. The figure reveals that with increase in the value of fuel utilization factor, there is an increment in the value of SOFC power but a decrement in the value of the residual heat load. This occurs so because with increase in the fuel utilization factor value, more power output is generated from the SOFC. However the waste heat of the burnt gases becomes very low due to the generation of more power from the SOFC which narrows down the residual heat available for process heating.

3.3 *Effect of Fuel Utilization Factor on Sustainability Index of the Plant*

Figure 6 shows the variation of fuel utilization factor with the sustainability index (*SI*) of the plant at different equivalence ratio values. From the figure, it is clearly understood that the sustainability index increases with the fuel utilization factor implying that the proposed CHP plant is environmentally sustainable. This happens

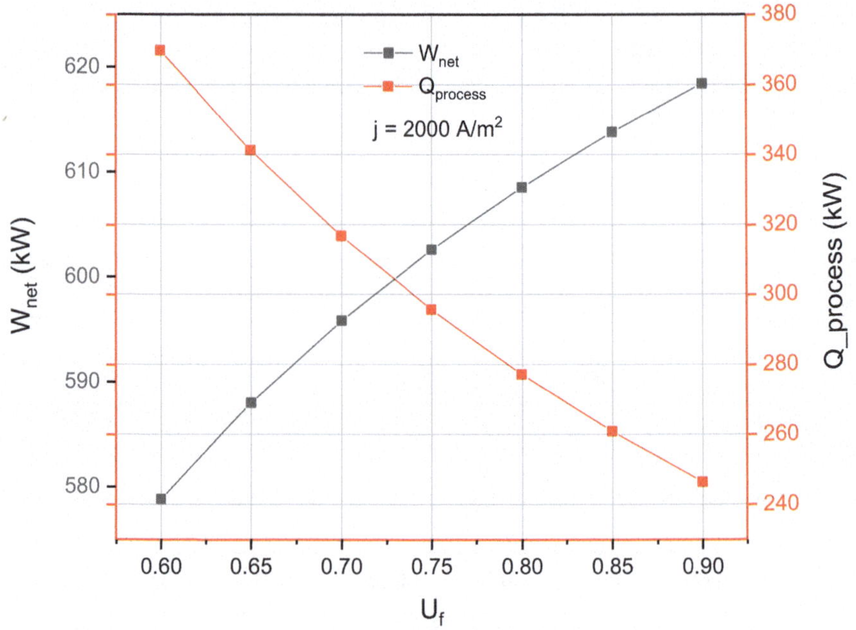

Fig. 5 Effect of U_f on SOFC power and heat load for process heating

so because when more fuel is utilized, there is more fuel consumption in the SOFC and hence the performance of the plant is enhanced which in turn increases the sustainability index of the plant.

3.4 Effect of Current Density on SOFC Voltage and SOFC Power

The influence of current density on the SOFC voltage and SOFC power can be observed in Fig. 7. It is observed that the cell voltage diminishes whereas the SOFC power increases with increase in the current density value. This is due to the fact that with increase in the current density, all the polarization losses, i.e. ohmic polarization, activation polarization and concentration polarization losses increase considerably, with activation polarization dominating the other two. Thus, a decrease in the net cell voltage is observed. On the other hand, with the increment in current density value, fuel consumption by the SOFC increases leading to high power generation at the SOFC.

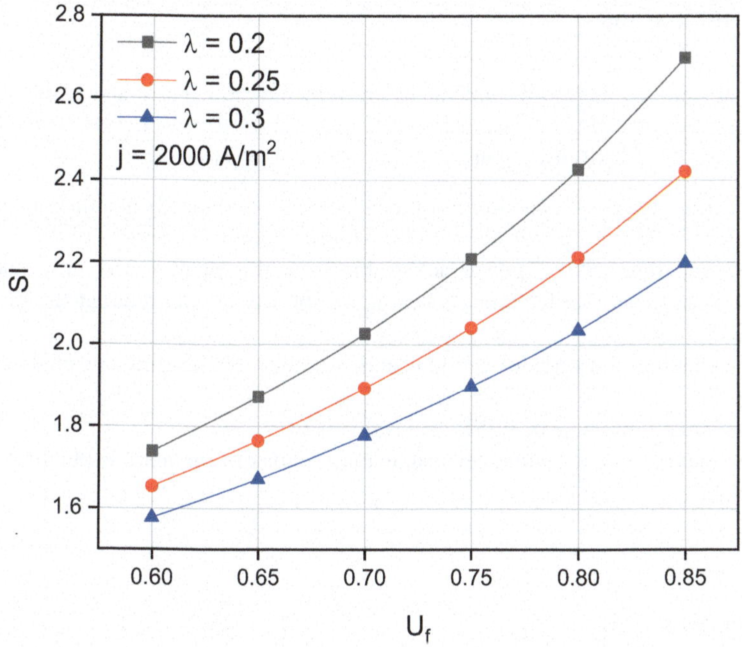

Fig. 6 Effect of U_f on sustainability index of the plant

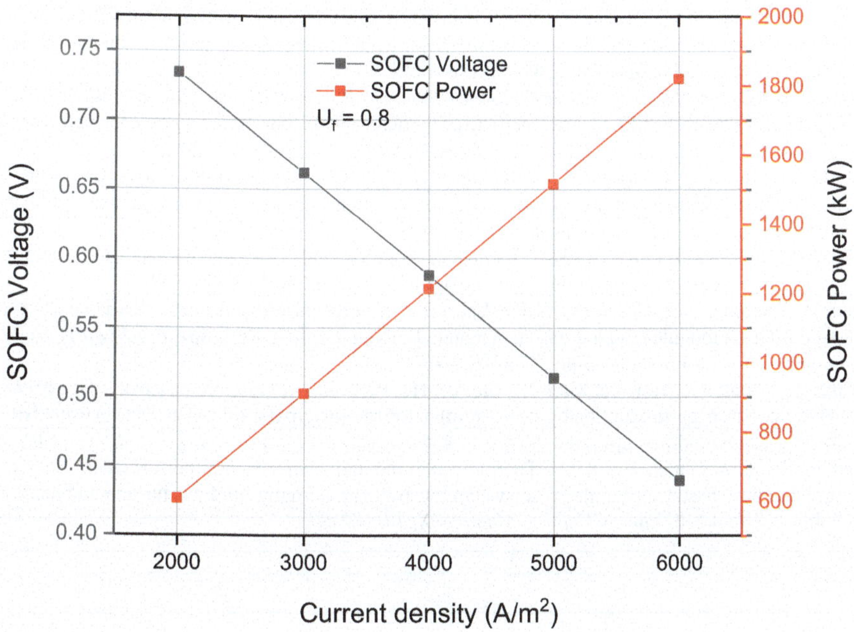

Fig. 7 Effect of current density on SOFC voltage and SOFC Power

4 Conclusions

From the detailed thermodynamic analyses of the combined heat and power (CHP) plant employing SOFC by gasification of municipal solid waste (MSW) using air as well as steam as gasifying agents, certain points can be concluded:

a. The desirable output conditions of the plant are found mostly at a fuel utilization factor of 0.9, current density of 2000 A/m^2 and equivalence ratio of 0.2.
b. The maximum SOFC power achievable from this plant is 1821 kW and the maximum heat rate for process heating is 360 kW at a current density of 6000 A/m^2.
c. The maximum values of thermal and exergetic efficiencies are 64.49% and 57.77%.
d. The maximum value of FESR achievable is 64%.
e. The maximum environmental sustainability index of the plant is obtained to be 2.368.

References

Awasthi, S. K., et al. Processing of municipal solid waste resources for a circular economy in China: An overview. *Fuel, 317*, 123478.

Baggio, P., Baratieri, M., Gasparella, A., & Longo, G. A. (2008). Energy and environmental analysis of an innovative system based on municipal solid waste (MSW) pyrolysis and combined cycle. *Applied Thermal Engineering, 28*(2–3), 136–144.

Horlock, J. H. (1987). *Cogeneration-Combined Heat and Power (CHP)*. Pergamon Press.

Mondal, P., & Ghosh, S. (2016). Externally fired biomass gasification-based combined cycle plant: Exergo-economic analysis. *International Journal of Exergy, 20*(4), 496–516.

Mondal, P., Samanta, S., Zaman Sk, A., & Ghosh, S. (2021). Municipal solid waste fired combined cycle plant: Techno-economic performance optimization using response surface methodology. *Energy Conversion and Management, 237*, 114–133.

Murphy, J. D., & Mckeogh, E. (2004). Technical, economic and environmental analysis of energy production from municipal solid waste. *Renewable Energy, 29*(7), 1043–1057.

Roy, D., Samanta, S., & Ghosh, S. (2018). Energetic and exergetic Analyses of a Solid Oxide Fuel Cell (SOFC) module coupled with an Organic Rankine Cycle (ORC). *Advance Fluid Thermal Engineering: Select Proceedings FLAME, 2019*, 13–24.

Wang, J., Cheng, G., You, Y., Xiao, B., Liu, S., He, P., Guo, D., Guo, X., & Zhang, G. (2012). Hydrogen-rich gas production by steam gasification of municipal solid waste (MSW) using NiO supported on modified dolomite. *International Journal of Hydrogen Energy, 37*, 6503–6510.

Zaman Sk, A., & Ghosh, S. (2022). Thermo-economic and environmental performance analyses of a biomass-based carbon negative system integrating externally fired gas turbine and molten carbonate fuel cell. *Energy Convers Manage: X, 14*, 100187.

Recovery Potential of Municipal Solid Waste for Hydrogen Generation

P. Agamuthu, S. C. Aboudi Mana, and N. Arumugam

Abstract By 2050, 3.40 billion tonnes of municipal solid waste (MSW) will be generated (Kaza et al. in What a waste 2.0: A global snapshot of solid waste management to 2050. World Bank Publications, 2018). In 2018, the United States produced approximately 811 kg per capita, totalling 292 billion tonnes, while according to Eurostat, 530 kg of MSW was produced per capita in 2021. Landfilling treatment was 50% in the United States and Europe, while 11.8% was combusted, 32% recycled or composted, and 6.2% was submitted to alternative treatments. Regarding the European situation, led by Austria, Denmark, and Luxembourg, of all the municipal wastes generated by the E.U. in 2020, 48% were recycled (considering recycling and composting), while 26% were submitted to incineration and landfilling. Incineration accounted for over 80% of energy recovery facilities in 2015, while the remaining have only functioned as final disposal units. There is a need to shift from conventional waste disposal methods to advanced technologies to manage these waste piles and the related health and environmental issues related to their disposal. Despite the up-to-date technologies, only 250 Waste-to-energy (WtE) plants have been built worldwide in the last 15 years. Here, we discussed some existing MSW technologies used in converting waste to energy, analysed the challenges through a look at the life cycle assessment (LCA) of waste-to-energy technologies, and identified the features of waste-to-energy in promoting the Circular economy.

P. Agamuthu (✉)
Jeffrey Sachs Center on Sustainable Development, Sunway University, Subang Jaya, Malaysia
e-mail: profagamuthu@gmail.com

S. C. A. Mana
Solid Waste Laboratory, Institute of Biological Sciences, Universiti Malaya, Kuala Lumpur, Malaysia

N. Arumugam
Kingsway International Secondary School, Batu Caves, Selangor, Malaysia

A. Anouzla and S. Souabi (eds.), *Generation of Energy from Municipal Solid Waste*,
https://doi.org/10.1007/978-3-031-74334-4_10

1 Introduction

Municipal solid waste (MSW) generation is a global threat rising rapidly. According to the World Bank (2020), an average individual generates nearly 0.74 kg of solid waste footprint every day. Waste disposal in developed countries and transitory countries are always higher than in underdeveloped countries, as the latter nations practiced better waste recycling (Agamuthu, 2001). Globally, landfills receive 70% of the total MSW, 19% are recycled and composted, and 11% are used for energy recovery (Kaza et al., 2018). Cost of MSW treatment and disposal which is skyrocketing and scarcity of lands for waste disposal has increased the environmental awareness and prompted the need to use effective method for waste management and reduction. Besides, the increasing world energy demand drives the need for new and sustainable renewable fuel to mitigate problems associated with greenhouse gas emissions, such as climate change and the emission of unpleasant odours and pollution of land and groundwater due to landfill management (Aziz et al., 2021a, 2021b; Hassan et al., 2022).

The composition of MSW varies significantly among different municipalities worldwide, but it typically comprises both biodegradable and non-biodegradable materials from organic and inorganic sources. MSW collected from households, offices, institutions, and commercial enterprises typically contains organic waste. The heterogeneity of the organic waste materials makes suitable feedstock for conversion into value-added products. Waste-to-energy (WtE) is an economically viable and environmentally sustainable solution to recover energy from waste resources in the form of fuel, heat, and electricity. Numerous biological and thermochemical technologies such as incineration, gasification, and pyrolysis have been used to convert MSW into usable solid, liquid, and gaseous fuels to supplement the amplifying energy demands. However, some of these technologies have high energy requirements and generate by-products such as bio-oils, gas, char, and corrosive pollutants, which negate the environmental benefit of energy production from MSW (Nanda & Berruti, 2021). Biological conversion is usually anaerobic digestion and composting that produce biogas (biomethane) and compost (decomposed organic matter). Some next-generation biological conversion processes applied to the MSW organic fraction and leachate have energy recovery potential.

Waste-derived hydrogen (H_2) from biological sources, commonly addressed as biohydrogen (Bio-H_2), is a clean energy source that could simultaneously decrease greenhouse gas (GHG) emissions and avert municipal solid waste (MSW) from being deposited in perpetual landfills. Organic wastes like industrial waste, municipal sewage sludge, solid waste, agricultural residues, poultry waste, and manure are used for bioenergy production. Recent research has shown the potential of biohydrogen to be recovered through dark and photo-fermentation, direct and indirect bio-photolysis, microbial electrolysis cells, as well as microbial electro-hydrogenesis cells (Allegue et al., 2020; Osman et al., 2020; Zhen et al., 2016). Hydrogen is a versatile energy carrier with the potential to be utilised in multiple industrial sectors, including energy generation, chemical production, transport sector, and iron and steel production (Osman et al., 2021). Hydrogen is a low-carbon fuel and one option for transitioning towards a low-carbon emission and promoting a circular economy.

Hydrogen is produced through various technologies using fossil and renewable sources because it is technologically mature and has high energy conversion efficiencies. Biohydrogen is considered one of the energy sources that have the potential to replace part of conventional fossil-based fuels. It produces no harmful greenhouse gases upon combustion but only water. Hydrogen is the most prevalent and invisible gas in nature. Its benefits include availability, renewability, and high energy content, which fuel many industries. It is seen as a possible substitute for fossil fuels in the green economy. It can be transported or kept in various valued forms, including solid and liquid. Recently, hydrogen production has gained global attention as one of the most promising and eco-friendly energy alternatives, so substantial efforts have been made to effectively and safely manage municipal solid waste in conjunction with its sustainable biohydrogen valorisation. Herein, we discussed MSW treatment technologies used in producing hydrogen derived from MSW, identified the challenges of waste to hydrogen technologies and provided a comprehensive look at life cycle assessment studies applied to MSW conversion into hydrogen production.

1.1 Objectives

a. Discuss the existing MSW practices and the technologies used to convert waste to biohydrogen.
b. Analyse the challenges in waste to biohydrogen technologies.
c. Identify the features of Waste to Energy in promoting the Circular economy using already applied life cycle assessment.

2 Advanced Technologies Adopted in the Conversion of MSW to Energy

2.1 Biological Conversion

In most cases, "biological treatment" refers to treating municipal solid waste that uses microorganisms to break down organic contaminants. Biological treatment methods include membrane bioreactors, anaerobic digestion, dark fermentation, activated sludge, and biofilters. The efficiency, cost, and environmental impact of these processes can vary. A few benefits of biological treatment include the ability to recover valuable products from waste, such as biomaterials, biogas, and biofertiliser, as well as the fact that it is more economical and environmentally friendly than chemical or physical techniques. One of the difficulties with biological treatment is that it might be impacted by poisonous or resistant substances, which necessitates careful optimisation and control of operational parameters, including pH, temperature, oxygen level, and nutritional balance.

During the pre-treatment step, the primary carbon microorganisms source is usually free and readily available sugars. Biological pre-treatment is typically a slow process that requires a more extended retention period. Since they compete with the native microorganisms throughout the pre-treatment process, pure biological agents for the pre-treatment of waste are typically challenging to optimise and maintain. Few studies use pure microbial enzymes to pretreat waste since enzymatic pre-treatment is costly and needs high enzyme concentrations to be effective. Raw enzymes made from biomass lysate are also utilised directly for pre-treatment to reduce cost (Parthiba Karthikeyan et al., 2018). The use of consortia or microbial communities to enable the biodegradation of complex compounds, the use of combined toxicity assays, and ultimately, the development of integrated treatments by combining two or more methodologies to achieve higher degradation percentages all contribute to the effectiveness of biological treatment (Ceretta et al., 2021). Biological methods involve direct bio-photolysis, indirect bio-photolysis, photo-fermentation, and biochemical systems to produce bio-hydrogen. These methods have advantages such as low energy consumption, CO_2 fixation and renewable feedstocks, but also face challenges such as low hydrogen yield, high cost and low efficiency.

a. Biofermentation (dark fermentation and photofermentation)

Biofermentation is typically used in conjunction with anaerobic digestion (AD). However, this process separates rate-faster methanogenesis during waste conversion to energy from slower mechanisms than AD. Two-stage methods have the following main benefits: a better stabilisation of the AD process under two different pH optimum conditions, the control of the methanogenic process from organic overloads, the reduction of digestion time and buffering requirements, and an increase in overall yield (carbon conversion up to 40%) of gaseous end products like H_2 and CH_4. In the first stage of dark fermentation, hydrogen is mainly created from readily available sugars (10% of total COD is used for H_2 generation). In contrast, the more complicated unfermented ingredients are further digested to form CH_4 (80% of COD is used for H_2 generation). This approach is often used as a pre-treatment to thermochemical conversion (Zhen et al., 2016).

Anaerobic digestion (AD) or dark fermentation (DF) of waste is frequently preceded by wet-type thermal pre-treatment. Improved hydrolysis allows for the solubilisation of more sugars and produces a more homogeneous waste for AD or DF feeding. In addition, it is quicker and easier than other pre-treatment techniques. By controlling pH, which also encourages the development of resistant substances (like melanoidins) beyond 150°C (with short retention time) or 100°C (with prolonged retention time) and may have an effect on the product hydrogen recovery, the efficiency of the pre-treatment of MSW may be further improved by the capacity of the microorganism to convert organic carbon into hydrogen.

The conversion of biomass-based materials to hydrogen via dark fermentation can be conducted in a one- or two-stage process. The one-stage anaerobic digestion is the simplest system that allows all steps (hydrolysis, acidogenesis, and acetogenesis) to occur in a single reactor. The multistage process uses two or more reactors for digestion to separate the hydrolysis/acidogenesis and acetogenesis phases (Fig. 1). Acidogenic bacteria produce organic acids and more quickly grow and reproduce than methanogenic bacteria. The multistage process is adopted to optimise every digestion step and achieve more stable operation, higher organic loading capacity, and a higher resistance toward toxicants and inhibiting substances. In the optimisation process, the most DF investigated temperature ranges for dark fermentation are performed at 35–37°C (mesophilic conditions). While the importance of environmental factors such as hydraulic retention time (HRT) and pH are also important factors. The type of waste materials is essential. The use of carbon sources containing monosaccharides, such as glucose and disaccharides, such as lactose or sucrose, is the focus of the dark fermentation process because of their biodegradable nature (Aziz et al., 2021a, 2021b). Due to its heterogeneity, the organic fraction of municipal solid waste is a complex substrate to be considered when converting an organic fraction into biohydrogen. The soluble and readily accessible sugars represent the main fraction of biomass that can be converted into hydrogen. Generally, organic fraction constitutes food waste, which contains a high biodegradable carbohydrates fraction with 85–95% volatile solids and 75–85% moisture content, making it a suitable substrate for DF. The presence of H_2 synthesising bacteria is a mixed culture of microorganisms originating from the organic fraction during the DF process. These cultures influence higher H_2 yields and are generally preferred to pure hydrogen producing seed inoculum or pure culture due to their economic benefits and practicability in environmental engineering. However, this approach necessitates the enrichment of mixed cultures to enhance biohydrogen production and inhibit hydrogen consumers such as methanogens and homoacetogens in the inocula (Ghimire et al., 2015).

Similar to anaerobic conversion, photofermentation is the fermentative conversion of organic substrate to biohydrogen by various photosynthetic bacteria through a series of biochemical reactions, including three phases. Nitrogenases found in photosynthetic bacteria enable hydrogen synthesis. Recent research interest in photo-fermentation hydrogen production revealed the benefits of abundant substrate supplies and extensive material suppliers. Under anaerobic conditions, purple non-sulfur photosynthetic bacteria carry out anaerobic photosynthesis using light as the energy source for synthesising hydrogen. The purple non-sulfur bacteria use the captured light energy to produce ATP and high energy electrons through reverse electron flow, which reduces ferredoxin (Fig. 2). Then, the ATP and reduced ferredoxin drive the proton reduction to hydrogen by using light as their energy source during anaerobic photosynthesis to create hydrogen. Through reverse electron flow, which lowers ferredoxin, the purple non-sulfur bacteria utilise the light energy they have captured to make ATP and high-energy electrons (Fig. 2). Thus, nitrogenase promotes proton reduction to hydrogen using ATP and decreased ferredoxin.

However, there is a need for a further comprehensive investigation to determine the best reactor configuration and improve augmenting microbial immobilisation, novel microbial strains, mitigation of inhibitory effects, and an integrated fermentation mechanism that can effectively convert biomass into hydrogen.

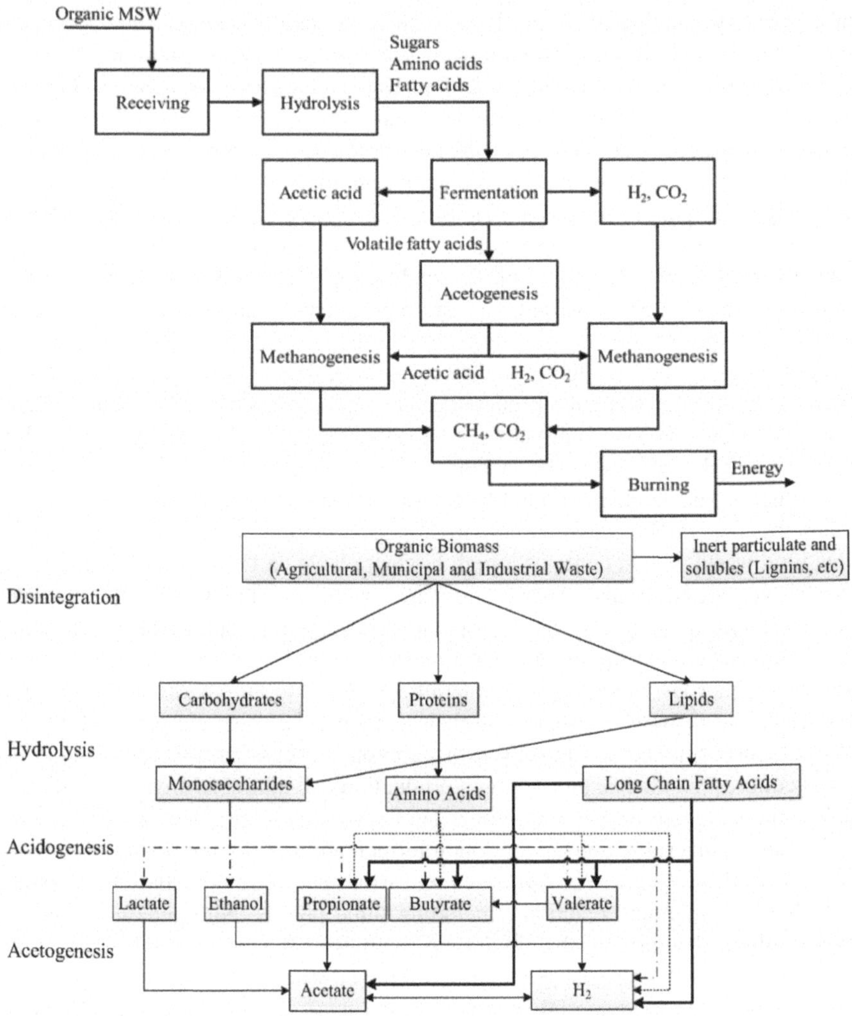

Fig. 1 Main reactions involved during anaerobic digestion of MSW for H_2 generation (Ghimire et al., 2015)

b. Biophotolysis (direct and indirect)

The prospect of generating clean hydrogen free from carbon as renewable energy from available natural resources such as sunlight and water has piqued interest in photobiological hydrogen generation by photosynthetic bacteria. Through biochemical processes, a photosynthetic bacteria will, with the help of a reducing agent such as nitrogenase or hydrogenase enzymes, use solar energy to break water into O_2 and produce photobiological by reducing protons. The growing organisms in the photobiological process provided by the hydrogenase enzyme are sunlight and CO_2.

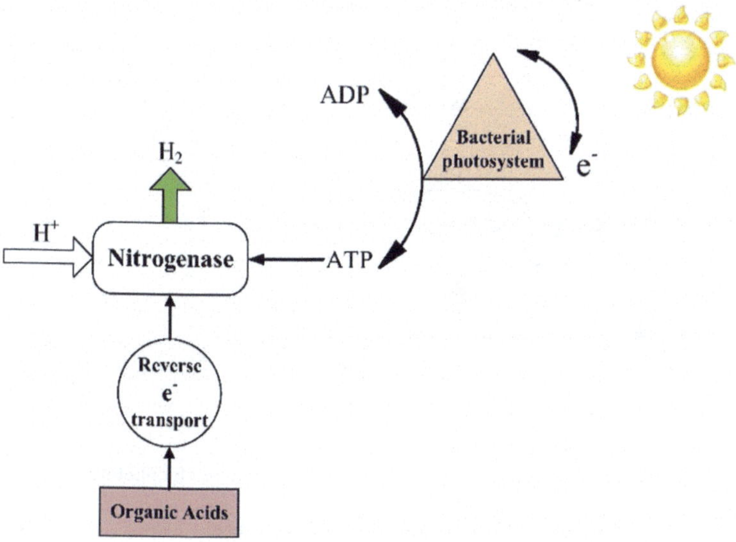

Fig. 2 Schematic process of photofermentation modified by Ghimire et al. (2015)

During photosynthesis, chlorophyll pigments in photosynthetic organisms known as thylakoid membranes absorb light and create molecular O_2. CO_2 is converted to triose phosphate via enzymatic activities with ATP and NADPH.

Direct and indirect biophotolysis are two types of biophotolysis used to produce H_2 (Fig. 2). In direct biophotolysis, water molecules are split using sunlight as an energy source in the photosystem. After transferring the residual electron e^- to the photosystem I, the hydrogenase helps to generate H_2 without releasing harmful GHG. This process also simultaneously releases O_2 into the atmosphere, as displayed

$$2H_2O + \text{light energy} \rightarrow 2H_2 + O_2$$

Microalgae perform indirect biophotolysis under anoxic conditions. In these conditions, microalgae can produce bio-H_2 through fermentation or respiration. Indirect biophotolysis is a two-step process:

(i) The first stage involves O_2 production and CO_2 fixing into chemical energy carbohydrates and lipids.
(ii) The second stage generates H_2 by the fermentation of stored carbohydrates.

Cyanobacteria and microalgae have shown the most significant attention among the microorganisms capable of producing H_2. Cyanobacteria have prospects to convert light energy to molecular hydrogen efficiently as an eco-friendly fuel (Fig. 3).

The requirement of the external electric supply for long-term stability and the simultaneous release of oxygen and hydrogen is a challenge in this approach. Cyanobacterial strains can produce 20-fold more hydrogen and optimise conditions for hydrogen production in the light than in the dark (Kossalbayev et al., 2020). They

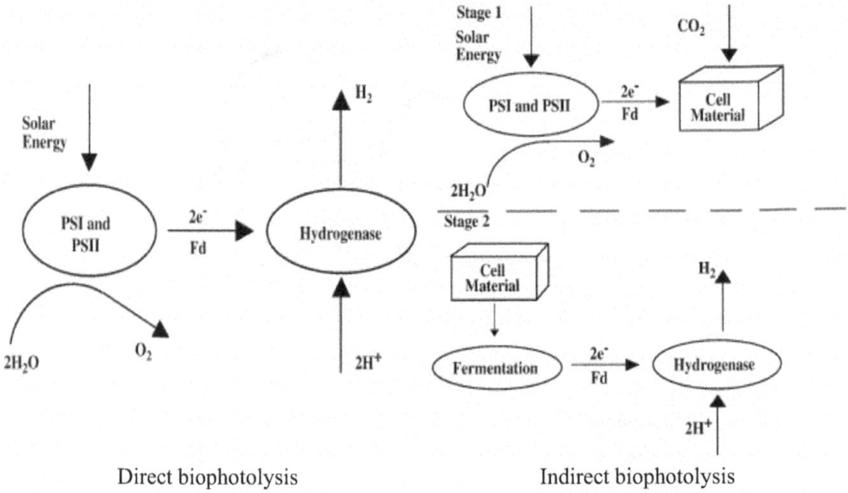

Fig. 3 Biophotolysis process of H_2 generation from MSW (Ni et al., 2006)

found that biomass optimisation in the first step of bacterial cultivation increased H_2 yield by 18.3%.

c. Microbial electrolysis cells.

Microbial electrolysis cell (Dolar et al., 2016) effectively enhances H_2 production. By oxidising organic material from various biomass sources, MEC uses microorganisms to catalyse biochemical reactions at the anode or cathode and produce protons and electrons. This method, which creates hydrogen through electrolysis, is typically used with the DF process. To fully convert the organic biomass, it is necessary to use the volatile fatty acids found in DF residues, essential by-products of the DF process. Two systems are interconnected by converting carbs to organic acids in the first stage (DF) and by-products to H_2 in the second stage during MEC. The recurrent DF products, acetate and butyrate, are usually substrates during MEC (Fig. 1). Different biochemical processes are undertaken by anaerobes, which generate H_2 from various bacteria. They convert glucose to pyruvate. Depending on the stoichiometric reactions, other products will yield H_2 generation. In some conditions, the production of acetic acid through metabolic pathways during stoichiometric reaction will produce H_2 with 4 mol for each mole of glucose (544 mL H_2/g hexose at 25 °C) or yield H_2 by 2 mol for a mole of glucose (272 mL H_2/g hexose at 25°C) when the final product is butyric acid. Under another condition, the metabolic pathways result in the synthesis of ethanol and acetate, which reduces the stoichiometric hydrogen yield to 2 mol of H_2 for a mole of glucose (272 mL H_2/g hexose at 25 °C) (Li & Fang, 2007). DF system cannot achieve beyond the highest yield of 4 mol H_2 per mole hexoses, and DF has a maximum output of 33% on sugars. To increase the H_2 generation, the conversion of by-products through MEC improves the biohydrogen yield and the economic viability.

An MEC technology needs an oxidising acetate under standard biological conditions (25 °C, 1 bar pressure and pH 7). Hydrogen can be produced at the cathode by applying a circuit voltage as low as 0.14 V. Some exoelectrogenic microorganisms (like the genera Geobacter, Shewanella and Pseudomonas sp.) will be capable of electron transfer to an electrode anode, then a minimum voltage of more than 250 mV is applied for the ohmic resistance and the electrode potential for hydrogen production via electrolysis of water at neutral pH conditions. This process will convert the DF by-products to hydrogen and increase the H_2 yield (Ghimire et al., 2015; Shao et al., 2019).

2.2 *Thermochemical Conversion*

a. pyrolysis.

In pyrolysis, organic materials are heated without oxygen to produce bio-oil, charcoal, and combustible rngases. The types and quality of the organic feedstock, reactor design, temperature, heating length, and other factors all affect the yield of each product differently. Similar to other thermochemical processes, pyrolysis has the potential to extract a significant amount of energy from MSW with lower NOx and SOx emissions due to the lack of oxygen. Pyrolysis processes occur before gasification at high temperatures, splitting the fuel without a catalyst (Fig. 4).

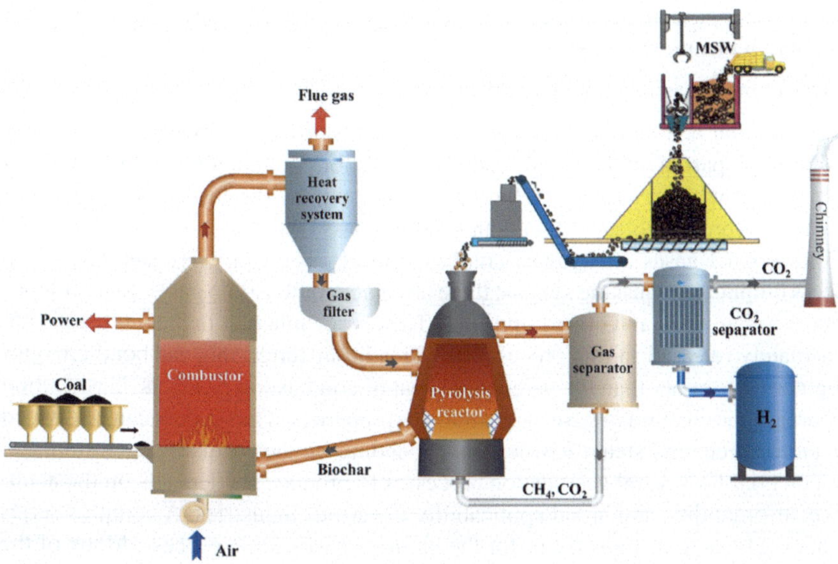

Fig. 4 Scheme of municipal solid waste pyrolysis adopted from Yan et al. (2020)

Based on the temperature at which it operates, this process, which turns biomass into bio-oil, can be divided into three categories: traditional (or slow), rapid, and flash pyrolysis. Conventional pyrolysis produces a substantial amount of charcoal because it is carried out at temperatures lower than 450°C. At medium temperatures (450–600°C), a high heating rate (about 300°C/min), and a brief residence time, fast pyrolysis can produce up to 75% weight of bio-oil. Rapid pyrolysis and flash pyrolysis are comparable processes. Flash pyrolysis is carried out at higher temperatures (over 600°C) and faster heating rates (>1000°C/s). However, because of the lower operating temperatures, quick and flash pyrolysis produces fewer gas yields than gasification. Oxygenated molecules (Ketones, phenolic compounds, aldehydes, alcohols, and carboxylic acids), water, and ash are condensed to create a complex liquid fraction similar to tar known as "bio-oil." There are two types of bio-oils: water-insoluble fractions and water-soluble fractions. A steam gasification device can be added to the process to boost the H_2 production yield from the soluble fraction. The production of H_2 from biomass pyrolysis is highly dependent on the catalyst type, process temperature, feedstock type, and residence time.

H_2 production yield is particularly upgraded by pyrolysis temperature increase and further improved by high heating rate and longer residence time. It is higher by fast pyrolysis at high temperatures than slow pyrolysis. The process kinetics generally produces a higher gas yield when the operating temperature is high, the heating rate is high, and the residence period is long enough. But for such processes to produce more H_2, sufficient optimisation is required. Even though longer residence time promotes thermal cracking of heavier hydrocarbons and correspondingly higher gas release rates, they also cause some H_2 to be consumed during the reaction and produce other gaseous substances like carbon oxides and methane. This results in a reduction in the amount of H_2.

b. Dry thermal gasification or gasification.

Gasification is a thermochemical process that transforms biomass into gaseous fuels through partial oxidation and reduction reactions at high temperatures and can occur with or without a catalyst. Hydrogen production by the gasification process of biomass can produce cleaner fuel than solid biomass (Fig. 5).

The product gases are syngas, composed of H_2, N_2, CO, CH_4, and CO_2. The process output is not just the syngas; there are other combustible gases, inert components, tar, char, dust, and other impurities. The energy inherent in the biomass is not immediately released in this process but instead transferred into the bond energies of lighter fuels rather than the perceptible heat of combustion products. Gasification encourages the creation of gases with multiple properties. The process can be divided into air, oxygen, and steam gasification depending on the oxidising agent utilised. The potential for several gasification processes to produce H_2 depends on the abundance, affordability, and broad applicability in various industries. A complex series of chemical reactions must occur for the biomass gasification process. Many of the techniques are endothermic, requiring heat to continue, while some reactions are exothermic, releasing heat as they proceed.

c. Hydrothermal gasification

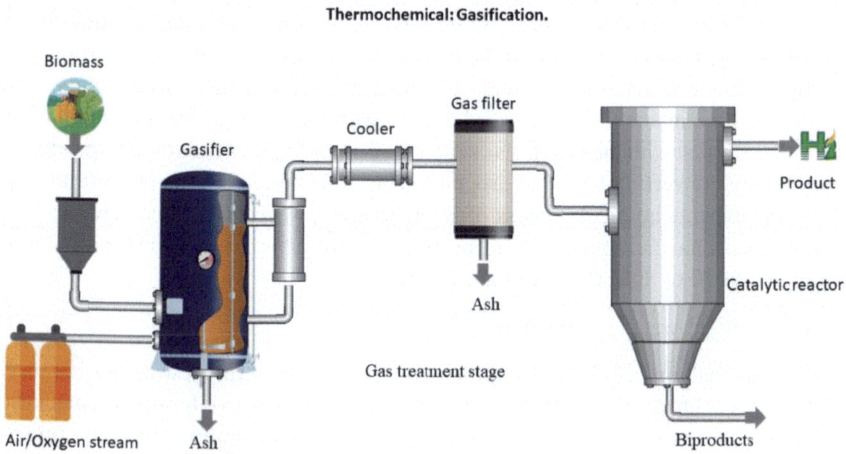

Fig. 5 Schematic diagram of gasification process (Eloffy et al., 2022)

Hydrothermal gasification occurs at higher pressures, including sub- and super-critical pressure (22.1 MPa). According to the reaction temperature, particularly the water conditions, hydrothermal gasification can be divided into three different types: aqueous phase reforming (temperature range of 215–265°C), (b) near-critical catalysed gasification (temperature range of 350–400°C), and (c) supercritical water gasification (temperature greater than 374°C). Under subcritical conditions, aqueous phase reforming is conducted, and multiple noble metal catalysts are used to promote hydrogen production. Although biomass is hydrolysed slowly, the feedstock concentration should generally be low (less than 1%). Higher molecular weight constituents (such as cellulose and lignin) are particularly difficult to convert with this approach, and the conversion is relatively restricted to a few simple constituents. Commonly, near-critical catalysed gasification is conducted to produce methane with the help of a catalyst. This catalyst enhances the conversion of syngas. Finally, the supercritical water gasification is generally carried out at a temperature higher than 500°C and pressure more elevated than the critical value (22.1 MPa).

1. Steam reforming

The steam reforming (SR) process is an attractive complementary technology that may be incorporated into current operations to minimise the amounts of tar and char and release syngas from the gasifier with high H_2 concentrations. Steam gasification produces significantly more H_2 than pyrolysis or air gasification, and the overall conversion efficiency of thermal energy to H_2 can reach up to 52% (Eloffy et al., 2022). However, during the SR process, a simultaneous purifying technique reduces the mass ratio of C to H_2 (C/H) in syngas.

The amount of light hydrocarbons and tar can cause pipe corrosion and clogging because polymerisation and condensation are reduced by adjusting the operational parameters of SR. The specific challenges encountered in this approach are high temperature, massive energy input, large CO_2 generation, and a requirement of catalyst regeneration, which can be nickel. This approach turns hydrocarbon into hydrogen gas. The reaction is usually endothermic and requires a heat supply. The reaction is done inside the pressurised reactor with the catalyst. Steam reforming is generally coupled with the water–gas shift at high temperatures to enrich the produced hydrogen by converting CO to hydrogen. Steam reforming is also a primordial step during the thermochemical process of H_2 generation. Most of the catalysts used are the same as for gasification. Steam reforming by gasification can yield H_2 production ranging from 70 to 80% (Eloffy et al., 2022).

2. Supercritical water gasification

The thermochemical alternative process known as supercritical water gasification (ScWG) has been developed to be more resilient for all types of biomass, especially wet biomass of moisture > 35%, such as wood and carbohydrates. The entire procedure resembles aqueous-phase reforming and is endothermic. However, high energy levels are needed to maintain these conditions, increasing costs and restricting scalability. At supercritical conditions, the hydrogen bond becomes significantly lower, and water behaves as an active reactant, promoting hydrolysis and water–gas shift reaction. It has high diffusivity, resulting in its excellent performance as a solvent; therefore, its use can replace the use of corrosive acid and organic solvents. Similarly, operating parameters in dry thermal gasification (temperature, feedstock-to-water ratio, and catalyst) strongly influence the conversion during hydrothermal gasification. For ScWG systems to be economically viable, hydrogen is mainly produced during water–gas shift reactions, and a small amount is produced during decomposition and pyrolysis. On the other hand, methanation and hydrogenation are oppositely reducing the hydrogen yield, as they convert the produced hydrogen to other compounds (M. Aziz et al., 2021a, 2021b).

d. Incineration

Incineration technology is being used widely in many developed and developing countries to tackle solid waste management issues (Allergrini et al., 2015) as the mass of MSW can be reduced up to 80 wt.% (Lam et al., 2010) as well as break down some hazardous and non-metallic organic wastes and get rid the growth of bacteria and viruses (Yu et al., 2013).

Municipal solid waste collected from the household area transported to an incineration facility and the waste dumped into the refuse pit for waste homogenization and increase the calorific value. Later wastes are transported to the charging hopper and then to the incineration chamber, where the waste will be incinerated at 850–1000 C. Incinerated waste (ash) then dropped into a quenching tank to cool down and later transferred to the ash pit via conveyor. The wet bottom ash generated is about 80–85% of the total solid residue generated during the incineration process (Cioffi et al., 2011). Bottom ash residue is a complex heterogeneous slag, which mainly consists of glass, magnetic and paramagnetic metals, minerals and ceramics (Chimenos et al., 2003).

In a typical incineration bottom ash (IBA), metals such as aluminium, zinck, iron and copper can be observed in 6–13 wt.% (Huber et al., 2021). However, the actual recovery rate of metals in bottom ash is limited by the smaller size of non-ferrous scraps which is dominant in finer fraction (Hu et al., 2011). Although the presence of metal may be seen as a drawback for bottom ash reuse purpose, its potential to produce hydrogen gas as a clean source of energy may be considered an advantage. Hydrogen from MSWI can be generated through metal (aluminium)—water reaction. Approximately 8.41–39.4 litter of hydrogen gas can be generated per kg of dry ash (Nithiya et al., 2018 and Saffarzadeh et al., 2016). Hydrogen generation from bottom ash is considered as sustainable and cheaper alternative solution for the traditional metal recovery in MSW IBA. Recovered hydrogen gas could provide an additional economic and environmental values if directed as clean fuel for electricity generation.

3 Challenges in the Waste to Hydrogen Technologies

The technologies used to transform MSW to Hydrogen might reduce our dependency on fossil fuels and offer sustainable options for environmentally responsible remediation of sites. However, these technologies are challenging. The syngas produced by gasification are highly rich in H_2, but MSW can yield energy-dense bio-oil by pyrolysis and liquefaction (Table 1). In addition, the operating parameters of these thermochemical processes, such as temperature, heating rate, pressure, reaction time, feed concentration, and catalyst type, determine the quality and the H_2 production efficiency. Hydrogen can be produced from biomass and organic solid waste fractions of MSW. However, those technologies are still in the research and development stage, and several problems, including those related to the economic and technological aspects, still need to be resolved. The rate of hydrogen synthesis from organic biomass channels from thermochemical processes is higher than that of biological ones. Besides gasification, pyrolysis and steam reforming are the most developed and prepared technologies. Thermochemical transformation produces hydrogen from organic material more quickly than biological methods. Using catalysts to improve the conversion rate to H_2 production is challenging due to the impossibility of regenerating the by-products and the increasing cost of using catalysts.

Alternatively, without using enzymes, biological conversion routes have the advantage of low temperature, pressure, and energy input, especially dark fermentation and photobiological processes. These biological conversion techniques suit liquid and organic-rich materials, including raw food waste.

These technologies need pre-treatment procedures. Some thermochemical processes include complex, high-temperature reactions, which make it difficult to manage operations, thus increasing reactor costs and the lack of infrastructure for stable supply and transmission of hydrogen. The advantages of low temperature, pressure, and energy in the treatment plant outweigh the biological conversion pathways. Storage of hydrogen is also considered challenging. Liquified hydrogen requires less storage volume than gaseous hydrogen, but energy loss and energy consumption

Table 1 Challenges of Hydrogen generation

Technologies	Challenges	References
Biological route		
Dark fermentation	Low production rate, low hydrogen yield (4–44 g H_2/kg-feedstock), production of a large quantity of by-products, the requirement for pre-treatment, low removal of chemical oxygen demand	Lepage et al. (2021)
Photobiological-process	Low production rate, low energy conversion efficiency, relatively low hydrogen yield (9e49 g-H2/kg-feedstock), the requirement of a large surface area	Dincer and Acar (2015)
Microbial electrolysis	The requirement for an electric external supply, expensive construction material, and long-term stability is questionable	Kadier et al. (2016)
Thermochemical route		
Dry thermal-gasification	Formation of tar and char (shorter catalyst life), the requirement for catalyst regeneration, high temperature, complex reaction leading to challenging operation control, high cost for reactor and apparatus, fluctuating and varying gas composition, CO2 generation	Shahbaz et al. (2017)
Hydrothermal gasification	Limited to the material with high moisture content, CO2 generation, the high energy input for pressure elevation, high reactor cost, problems related to corrosion and plugging, difficulty in recovering the catalyst	Abdalla et al. (2018), Darmawan et al. (2018)
Pyrolysis	Formation of char and tar, CO2 generation, fluctuating and varying gas composition, high cost	Zhao et al. (2010)
Steam reforming	High temperature, massive energy input, large CO2 generation, the requirement of catalyst regeneration	Dincer and Acar (2015) and Lepage et al. (2021)
Incineration Alumimum-water based reaction	Considerable amount of H_2 escaped during the quenching process, requirement of suitable storage method and the amount of H_2 generation depends on the MSW composition	Nithiya et al. (2018), Saffarzadeh et al. (2016)

during the liquefying processes are disadvantageous (Cho et al., 2023). Due to the numerous and varied sourcing points, MSW introduces a higher concentration and wider variety of pollutants than biomass.

Regarding climate change, using MSW as feedstock provides a plentiful and affordable source of consistent hydrogen and eliminates the use of present disposal techniques, which are well known to increase greenhouse gas (GHG) emissions and significantly cause water and land contamination. The main issue is that more advanced applications (such as fuel cells for transportation and catalytic processes for the synthesis of bio-H_2) have very low tolerances for H_2 impurities (Sharma et al., 2020).

The MEC technologies are still under research and development. One of the challenges of MECs is suppressing the methanogenic activity during the electro-hydrogenesis with mixed cultures, as it negatively affects the H_2 production rate. The microbial community of hydrogen producers and innovative substrates needs to be explored. Using dark fermentation residues in fermentation and electrochemical systems as a secondary step could pave the way towards sustainable biohydrogen production in increasing H_2 yield (Eloffy et al., 2022). However, DF process development at the industrial scale is limited by its lower hydrogen yield compared to its theoretical application and the estimated costs associated with H_2 production. Additionally, the ancillary processes (such as dilution, centrifugation, and methanogen inhibition) are recommended for further study to maximise the performance. The need for expensive catalysts, methane production, and low hydrogen purity are some inconveniences of the MEC technologies.

Most of the challenges these technologies face for their implementation are related to the lack of large-scale applications, the storage facilities for hydrogen generation, the fuel cell cost and the hydrogen purity.

4 Comparative Look of LCA in MSW to Hydrogen Technology Contribution in the Circular Economy

The life cycle assessment (LCA) of hydrogen production has received increasing attention as a high-efficiency energy generation produced from fossil and non-fossil sources. In a circular economy, integrating H_2 as renewable energy with the continuous recycling of MSW could maximise the waste value, increasing economic growth from resource consumption and waste generation. The three guiding concepts of circular economy are to protect and improve resource utilisation in ways that regenerate natural systems; maximisation of resource, material, and component output and value through reuse and recycling to retain them in the economic system; building a procedure to remove the adverse effects on the environment. The technologies of waste to energy processes support the overarching goals of a circular economy by transforming waste into useful forms of energy, as shown in Fig. 6.

Most hydrogen production technologies mentioned earlier are considered low greenhouse gas (GHG) emissions associated with H_2 production. China, the largest hydrogen producer in the world, established criteria for classifying hydrogen by setting emission thresholds for each production process. The emissions threshold for low-carbon and clean hydrogen is set at 14.51 kg CO_2 eq/kg H_2 and 4.9 kg CO_2 eq/kg H_2, respectively, while the threshold for green hydrogen is below 4.9 kg CO_2 eq/kg H_2, based on the GHG emission from coal gasification, which is China main production route (about 63% of total production in 2020) (Khan, 2020). Hydrogen is promoted as an ideal energy/fuel source for diverse applications. LCA of hydrogen

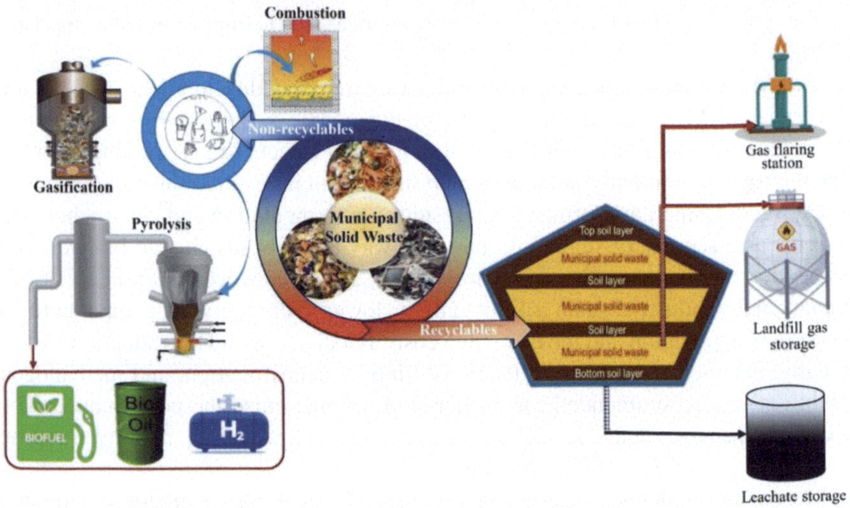

Fig. 6 MSW management using the circular economy framework for H_2 production (Yaashikaa et al., 2020)

(H_2) production from most literature evaluates its environmental assessment on GHG emission to support the climate change target and the decarbonisation associated with carbon capture and storage (Amaya-Santos et al., 2021; Battista et al., 2017; Cvetković et al., 2021). When MSW incineration loads are considered, the benefits of reducing climate change are increased by an additional 90% (Amaya-Santos et al., 2021). The production of hydrogen from MSW is considered bio-hydrogen (Bio-H_2), and to identify the environmental burden of the production of Bio-H_2, the life cycle assessment revealed that the most GHG emissions are from the syngas generation and syngas cleaning, which could impact climate change, acidification, eutrophication fresh water, ecotoxicity freshwater and photochemical ozone formation. Those two operational units are considered hotspots with environmental impacts, which are areas for future process improvement. The material recovery during pre-treatment for refuse derived fuel and production of net electricity improves the reduction of climate change and photochemical ozone formation during the waste to hydrogen process. Depending on the technologies used, reforming processes, gas cleaning, liquefaction (CO_2) and compression (H_2) attribute burden to Bio-H_2 generation. Contributions to acidification, eutrophication and ecotoxicity are dominated during the gas cleaning process (Amaya-Santos et al., 2021). Life cycle assessments are performed for different scenarios and configurations to evaluate their environmental footprint. In the waste to energy paradigm, global warming is the critical impact assessment done by life cycle assessment. Factors such as a high dependence upon the reactor source for heating and electricity greatly influence clean H_2 production. All the above technologies proved to be performing optimally under different weightings of environmental, economic, and social elements, suggesting the versatility of the waste-to-hydrogen concept and its suitability to be applied in various decision-making contexts (Wijayasekera et al., 2022).

The life cycle assessment typically depends on the research goals. The innovative application of auto thermal reforming (ATR) technology on food waste in comparison to steam reforming (SR) and internal combustion engines coupled with electrolysers (ICE + E) revealed that the life cycle assessment of hydrogen generation was to compare the three systems for producing high-purity bio-hydrogen from biogas in terms of their effects on the environment, energy needs, and efficiency. ATR demonstrated a marginally higher potential than steam reforming in energetic and environmental behaviour. ATR was twice as effective as ICE + E. However, ATR and SR required high electric consumption, which significantly increased greenhouse gas emissions and had a negative global impact. At the same time, the liquid electrolyte in the ICE + E required a significant amount of energy to support the electrochemical processes (Battista et al., 2017).

Life cycle assessment is usually helpful in assessing the environmental performance when considering bio-hydrogen generation. Due to methane emissions, steam reforming generally performs better in terms of environmental performance than other available technologies, primarily when organic substrates are used as reagents for Bio-H_2 production. Life cycle analysis evaluated steam reforming to convert alcoholic waste to hydrogen from the anaerobic sludge of palm oil mill effluent. The sensitivity analysis of life cycle assessment revealed that wastewater treatment could produce 33% more alcoholic waste than incineration to produce Bio-H_2 (Cortés et al., 2019).

On the other hand, steam reforming has poor results in life cycle assessment on global warming potential (GWP) and Photochemical Oxidant (POF) due to methane emissions. The hydrogen production process when biogas is used as feedstock during steam reforming has a negative balance; it uses large amounts of natural gas to reach the temperature of the chemical reactions of the steam-reforming process, and it is unsustainable from an energy point of view (Cvetković et al., 2021). Gasification and landfill bioreactor technology are the methods with the highest and lowest H_2 outputs, respectively. Gasification coupled with pyrolysis increased the H_2 output by 20% compared to the standalone pyrolysis process, even with an 85% reduction in final biochar output (Wijayasekera et al., 2022). One lacking LCA aspect is the life cycle cost during H_2 post-production. The storage and transportation of H_2 involve specific logistics depending on the H_2 form generated, and high costs accompany these processes.

5 Conclusion

Bio-hydrogen is regarded as one of the expanding environmentally friendly energy sources in recent years due to its advantages in providing clean and sustainable energy. A mechanistic understanding of the technologies used to produce hydrogen from municipal solid waste was discussed. As a result, the current analysis has carefully considered recent developments in MSW-sourced H_2 generation routes (thermochemical, biological, and electrochemical). Direct or indirect biophotolysis, photo- and dark-fermentation are examples of biological processes that can produce hydrogen from MSW, as can thermochemical techniques like pyrolysis, liquefaction, and gasification. The biological approaches outperformed thermochemical and electrochemical methods regarding environmental performance for the most promising identified hydrogen conversion pathways. Dark and photo-fermentation by bacteria, cyanobacteria, microalgae, and stringent anaerobes are the most widely used methods for biological hydrogen production. Life cycle assessment on the potential Bio-H_2 recovery mainly investigated the environmental impacts of these technologies, which aligned with the aspect of the circular economy.

Further studies are still needed in all aspects of the H_2 recovery. Also, hydrogen transportation is another factor impacting global warming in life cycle assessment. The comparative analysis related to hydrogen transportation from production plants is generated from different raw materials sources at a laboratory scale, and more studies on a larger scale will be needed in the future to determine a more accurate estimate of the actual environmental impact of hydrogen transportation. Identifying the technical barriers associated with large-scale commercial applications requires studies on continuous operation. It is essential to conduct in-depth studies on the toxicology of residues. Research on recycling residuals from hydrogen generation for further use might be able to solve this problem in the future.

References

Abdalla, A. M., Hossain, S., Nisfindy, O. B., Azad, A. T., Dawood, M., & Azad, A. K. (2018). Hydrogen production, storage, transportation and key challenges with applications: A review. *Energy Conversion and Management, 165*, 602–627. https://doi.org/10.1016/j.enconman.2018.03.088

Agamuthu, P. (2001). *Solid waste: Principles and management* (pp. 174–179). University of Malaya Press.

Allegrini, E., Vadenbo, C., Boldrin, A., & Astrup, T. F. (2015). Life cycle assessment of resource recovery from municipal solid waste incineration bottom ash. *Journal of Environmental Management, 151*, 132–143.

Allegue, L. D., Puyol, D., & Melero, J. A. (2020). Novel approach for the treatment of the organic fraction of municipal solid waste: Coupling thermal hydrolysis with anaerobic digestion and photo-fermentation. *Science of the Total Environment, 714*, 136845. https://doi.org/10.1016/j.scitotenv.2020.136845

Amaya-Santos, G., Chari, S., Sebastiani, A., Grimaldi, F., Lettieri, P., & Materazzi, M. (2021). Biohydrogen: A life cycle assessment and comparison with alternative low-carbon production routes in UK. *Journal of Cleaner Production, 319*. https://doi.org/10.1016/j.jclepro.2021.128886

Aziz, A., Agamuthu, P., Hassan, A., Auta, H., Fauziah, S. J. E. T., & Innovation. (2021). Green coagulant from Dillenia indica for removal of bis (2-ethylhexyl) phthalate and phenol, 4, 4'-(1-methylethylidene) bis-from landfill leachate. *24*, 102061. https://doi.org/10.1016/j.eti.2021.102061

Aziz, M., Darmawan, A., & Juangsa, F. B. (2021b). Hydrogen production from biomasses and wastes: A technological review. *International Journal of Hydrogen Energy, 46*(68), 33756–33781. https://doi.org/10.1016/j.ijhydene.2021.07.189

Bank, W. (2020). *Solid waste management*. www.worldbank.org/en/topic/urbandevelopment/brief/solid-waste-management

Battista, F., Montenegro Camacho, Y. S., Hernández, S., Bensaid, S., Herrmann, A., Krause, H., & Fino, D. (2017). LCA evaluation for the hydrogen production from biogas through the innovative BioRobur project concept. *International Journal of Hydrogen Energy, 42*(19), 14030–14043. https://doi.org/10.1016/j.ijhydene.2016.12.065

Ceretta, M. B., Nercessian, D., & Wolski, E. A. (2021). Current trends on role of biological treatment in integrated treatment technologies of textile wastewater. *Frontiers in Microbiology, 12*, 651025. https://doi.org/10.3389/fmicb.2021.651025

Chimenos, J. M., Fernandez, A. I., Miralles, L., Segarra, M., & Espiell, F. (2003). Short term natural weathering of MSWI bottom ash as a function of particle size. *Waste Management, 23*, 887–895.

Cioffi, R., Colangelo, F., Montagnaro, F., & Santoro, L. (2011). Manufacture of artificial aggregate using MSWI bottom ash. *Waste Management, 31*, 281–288.

Cho, H. H., Strezov, V., & Evans, T. J. (2023). A review on global warming potential, challenges and opportunities of renewable hydrogen production technologies. *Sustainable Materials and Technologies, 35*. https://doi.org/10.1016/j.susmat.2023.e00567

Cortés, A., Feijoo, G., Chica, A., Da Costa-Serra, J. F., & Moreira, M. T. (2019). Environmental implications of biohydrogen based energy production from steam reforming of alcoholic waste. *Industrial Crops and Products, 138*, 111465. https://doi.org/10.1016/j.indcrop.2019.111465

Cvetković, S. M., Radoičić, T. K., Kijevčanin, M., & Novaković, J. G. (2021). Life cycle energy assessment of biohydrogen production via biogas steam reforming: Case study of biogas plant on a farm in Serbia. *International Journal of Hydrogen Energy, 46*(27), 14130–14137. https://doi.org/10.1016/j.ijhydene.2021.01.181

Darmawan, A., Ajiwibowo, M. W., Yoshikawa, K., Aziz, M., & Tokimatsu, K. (2018). Energy-efficient recovery of black liquor through gasification and syngas chemical looping. *Applied Energy, 219*, 290–298. https://doi.org/10.1016/j.apenergy.2018.03.033

Dincer, I., & Acar, C. (2015). Review and evaluation of hydrogen production methods for better sustainability. *International Journal of Hydrogen Energy, 40*(34), 11094–11111. https://doi.org/10.1016/j.ijhydene.2014.12.035

Dolar, D., Kosutic, K., & Strmecky, T. (2016). Hybrid processes for treatment of landfill leachate: Coagulation/UF/NF-RO and adsorption/UF/NF-RO. *Separation and Purification Technology, 168*, 39–46. https://doi.org/10.1016/j.seppur.2016.05.016

Eloffy, M. G., Elgarahy, A. M., Saber, A. N., Hammad, A., El-Sherif, D. M., Shehata, M., ... Elwakeel, K. Z. (2022). Biomass-to-sustainable biohydrogen: Insights into the production routes, and technical challenges. *Chemical Engineering Journal Advances, 12*. https://doi.org/10.1016/j.ceja.2022.100410

Ghimire, A., Frunzo, L., Pirozzi, F., Trably, E., Escudie, R., Lens, P. N. L., & Esposito, G. (2015). A review on dark fermentative biohydrogen production from organic biomass: Process parameters and use of by-products. *Applied Energy, 144*, 73–95. https://doi.org/10.1016/j.apenergy.2015.01.045

Hassan, A., Pariatamby, A., Ossai, I. C., Ahmed, A., Muda, M. A., Barasarathi, J., & Hamid, F. S. J. A. S. E. (2022). Synergistic association of endophytic fungi enhances tolerance, growth, and heavy metal uptake of Alocasia calidora in landfill contaminated soil. *Applied Soil Ecology, 170*, 104307. https://doi.org/10.1016/j.apsoil.2021.104307

Hu, Y., Bakker, M. C. M., & de Heij, P. G. (2011). Recovery and distribution of incinerated aluminum packaging waste. *Waste Management, 31*, 2422–2430.

Huber, F., Korotenko, E., Syc, M., & Fellner, J. (2021). Material and chemical composition of municipal solid waste incineration bottom ash fractions with different densities. *Journal of Material Cycles and Waste Management, 23*, 394–401.

Kadier, A., Kalil, M. S., Abdeshahian, P., Chandrasekhar, K., Mohamed, A., Azman, N. F., & Hamid, A. A. (2016). Recent advances and emerging challenges in microbial electrolysis cells (MECs) for microbial production of hydrogen and value-added chemicals. *Renewable and Sustainable Energy Reviews, 61*, 501–525. https://doi.org/10.1016/j.rser.2016.04.017

Kaza, S., Yao, L., Bhada-Tata, P., & Van Woerden, F. (2018). *What a waste 2.0: A global snapshot of solid waste management to 2050*. World Bank Publications.

Khan, I. (2020). Waste to biogas through anaerobic digestion: Hydrogen production potential in the developing world—A case of Bangladesh. *International Journal of Hydrogen Energy, 45*(32), 15951–15962. https://doi.org/10.1016/j.ijhydene.2020.04.038

Kossalbayev, B. D., Tomo, T., Zayadan, B. K., Sadvakasova, A. K., Bolatkhan, K., Alwasel, S., & Allakhverdiev, S. I. (2020). Determination of the potential of cyanobacterial strains for hydrogen production. *International Journal of Hydrogen Energy, 45*(4), 2627–2639.

Lam, C. H. K., Alvin, W. M. I., Barford, J. P., & McKay, G. (2010). Use of incineration MSW Ash: A review. *Sustainability, 2*, 1943–1968.

Lepage, T., Kammoun, M., Schmetz, Q., & Richel, A. (2021). Biomass-to-hydrogen: A review of main routes production, processes evaluation and techno-economical assessment. *Biomass and Bioenergy, 144*, 105920. https://doi.org/10.1016/j.biombioe.2020.105920

Li, C., & Fang, H. H. P. (2007). Fermentative hydrogen production from wastewater and solid wastes by mixed cultures. *Critical Reviews in Environmental Science and Technology, 37*(Krook et al.), 1–39. https://doi.org/10.1080/10643380600729071

Nanda, S., & Berruti, F. (2021). A technical review of bioenergy and resource recovery from municipal solid waste. *Journal of Hazardous Materials, 403*, 123970. https://doi.org/10.1016/j.jhazmat.2020.123970

Ni, M., Leung, D. Y. C., Leung, M. K. H., & Sumathy, K. (2006). An overview of hydrogen production from biomass. *Fuel Processing Technology, 87*(5), 461–472. https://doi.org/10.1016/j.fuproc.2005.11.003

Nithiya, A., Saffarzadeh, A., & Shimaoka, T. (2018). Hydrogen gas generation from metal aluminum-water interaction in municipal solid waste incineration (MSWI) bottom ash. *Waste Management, 73*, 342–350.

Osman, A. I., Deka, T. J., Baruah, D. C., & Rooney, D. W. (2020). Critical challenges in biohydrogen production processes from the organic feedstocks. *Biomass Conversion and Biorefinery, 13*(10), 8383–8401. https://doi.org/10.1007/s13399-020-00965-x

Osman, A. I., Mehta, N., Elgarahy, A. M., Hefny, M., Al-Hinai, A., Al-Muhtaseb, A. a. H., Rooney, D. W. (2021). Hydrogen production, storage, utilisation and environmental impacts: A review. *Environmental Chemistry Letters, 20*, 153–188.https://doi.org/10.1007/s10311-021-01322-8

Parthiba Karthikeyan, O., Trably, E., Mehariya, S., Bernet, N., Wong, J. W. C., & Carrere, H. (2018). Pre-treatment of food waste for methane and hydrogen recovery: A review. *Bioresource Technology, 249*, 1025–1039. https://doi.org/10.1016/j.biortech.2017.09.105

Saffarzadeh, A., Nithiya, A., & Shimaoka, T. (2016). Aluminum and aluminum alloys in municipal solid waste incinearion (MSWI) bottom ash: A potential source for the production of hydrogen gas. *International Journal of Hydrogen Energy, 41*(2), 820–831.

Shahbaz, M., Yusup, S., Inayat, A., Patrick, D. O., & Ammar, M. (2017). The influence of catalysts in biomass steam gasification and catalytic potential of coal bottom ash in biomass steam gasification: A review. *Renewable and Sustainable Energy Reviews, 73*, 468–476. https://doi.org/10.1016/j.rser.2017.01.153

Shao, Q., Li, J., Yang, S., & Sun, H. (2019). Effects of different substrates on microbial electrolysis cell (Dolar et al.) anodic membrane: Biodiversity and hydrogen production performance. *Water Science and Technology, 79*(6), 1123–1133. https://doi.org/10.2166/wst.2019.107

Sharma, S., Basu, S., Shetti, N. P., & Aminabhavi, T. M. (2020). Waste-to-energy nexus for circular economy and environmental protection: Recent trends in hydrogen energy. *Science of the Total Environment, 713*, 136633.

Wijayasekera, S. C., Hewage, K., Hettiaratchi, P., Siddiqui, O., Razi, F., Pokhrel, D., & Sadiq, R. (2022). Sustainability of waste-to-hydrogen conversion pathways: A life cycle thinking-based assessment. *Energy Conversion and Management, 270*. https://doi.org/10.1016/j.enconman.2022.116218

Yaashikaa, P., Kumar, P. S., Saravanan, A., Varjani, S., & Ramamurthy, R. (2020). Bioconversion of municipal solid waste into bio-based products: A review on valorisation and sustainable approach for circular bioeconomy. *Science of the Total Environment, 748*, 141312.

Yan, M., Zhang, S., Wibowo, H., Grisdanurak, N., Cai, Y., Zhou, X., Antoni. (2020). Biochar and pyrolytic gas properties from pyrolysis of simulated municipal solid waste (SMSW) under pyrolytic gas atmosphere. *Waste Disposal & Sustainable Energy, 2*(Krook et al.), 37–46. https://doi.org/10.1007/s42768-019-00030-y

Yu, J., Sun, L., Xiang, J., Jin, L., Hu, S., Su, S., & Qiu, J. (2013). Physical and chemical characterization of ashes from a municipal solid waste incinerator in China. *Waste Management & Research, 31*(7), 663–673.

Zhao, B., Zhang, X., Sun, L., Meng, G., Chen, L., & Xiaolu, Y. (2010). Hydrogen production from biomass combining pyrolysis and the secondary decomposition. *International Journal of Hydrogen Energy, 35*(7), 2606–2611. https://doi.org/10.1016/j.ijhydene.2009.04.011

Zhen, G., Kobayashi, T., Lu, X., Kumar, G., Hu, Y., Bakonyi, P., & Xu, K. (2016). Recovery of biohydrogen in a single-chamber microbial electrohydrogenesis cell using liquid fraction of pressed municipal solid waste (LPW) as substrate. *International Journal of Hydrogen Energy, 41*(40), 17896–17906. https://doi.org/10.1016/j.ijhydene.2016.07.112

Estimation of Methane Gas Emissions from Municipal Landfill Sites: A Comparative Review on the Existing Models

Mohammad Toha, Sadia Sikder, and Md. Mostafizur Rahman

Abstract Due to urbanization, increasing population growth, and inadequate landfill management, a massive amount of waste has been disposed of in landfills and generated a tremendous amount of methane from the landfills. Moreover, landfills have already been considered a significant source of methane and are becoming methane sanctuaries, especially in developing nations. Where the world is witnessing the devastating effects of climate change and global warming, producing this massive amount of methane gas is undoubtedly cause for concern in the global community, as methane's global warming potential is 25 times greater than carbon dioxide. On the other hand, methane is a very light gas that fluctuates quickly, making it difficult to estimate. Thus, measuring how much methane is generated from landfills is essential to taking the necessary steps and raising awareness. Thus, this book chapter emphasizes the overview of various methane estimation models from landfills and their comparative assessment. Lastly, the overview of these various methane estimation models will assist policymakers in adopting new policies and encourage innovation regarding these issues.

1 Introduction

Since industrialization, a rapidly increasing trend of greenhouse gas emissions has already been seen in our globe due to the large population, industrial activities, urbanization, and other factors (Chandra & Ganguly, 2023; Rafiq et al., 2018). In recent years, a large amount of methane has been emitted from landfill sites due to the tremendous amount of waste and inadequate management programs (Liu et al.,

M. Toha · S. Sikder · Md. Mostafizur Rahman (✉)
Department of Environmental Science, Bangladesh University of Professionals, Mirpur, Cantonment, Dhaka 1216, Bangladesh
e-mail: rahmanmm@juniv.edu

Md. Mostafizur Rahman
Laboratory of Environmental Health and Ecotoxicology, Department of Environmental Sciences, Jahangirnagar University, Savar, Dhaka 1342, Bangladesh

A. Anouzla and S. Souabi (eds.), *Generation of Energy from Municipal Solid Waste*,
https://doi.org/10.1007/978-3-031-74334-4_11

2023). Moreover, landfills are a sacred hub of methane gases responsible for 20% of global warming (Kumar et al., 2016). Air pollutants emitted from landfills contribute to the emission of greenhouse gases (GHGs) into the atmosphere and pose significant health risks to humans. In recent decades, greenhouse gases from human activities have surpassed those of natural origin (Kumar & Sharma, 2014). The waste sector is a significant contributor to greenhouse gas (GHG) emissions, accounting for roughly 5% of the global greenhouse budget. This 5% consists of methane and carbon dioxide emissions. So, knowing how and why methane gas is produced in landfills is a significant concern.

Generally, the anaerobic breakdown of organic waste in landfills, with the help of bacteria that thrive in these circumstances, is the primary source of methane gas. All organic waste, such as food and garden scraps, textiles, and paper, decomposes more quickly in the presence of oxygen in landfills (Amoatey et al., 2023; Kumar et al., 2004). Methane and carbon dioxide are combined in this method to make landfill gas. According to Kumar et al. (2016), landfill gas consists mainly of methane and carbon dioxide. It is produced in three phases that occur quickly, one after another, before reaching a stable state. Since it might take decades for most of the organic material in the trash to decay, these odours will likely persist for some time. Landfill gas not absorbed by the cover or oxidized by the bacteria is discharged into the environment (Chakraborty et al., 2011; Silva et al., 2020).

The transport processes that lead to methane migration and emission are affected by atmospheric conditions (barometric pressure, precipitation, temperature, wind), soil/cover conditions (cracks, fissures, permeability, diffusivity, porosity, moisture content, organic content, CH4 oxidation capacity), and waste and landfill conditions (gas production, temperature, and wind) (Bingemer & Crutzen, 1987; Osra et al., 2021; Rafiq et al., 2018). Moreover, soil's physical and chemical characteristics (composition, depth, moisture, temperature, aeration status) greatly influence gas movement and microbial activity throughout space and time (Kumar et al., 2023a, 2023b). According to existing studies, the concentration level of methane is being raised by 22.16 ppb, which contains almost (1–2) % annually (Anh et al., 2021; Rafiq et al., 2018). Furthermore, due to anthropogenic activities, 30% and 13% of methane gas have already been generated from European and Chinese landfills (Toha & Rahman, 2023), which has a tremendous detrimental impact on the nearby environment and human health. On the other hand, most of the developing nation follows a sanitary landfill system with an inadequate management program (Jeong et al., 2019; Kormi et al., 2017). As a result, these landfills have already generated a tremendous amount of GHGs. This worrying greenhouse gas emissions and atmospheric accumulation throughout the globe have prompted several countries to commit to reducing their contributions. However, while the developed nations have already measured the estimation of methane from landfills, at the same time, developing nations still need to be able to quantify the estimation of methane from their landfills due to inadequate data and technological problems. Thus, it is crucial to determine the methane emission to overcome this situation as it has a tremendous global warming potential and is responsible for climate change.

Standard methods of methane emission estimation are yet to be discovered from landfills. However, several models and methods have already been developed in methane emission estimation in recent years. The IPCC ZODM, IPCC FODM, MTM, and Land-GEM-V-3.02 are some of the renowned models for estimating methane from landfills (Saltykov & Trifonov, 2023; Toha & Rahman, 2023). In addition, In-situ methods are used to estimate methane from direct landfill environments (Anh et al., 2021; Hettiarachchi et al., 2023; Liu et al., 2023). Unlike In-situ methods, sensor-based models are utilized to estimate methane indirectly (Manheim et al., 2023; Ma et al., 2023; Mellqvist et al., 2023; Njoku & Edokpayi, 2023; Papale et al., 2023; Pliszka et al., 2023). As standard methods are yet to be discovered, inventing more reliable methane emission estimation procedures is essential. As the GWP of methane is more than 25 times that of carbon dioxide, is responsible for climate change, and has a detrimental impact on human health, it is high time to measure the amount of methane emission from landfills to detect the level of impact and mitigate these harmful gases.

Therefore, this chapter will discuss the production procedure of methane, the mobility of methane in landfills, and contributing factors. This chapter will also discuss the mechanisms of existing methane emission models and their comparative assessment. Lastly, with a comprehensive understanding of greenhouse gas (GHG) emissions from landfills and their estimating procedures, policymakers and stakeholders can make informed decisions to reduce emissions, implement sustainable waste management practices, and work toward a greener and more resilient urban environment.

2 Sources of Methane

Methane (CH4) and other trace gases in the atmosphere are commonly regarded as substantially contributing to the so-called "greenhouse effect." There are six primary environmental sources of methane: (1) natural wetland anaerobic decomposition, (2) paddy rice fields, (3) livestock production systems (including intrinsic fermentation and animal waste), (4) biomass burning (including forest fires, charcoal combustion, and firewood burning), and (6) fossil methane emission from the exploration and transportation of fossil fuels (Troiano et al., 2023; Yu et al., 2023). The increased methane emissions from these regions may be directly attributed to human activity. Methane emissions have risen since the advent of fossil fuels and the global proliferation of paddy rice cultivation. Although it's atmospheric concentration is relatively low (about 1.7 ppmv), this radioactively and chemically reactive gas has been increasing at around 1% per year (He et al., 2023; Kumar et al., 2016). Two to six times more methane exists now than before the industrial revolution. The complex and poorly understood function methane plays in atmospheric chemistry is the photochemical oxidation of methane by the hydroxyl radical (OH), which produces carbon monoxide, carbon dioxide, water, and methane (Rafiq et al., 2018). Methane is mainly discharged into the atmosphere by this oxidizing reaction with OH. Ozone is one of the additional trace gases that may be employed. Methane pollution causes many problems for the planet's thermal equilibrium. Its oxidation triggers the emission of additional climate-altering gases (including carbon dioxide and water vapour) and affects the half-life of many other gases (Yu et al., 2023).

3 Production Process of Methane

Landfill gas is a mixture of various types of gases. Generally, landfill gas contains mainly methane and carbon dioxide. In addition, other gases like nitrogen, oxygen, ammonia, hydrogen, and nonmethane organic compounds (NMOC) are found in landfills. The landfill gas components with their percentage are depicted in Fig. 1. Bacterial decomposition, volatilization, and chemical reactions significantly impact the methane production process. Bacterial decomposition can break down the organic wastes and create methane and carbon dioxide. On the other hand, NMOC is generally created due to the volatilization and chemical reactions process.

The bacterial decomposition follows four phases to generate methane from organic wastes in landfills. The phases are given below with descriptive information.

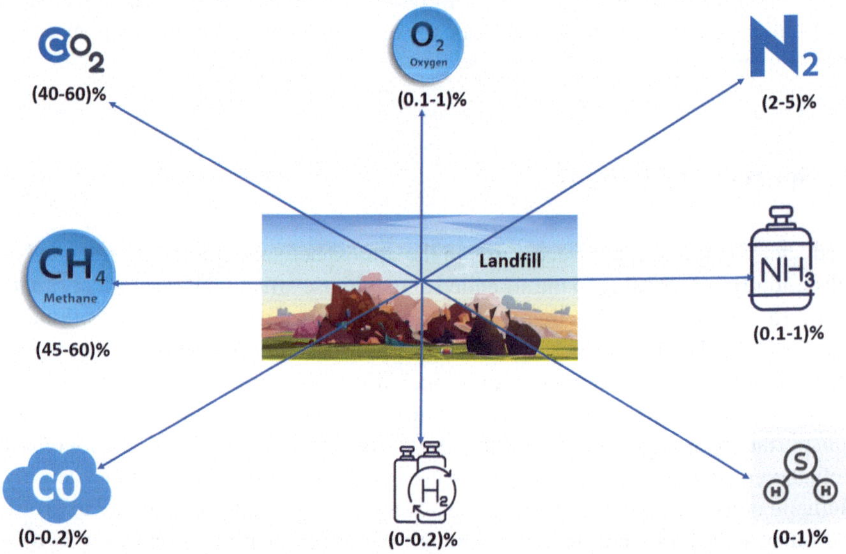

Fig. 1 Landfill gas components (percent by volume)

3.1 Phase I (Aerobic Stage)

In this phase, carbon dioxide and nitrogen gas are created due to aerobic bacteria' breakdown of organic wastes. Initially, the nitrogen gas level is high in this phase, but later, it declines when landfills go through the other phases (Kissas et al., 2022). As this phase depends on aerobic bacteria, the phase continues until the oxygen level ends (Wang et al., 2023).

3.2 Phase II (Anaerobic Non-Methanogenic Stage)

The gas production pattern shifts to Phase II once oxygen is depleted, during which anaerobic activity predominates. Around this time, there is a noticeable increase in hydrogen production and a carbon dioxide bloom (a peak in carbon dioxide concentration) (Scheutz et al., 2023). However, some nitrogen production may exist via denitrification, and nitrogen is displaced during this process. It has yet to begin to produce methane. A lack of dissolved carbon dioxide, which acts as a hydrogen acceptor, may account for the delay in methane synthesis following the establishment of anaerobiosis (Shafy et al., 2023). In this phase, the landfill contains a highly acidic situation as the production of aerobic bacteria is converted into various acids and alcohols. After that, these acids mix with the moisture content of landfills and create a favourable situation to dissolve nutrients.

Consequently, bacteria can get more nutrients (nitrogen and phosphorus) to grow and produce carbon dioxide and hydrogen. Two sanitary landfill sites in California were found to have elevated carbon dioxide levels. The Azusa facility reached 70% volume of carbon dioxide after 11 days, whereas the Calabasas location reached 50% after 23 days. According to Beluche (1968), after 40 days in a sanitary landfill test cell, the carbon dioxide content reached its high of 90%. Lin (1966) noted hydrogen concentrations in the range of 20% by volume during the second phase.

3.3 Phase III (Anaerobic Methanogenic Unsteady Phase)

In this phase, the production of the second phase (organic acids) is consumed by certain anaerobic bacteria. As a result, a neutral environment is created due to the symbiotic relationships between methane and acid-producing bacteria (ATSDR, 2019). Since Methanobacterium can rapidly absorb hydrogen, the beginning of this Phase is devoid of it. The levels of carbon dioxide and nitrogen both reach their final minimums. According to a research study, it will take around six months to 500 days to complete Phases I, II, and III. (ATSDR, 2019; Wang et al., 2023).

3.4 Phase IV (Anaerobic Methanogenic Steady Phase)

For the given environmental conditions, gas composition and the production rate are at their maximum during Phase IV. Neither does it take into consideration long-term alterations owing to variables like food depletion nor the buildup of harmful compounds, both of which are possible outcomes of environmental change that might lead to sudden shifts in gas output (Chandra & Ganguly, 2023; Scheutz et al., 2023). In this stage, landfill gas is expected to comprise 50–70% methane and 30–50% carbon dioxide. The data provided by Kotze et al. (1969) shows that the gas composition of gases generated during the digestion of sewage sludges is 66% methane and 34% carbon dioxide. Therefore, this fits in nicely with the other literature. It is also possible for there to be traces of nitrogen gas and hydrogen sulfide. Methane production is likely delayed if the concentration is less than 50% by volume (Kotze et al., 1969). Changes in gas concentrations can be used as an indicator of climatic shifts (Kriipsalu et al., 2023).

4 Factors Affecting Landfill Gas Production

Numerous well-established factors affect gas production in various media. Some evidence suggests that they could be used in a landfill scenario, which has been a long-held theory. Some of these factors, along with their effects, are stated below in detail:

4.1 Rate of Infiltration

The infiltration rate significantly impacts the production of landfill gas as it can reduce the alkalinity and pH of leachate. Consequently, the level of COD, BOD, and VDS are increased due to reducing these factors.

4.2 Moisture

The decomposition of municipal solid waste requires water. There are numerous ways that water can enter a landfill, but the most common are garbage and precipitation (Rafiq et al., 2018). The optimal moisture content for gas production varies between 60 and 80% wet weight, and gas production rates increase as refuse moisture content increases. The proportion of carbon and hydrogen in the resultant gases increases due to the rising relative humidity trend. Generally, moisture content helps to grow bacteria in landfills by transporting nutrients. As a result, gas production is relatively high in rainy seasons as heavy rainfall occurs (Jain et al., 2023).

4.3 Temperature and pH

Changes in pH throughout a landfill's lifespan significantly impact the composition of microorganisms and the gases they emit (ATSDR, 2019). The acidity of the onset of anaerobic decomposition causes the pH of a landfill to decrease once the aerobic phases are complete. Methane-producing microorganisms cannot convert carbon dioxide to methane in an acidic environment. Therefore, methane production is modest at corrosive pH (ATSDR, 2019). Temperature directly correlates with bacterial activity, with increased gas production at higher temperatures (ATSDR, 2019).

The landfill environment regulates temperature through insulation (70–130°C) and thus allows for relatively stable bacterial colonies. Temperature significantly affects the diversity and activity of anaerobic microbes and bacteria. Most of the chemical reactions that lead to the formation of acetate, which in turn fuels methane-producing microbes in the landfill, are temperature-dependent, with a higher reaction rate at higher temperatures. The thermophilic, mesophilic, and psychrophilic temperature regimes have characterized the effect of heat on methane production. Temperatures over 50°C are optimum for methane production during sewage sludge digestion, according to Kotze et al. (1969). The three categories had optimal temperatures, but the mesophilic range thrived at approximately 37°C. Thus, the presence of high temperatures can increase the gas production rate.

Consequently, seasonal patterns have a significant influence on it. On the other hand, acidic or alkaline substances within the waste can cause pH levels to fluctuate wildly. The accumulation of organic acids may decrease pH due to the inhibition of methane production. Because of them, the alkalinity of the water would decrease.

4.4 Waste Composition

Various things are harmful or inhibitory to gas-generating microbes, and some of them may end up in landfills that are supposed to be clean. In this situation, the gas output heavily depends on the garbage's composition. However, it is also possible that gas generation may occur under the current circumstances, but nutrient deficits in the garbage are slowing things down by hindering the microbial population (Kumar et al., 2004; Rafiq et al., 2018). According to various studies, adjusting the proportions of N, P, and K in the garbage may affect the gas production pace. Moreover, carbonaceous compounds release more gas over time than proteinaceous ones. The methane generation during sewage sludge digestion is maximized at a C: N ratio of 16:1 (Kotze et al., 1969). Thus, gas production rates will depend on the organic substance being digested; therefore, such ratios should be employed with care. Decomposing paper, which has a C: N ratio of around 400:1, will result in a different gas stream than grass, which has a C: N ratio of about 16:1. Decomposition times are also influenced by the refuse's physical state, especially its size and surface area. Therefore, the composition of the waste significantly influences the patterns of gas generation.

4.5 Oxygen Content

Bacteria can only produce methane if the atmosphere is completely depleted of oxygen. When more oxygen is available, aerobic microorganisms may continue Phase I decomposition for an extended period (Bingemer & Crutzen, 1987; Chakraborty et al., 2011). If garbage is only partially buried or is habitually disturbed, oxygen-dependent microbes may proliferate for extended periods, consuming carbon dioxide and water. However, if the trash is exceptionally compacted, methane-producing anaerobic bacteria will rapidly replace aerobic bacteria in Phase III, and methane production will commence sooner. Methane gas production is inhibited by oxygen in the landfill and cannot begin until aerobic microorganisms have consumed all oxygen. The surface soils of the landfill are exposed to atmospheric oxygen at barometric peaks, which may influence microbial activity (ATSDR, 2019). Waste from Phase IV may be transferred back to Phase I for a limited duration if all the oxygen is consumed.

5 Landfill Gas Moving Procedure and the Affecting Factors

As landfill gas emissions escape to the surface, they are typically reversed. As a result of their tendency to expand until they occupy the available space, gases "migrate" or travel through the compacted waste and compacted sediments that cover a landfill. Lighter-than-air landfill gases, such as methane, are naturally inclined towards rising, often by permeating the landfill's upper stratum. A substantial layer of detritus or material (such as soil and daily sealing) may be deposited on top of refuse to prevent the escape of methane gas from landfills. When its upward path is halted, landfill gas may migrate laterally to other areas of the landfill or to sites outside the landfill to continue its ascent. The molecules are attracted to paths with the lowest resistance. Carbon dioxide and other gases with a lower specific gravity prefer congregating underground, particularly in utility corridors. Diffusion (concentration), pressure, and permeability are three possible influences on the movement of landfill gas.

5.1 *Diffusion*

When discussing gases, "diffusion" refers to their tendency to disperse uniformly throughout a confined area. The concentration of gases in garbage progressively decreases as they disperse throughout the garbage. Since gas concentrations are more significant in landfills, landfill gases migrate outward and accumulate in regions with lesser gas concentrations (ATSDR, 2019).

5.2 *Atmospheric Pressure*

When gases accumulate in a landfill and are forced against compacted garbage or soil covers, high-pressure zones where gas passage is restricted and low-pressure zones where it is not being produced due to the pressure gradient in a landfill, gases migrate from high-pressure to low-pressure regions. Convection is the movement of gases from regions of high pressure to low-pressure areas (Bingemer & Crutzen, 1987). As the quantity of gas produced rises, the pressure under a landfill can occasionally exceed that of the surrounding air or the air within it. When landfill pressure rises, gases escape into the atmosphere.

5.3 *Permeability*

Gases pursue the path with the least resistance. The permeability of debris and soils quantifies the ease with which gases and liquids can travel through their interconnected spaces or pores (Chakraborty et al., 2011). Wet clay is typically much less permeable than powdery soils because it has fewer interconnected pores. Gases migrating underground prefer permeable materials (such as sand or gravel) to those with limited permeability (such as clay or sediment) (Rafiq et al., 2018). Due to its limited permeability, clay is frequently used as a landfill cover. Therefore, gases in a covered landfill may travel horizontally rather than vertically.

6 Methane Estimation Models

Estimating methane from landfills is essential to discovering the characteristics and patterns of greenhouse gases (GHGs) in regional areas. Moreover, methane has a significant impact on human health, global warming, and the atmospheric air quality system, so making a comprehensive emission inventory is necessary to determine the probable levels of methane pollution and its impact. The quantification of methane from a landfill vastly depends on atmospheric factors, geographical locations, and

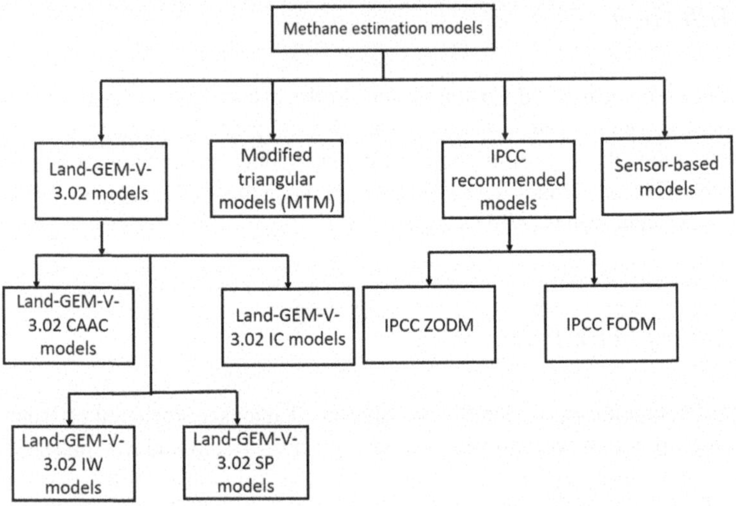

Fig. 2 Methane emission estimation models

temporal dimensions (Adhikary & Sengupta, 2023, Njoku & Edokpayi, 2023; Zhang et al., 2023). Various models and in-situ measurement techniques are widely recognized to estimate methane emissions from landfills in developed nations (Kumar et al., 2023a, 2023b, Kim et al., 2023; Silva et al., 2020). Problematically, due to a lack of resources and technical barriers, developing nations need to catch up in this kind of study. As a result, they silently suffer the harmful effects of methane. Thus, various methane estimation models have been discussed to create awareness and encourage scientific studies. All of these are highlighted below (Fig. 2).

6.1 IPCC Recommended Models

The IPCC recommended two specific models for estimating methane gas emissions from landfills. The IPCC zero order decay model (ZODM) is a simple default value model that follows the mass balance equation. On the other hand, the IPCC first order decay model (FODM) follows an anaerobic degradation process and focuses on time-dependent factors (Kim et al., 2023; Kumar et al., 2004, 2023). As a result, it gives a more accurate value than ZODM. However, ZODM is generally utilized when the availability of field data is lacking. The procedure for choosing IPCC-recommended models is depicted in Fig. 3.

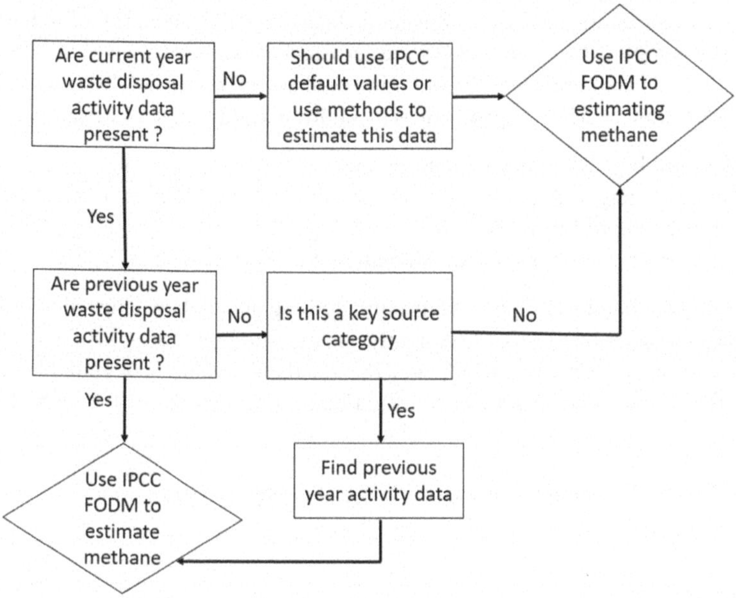

Fig. 3 Choosing procedures of IPCC recommended models

6.2 *IPCC Zero Order Decay Model (ZODM)*

The ZODM model was first invented by Bingemer and Crutzen in 1987. After that, the IPCC recommended this model for estimating methane gas emissions from landfills in 1996. This ZODM follows simple mass conservation techniques where the IPCC sets maximum parameters. This model assumes that the amount of methane directly depends on the number of wastes disposed of in landfills in a particular year. This model does not focus on the degradation rate of wastes, so the assumption of this model is overestimated and has several uncertainties. However, as most developing nations do not have any field data on methane correction factor, oxidation factor, degradable organic carbon value, and recovery rate, this ZODM model could be a good choice for estimating methane from landfills by focusing on Eqs. (1) and (2), ZODM calculates the amount of methane emitted in a given year and the value of DOC.

$$QCH4 = MSWT * MSWF * MCF * DOC * DOCf * F * \left(\frac{16}{12} - R\right) * (1 - OX) \quad (1)$$

In this model, QCH_4, MSWT, R, OX, MSWF, MCF, CF, and F represent the estimation of methane generation from a particular year, the total amount of waste generation, the recovery rate, the oxidation factor, the municipal solid waste fraction, the methane correction factor, the conversion factor, and the methane fraction, respectively. The default values of R and OX are zero, and the value of CF is 1.33. The value of MCF is a crucial parameter for estimating methane in this model (Toha & Rahman, 2023). The value of MCF depends on the condition of landfills. For example:

- MCF could be 1 for managed landfill sites
- If the waste height at an open dump exceeds 5 m, the MCF value must be 0.80. Otherwise, it could be 0.40
- For unspecified landfill sites, the default of MCF could be 0.60

On the other hand, DOC_f means dissimilated organic carbon. The value of DOC_f depends on temperature (T). Consequently, the measurement formula of dissimilated organic carbon is ($DOC_f = 0.014\ T + 0.28$). Lastly, the value of degradable organic carbon (DOC) can vary from city to city. Equation (2) measures the DOC for any landfills.

$$\boldsymbol{DOC} = 0.40\boldsymbol{A} + 0.17\boldsymbol{B} + 0.15\boldsymbol{C} + 0.30\boldsymbol{D} \tag{2}$$

Here,

A=paper + rags
B=leaves + hay + straw
C= stands for fruits and vegetable
D=wood

6.3 IPCC First Order Decay Model (FODM)

The IPCC recommended the first-order decay model (FODM) in 2006. This model assumes that organic waste takes several months to a year to generate landfill gas (LFG). The main components of this LFG are methane and carbon dioxide. The LFG is estimated by first-order kinetics. The types of municipal solid waste in landfills are depicted in Fig. 4.

In FODM mechanisms, microbiological and biochemical processes significantly influence methane gas generation. In the first few years, this model's methane generation peaks. On the other hand, the amount of methane generated decreases with increasing time. This FODM focuses on three segments: Tier I, II, and III (Kumar & Sharma, 2014). Tier I is used when a country does not have specific values. In this situation, default activity data generally estimates methane gas generation from landfills. On the other hand, Tier II is used when any country has at least ten years of data. Moreover, Tier III is usually used for methane estimation in the presence of half-life data, methane generation potential, and degradable organic carbon (Anh et al., 2021; Sohoo et al., 2021).

Generally, Eqs. (3), (4), (5), and (6) are used to determine the methane generation, the equation correction factor, the methane generation constant, and the methane generation potential of a particular landfill.

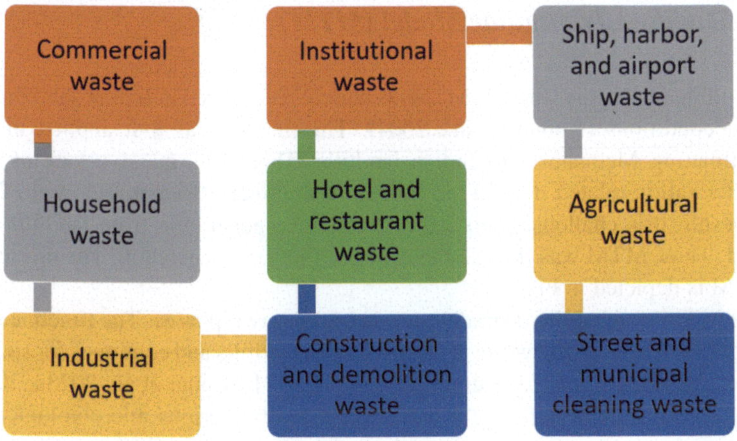

Fig. 4 Various types of municipal solid waste in landfills

$$Q(T, X) = A * K * MSWTx * MSWFx * MCFx * Lo(X) * e^{-K(T-X)} \tag{3}$$

a). A represents the equation correction factor. It is determined by the below formula (Eq. 4)

$$A = \frac{(1 - e^{-k})}{k} \tag{4}$$

b). K means methane generation constant. It is determined by

$$K = \frac{In(2)}{t.5} \tag{5}$$

$t_{.5}$ means the half-life of the waste. The value of half-life depends on the degradability of waste. The high degradable waste has less half-life (3-5) years rather than the slowly degradable waste (10–12) years. On the other hand, the value of methane generation potential (Lo) is determined by Eq. (6).

$$Lo = MCF * DOC * DOCf * F * \frac{16}{12} \tag{6}$$

c) (T-X) means the subtraction of Current year of estimation and the year of first waste disposal.

6.4 Modified Triangular Model (MTM)

The modified triangular model (MTM) is widely recognized in Indian sub-continent landfill conditions (Kumar et al., 2004). This model was first applied in 2004 for estimating Methane from Indian landfills. There is a massive lack of waste characterization-related data in the Indian subcontinent (Ramprasad et al., 2022). As a result, it is challenging to estimate methane generation from landfills using FODM. Thus, MTM was invented to estimate methane generation. The mechanism of MTM is depicted in Fig. 5.

The MTM divided the degradation process into two phases. The timeline of the first phase starts when the waste is disposed of in landfills and continues for six years (Chandra & Ganguly, 2023; Kumar & Sharma, 2014; Kumar et al., 2023a, 2023b). During this first phase, organic waste is degraded by various microbiological and biochemical processes, producing methane gas. As the degradation rate is high in this phase, most methane is predicted to be generated. On the other hand, the second phase starts just after the end of the first phase and shows a decreasing trend for the rest of the time. Moreover, this model assumes that the total duration of these two phases is 15 years (Kumar et al., 2016; Saltykov & Trifonov, 2023). So, no methane gas will be generated after the end of the second phase. According to this model, the first and second phases show an increasing and decreasing trend, which looks triangular. This triangle shape's size equals the amount of methane generated by ZODM (Eq. 7) (Kumar et al., 2004, 2016). After that, the peak year of methane generation is determined by converting the triangle value into cubic meters per year

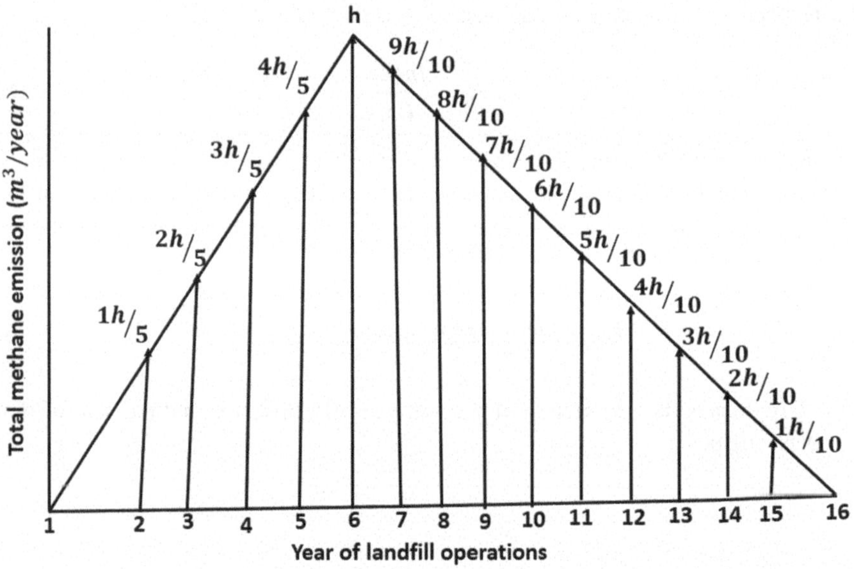

Fig. 5 Mechanism of modified triangular method (MTM)

(Eq. 8). Lastly, the multiplication of the volume of gas and the base of the triangle would equal this peak methane generation value (h) (Eq. 9). The following equations are given below:

$$\text{a) The area of triangle} = \text{methane emission estimation through ZODM} \quad (7)$$

$$\text{b) } V = \frac{x * 10^6}{D} \quad (8)$$

Here,V = Volume of gas $(M^3/year)$
X = Emissions of methane through ZODM
D = Density of methane gas

$$\text{c) } h = \frac{V * 2}{15} \quad (9)$$

Here, the value of h represents the peak generation value.

6.5 *Land-GEM-V-3.02 Model*

The land-GEM-V-3.02 model was one of the most renowned models first explored by USEPA in 2005 to quantify methane generation from specific landfill sites (Alexander et al., 2005; Amoatey et al., 2023). Like FODM, the first-order decay equation is used in this Land-GEM-V-3.02 model (Dimishkovska et al., 2019). This model assumes that methane generation is initially high in landfill conditions, but it would be decreased significantly with the increasing time of waste in landfills. This model estimates and predicts methane generation by using Eq. (10).

$$QCH4 = \sum_{i=0}^{n} \sum_{j=.1}^{n} K * L * \frac{Mi}{10} * e^{-Ktij} \quad (10)$$

In this Land-GEM-V-3.02 models, Lo, i, n, j, K, Mi, and tij contains methane generation potential $\left(m^3/Mg\right)$, , one year time increment, subtraction of current year of estimation and the first waste disposal year, time increment for 0.10 years, methane generation rate constant $(year^{-1})$, amount of waste deposited in the *i*th year (Mg), and shows age of the *j*th section of waste mass M_i accepted in the *i*th year, respectively (Toha & Rahman, 2023).

The value of the methane generation constant (K) is highly influenced by moisture content, temperature, and PH (Sohoo et al., 2021). Due to these factors, the cellulose and lignin of organic wastes break down and produce methane (Rafiq et al., 2018). In this production process, water contents and temperature can accelerate the development of bacteria by providing more nutrients (BP et al., 2023; Sohoo

Table 1 The default methane generation constant (K) and methane generation potential (Lo) values of Land-GEM-V-3.02 model

Type	K ($year^{-1}$)	Lo (m^3/Mg)
Clean Air Act Conventional	0.05	170
Inventory Conventional	0.04	100
Inventory Wet	0.70	96
Site-Specific	0.20	96.22

et al., 2021). As a result, more bacteria can participate in the degradation process and increase the gas generation rate. The value of K varies from one geographical area to another. However, the general range of K value ranges from 0.01 to 0.21, where the K value of 0.04 is practically found most worldwide (Osra et al., 2021). Moreover, in highly degradable conditions, the range of K values are 0.30 and 0.50, indicating a geographically moist area. The default K and Lo values are shown in Table 1.

However, the value of K for specific area is determined by (Eq. 11), respectively.

$$K = 3.20 \times 10^{-5}(\boldsymbol{annualmeanrainfall}) + 0.01 \tag{11}$$

The amount of cellulose has also significantly impacted the methane production potential capacity (Lo) (Fallahizadeh et al., 2019). In the Land-GEM-V-3.02 model, five different values of Lo are usually utilized to determine the amount of methane gas generated from landfills based on the geographical area.

6.6 Sensor Based Methane Estimation Models

There are various sensor-based techniques, especially laser absorption spectroscopy, optical gas imaging, and satellite imaging, to determine methane concentration from a particular place or point (Kim et al., 2023; Kwasny & Bombalska, 2023; Li et al., 2023). The measuring procedure depends on the sensor type, placement of the sensor, and the climatic variables (Ma et al., 2023; Manheim et al., 2023; Mellqvist et al., 2023; Monster et al., 2019). Various types of sensor-based methane estimation techniques are shown in Fig. 6.

The placement of sensors identifies the point from which methane concentration and the data type would be measured for estimating methane flow rates. Existing placement instruments can cover from 10 cm to 5000 km area (Papale et al., 2023; Pliszka et al., 2023; Sherwin et al., 2023). The placement of sensors with area coverage is shown in Fig. 7. Unlike in-situ measurements, some sensor-based technologies detect methane concentration without the interactions of methane plumes. On the other hand, some types of sensor-based technologies can use methane plumes where methane molecules must show physical and chemical interaction with sensors.

However, most of the sensor-based technologies formed based on detecting electromagnetic signals (Zhang et al., 2023). Generally, various molecules can absorb

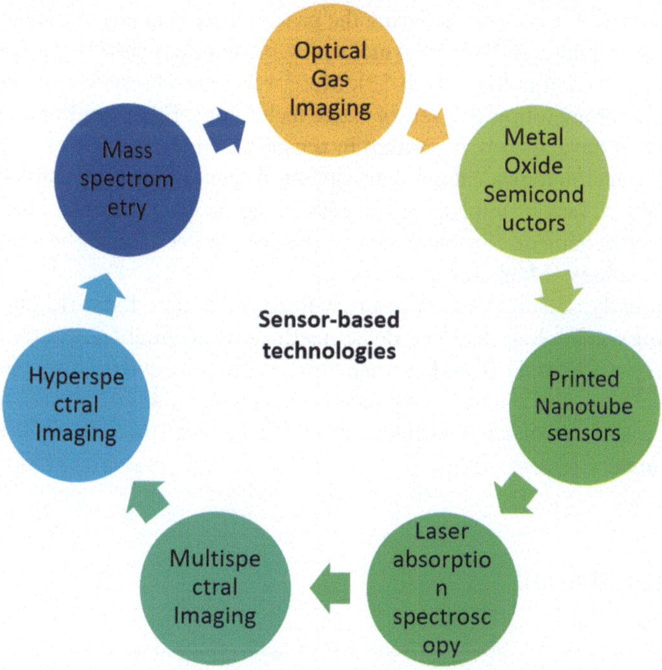

Fig. 6 Sensor based technologies for estimating methane gas emissions from landfills

Fig. 7 The placement of sensors with area coverage

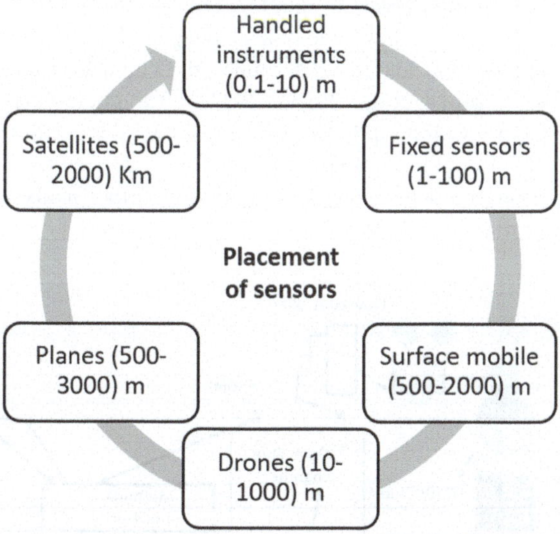

particular types of wavelengths during the passing period of electromagnetic radiation in the atmosphere. As with other molecules, methane can absorb an infrared range of (0.75–14 μm) (Zifarelli et al., 2023). Moreover, laser absorption and optical gas imaging technology follow this sensor-based technology. Ethane and carbon isotopes 12, 13, and 14 have already been used in sensor-based technology. However, these sensor-based technologies cannot detect methane concentration accurately like in-situ methods as various atmospheric variables (aerosols, cloud) can interfere with the sensor's mechanisms. Problematically, this interference cannot be eradicated by present technology (Manheim et al., 2023).

Consequently, scientists have already worked on modern Lidar (Light Detection and Ranging) technology that can detect methane concentration accurately worldwide. Lastly, AERMOD is widely renowned as an atmospheric dispersion model based on the Gaussian plume. Using meteorological data, this dispersion model can identify the sources of gas and their dispersion pattern (more than 50 km) for a particular area.

7 In-Situ Method

In-situ is the most accurate direct field-based measurement method for determining the concentration of methane (Fig. 8).

This method uses various sizes of stainless steel chambers to investigate methane concentration levels in landfills (Kormi et al., 2017; Anh et al., 2021). This steel chamber has to have two holes (Jeong et al., 2019). One hole is used to collect LFG samples, and another determines the chamber's temperature (Chakraborty et al., 2011; Kumar et al., 2023a, 2023b). The chamber is generally implanted at least 10 cm into the top layers to control lateral emissions (Chandrasekaran & Busetty, 2023; Liu et al., 2023). Lastly, syringes are used to collect gas samples at a specific period and sent to the laboratory for further analysis (Hettiarachchi et al., 2023; Kormi et al., 2017).

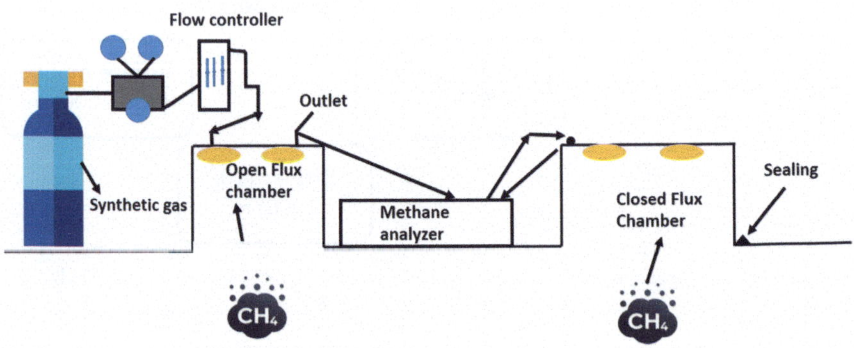

Fig. 8 The mechanism of In-situ methods

This method follows the Eq. (12):

$$\textit{ConcentrationofMethane} = \frac{\Delta C}{\Delta t} * h \tag{12}$$

where , $\frac{\Delta C}{\Delta t}$ represents the ratio of concentration gradient to sampling intervals and 'h' represents the height of the enclosure above the soil surface.

8 Comparative Assessment and Data Validation of Methane Estimation Models

According to various research, it has been found that ZODM, FODM, and MTM tremendously overestimated the methane gas emissions (Kumar & Sharma, 2014; Toha & Rahman, 2023). On the other hand, Land-GEM-V-3.02 are measured ideally compared to other models, and their estimated value is very near to in-situ methods (Amirmahani et al., 2020; Kumar et al., 2004; Rafiq et al., 2018). Thus, Land-GEM-V-3.02 models have already been used in various landfills around the globe. Table 2. represents the methane estimation from various landfills by Land-GEM-V-3.02 models.

Lastly, these model's data can be compared to in-situ methods for data validation.

Table. 2. Methane emission estimation by Land-GEM-V-3.02

Name of landfills	Country	Methane emission (Gg/y)	References
Amin-Bazar, Dhaka	Bangladesh	25.95	Toha and Rahman (2023)
Matuail, Dhaka	Bangladesh	24.97	Toha and Rahman (2023)
Faisalabad	Pakistan	1.84	Rafiq et al. (2018)
Dhapa	India	17.14	Adhikary and Sengupta (2021)
Yasuj	Iran	1.93	Fallahizadeh et al. (2019),
Kakia	Saudi-Arabia	22	Osra et al. (2021)
Aksaray	Turkey	7.85	Kale and Gokcek (2020)
Kerman	Iran	5015 m^3/year	Amirmahani et al. (2020)
Nam Binh Duong	Vietnam	10,053 Mg/year	Anh et al. (2021)
Phnom Penh	Combodia	18 M kg/year	Pheakdey et al. (2023)
Tirupati	India	39, 04, 794 m^3/year	Ramprasad et al. (2022)
Jakarta, Medan	Indonesia	875,130 tons/year	Andriani et al. (2019)
Kanpur	India	233.44×106 m^3/year	Chandra and Ganguly (2023)

9 Conclusions

Landfills are a sacred source of methane emissions associated with other anthropogenic activities. Consequently, over time, incorrect disposal of solid waste and mismanagement of landfill procedures have resulted in the production of methane, which can destroy our ecosystems and cause other serious issues. As a result, this issue has already been taken seriously in the scientific community. Thus, it is vital to find the concentration of methane level in landfills. Consequently, various models and methods mechanisms, their application procedure, advantages, and limitations have been discussed. According to various research, IPCC ZODM, IPCC FODM, and MTM models overestimated methane gas compared to Land-GEM-V-3.02 models from landfills. On the other hand, Land-GEM-V-3.02 model estimation is very similar to field measurement values. It indicates that the Land-GEM-V-3.02 model is comparatively better to use in landfills for quantifying methane gas. Various types of sensor technology also have limitations as they follow passive measurements. Thus, establishing active measurement techniques (LIDAR) is essential for better estimating methane from a particular place. In developing nations, most methane gas has been generated from landfills due to the inappropriate management of municipal solid wastes. As a result, organic wastes are rapidly degraded, generating this greenhouse gas. In these circumstances, properly managing landfilling and integrated solid waste management programs demand time for tackling global warming and climate change. According to recent research, soil cover can reduce more than 50% of methane gas from landfills. So, a proper soil management program and waste-to-energy recovery technology will help maintain a sustainable landfill management program. Lastly, as developing nations are facing most of the catastrophic effects of greenhouse gases, policymakers should focus on proper management programs and indexing the contamination level of methane. Furthermore, more research is needed to establish cost-effective technology that would help reduce the deterioration of our Mother Nature and combat climate change.

References

Alexander, A., Burklin, C., & Singleton, A. (2005, May). Landfill gas emissions model. United States Environmental Protection Agency, Version 3.02 user's guide, U.S. Environ. Prot. Agency Off. Res. Dev, no., 48, https://www3.epa.gov/ttncatc1/dir1/landgem-v302-guide.pdf

Abdel-Shafy, H. I., Ibrahim, A. M., Al-Sulaiman, A. M., & Okasha, R. A. (2023). Landfill leachate: Sources, nature, organic composition, and treatment: An environmental overview. *Ain Shams Engineering Journal, 15*(1), 102293.

Kumar, A., & Sharma, M. P. (2014). Estimation of GHG emission and energy recovery potential from MSW landfill sites. *Sustainable Energy Technologies and Assessments, 5*, 50–61. https://doi.org/10.1016/j.seta.2013.11.004

Rafiq, A., Rasheed, A., Arslan, C., Tallat, U., & Siddique, M. (2018). Estimation of greenhouse gas emissions from Muhammad Wala open dumping site of Faisalabad, Pakistan. *Geology, Ecology, and Landscapes, 2*(1), 45–50. https://doi.org/10.1080/24749508.2018.1452463

Amirmahani, N., Sadeghi, S., Yazdanpanah, G., Tayebiyan, A., Nasiri, A., & Malakootian, M. (2020). Estimating methane gas generation rate from Kerman city landfill using LandGEM software. *International Journal of Environment and Waste Management, 26*(4), 520–530.

Amoatey, N. K., Darmey, J., & Tabbicca, K. E. (2023). Estimation of the methane generation potential of the tamale landfill site using LandGEM. *Ghana Mining Journal, 23*(1), 27–33.

Andriani, D., & Atmaja, T. D. (2019). The potentials of landfill gas production: A review on municipal solid waste management in Indonesia. *Journal of Material Cycles and Waste Management, 21*, 1572–1586.

Anh, L. H., Thanh Truc, N. T., Tuyen, N. T. K., Bang, H. Q., Son, N. P., Schneider, P., & Moustakas, K. (2021). Site-specific determination of methane generation potential and and estimation of landfill gas emissions from municipal solid waste landfill: a case study in Nam Binh Duong, Vietnam.

ATSDR. (2019). *Landfill gas primer—Chapter 2: Landfill gas basics*. https://www.atsdr.cdc.gov/HAC/landfill/html/ch2.html#:~:text=Three%20processes%E2%80%94bacteria%20decomposition%2C%20volatilization.

Beluche, R. (1968). Ph.D. Thesis, University of Southern California.

BP, N., Tabaroei, A., & Garg, A. (2023). Methane emission and carbon sequestration potential from municipal solid waste landfill, India. *Sustainability, 15*(9), 7125.

Kale, C., & Gökçek, M. (2020). A techno-economic assessment of landfill gas emissions and energy recovery potential of different landfill areas in Turkey. *Journal of Cleaner Production, 75*, 122946. https://doi.org/10.1016/j.jclepro.2020.122946

Chandra, S., & Ganguly, R. (2023). Assessment of landfill gases by LandGEM and energy recovery potential from municipal solid waste of Kanpur city, India. *Heliyon, 9*(4).

Chandrasekaran, R., & Busetty, S. (2023). Estimation of biogas generation rate and carbon sequestration potential from two landfill sites in southern India. *Environmental Science and Pollution Research*, 1–12.

da Silva, N. F., Schoeler, G. P., Lourenco, V. A., de Souza, P. L., Caballero, C. B., Salamoni, R. H., & Romani, R. F. (2020). First order models to estimate methane generation in landfill: A case study in south Brazil. *Journal of Environmental Chemical Engineering, 8*(4), 104053.

Dimishkovska, B., Berisha, A., & Lisichkov, K. (2019). Estimation of methane emissions from Mirash municipal solid waste sanitary landfill, differences between IPPC 2006 and LandGEM method. *Journal of Ecological Engineering, 20*(5).

Environmental Protection Agency United States, global mitigation of non-CO2 Greenhouse gases: 2010–2030. https://www.epa.gov/sites/default/files/2016-06/documents/mac_report_2013.pdf. Accessed September 2013.

European Environment Agency Denmark, Annual European Union greenhouse gas inventory 1990–2012 and inventory report 2014, Submission to the UNFCCC Secretariat, https://www.eea.europa.eu/publications/european-union-greenhouse-gas-inventory-2014/fileno. Accessed 15 April 2014.

Osra, F. A., Ozcan, H. K., Alzahrani, J. S., & Alsoufi, M. S. (2021). Municipal solid waste characterization and landfill gas generation in kakia landfill, makkah. *Sustainability, 13*, 1–13. https://doi.org/10.3390/su13031462

Bingemer, H. G., Crutzen P. J. (1987). The production of methane from solid wastes. *Journal of Geophysical Research, 92* (2181–2187). https://doi.org/10.1029/JD092iD02p02181

He, L., Groom, J. D., Wilson, E. H., Fernandez, J., Konopka, M. C., Beck, D. A., & Lidstrom, M. E. (2023). A methanotrophic bacterium to enable methane removal for climate mitigation. *Proceedings of the National Academy of Sciences, 120*(35), e2310046120.

Hettiarachchi, H., Irandoost, E., Hettiaratchi, J. P., & Pokhrel, D. (2023). A Field-Verified model to estimate landfill methane flux using surface methane concentration measurements under calm wind conditions. *Journal of Hazardous, Toxic, and Radioactive Waste, 27*(4), 04023019.

Sohoo, I., Ritzkowski, M., Sohu, Z. A., Cinar, S. Ö., Chong, Z. K., & Kuchta, K. (2021). Estimation of methane production and electrical energy generation from municipal solid waste disposal sites in Pakistan. *Energies, 14*, 9. https://doi.org/10.3390/en14092444

Jain, P., Kanneganti, A., Townsend, T., Krause, M., & Tolaymat, T. (2023). Isothermal dual-phase flow modeling to assess the impact of gas collection on geotechnical and hydraulic performance of landfills. *Journal of Geotechnical and Geoenvironmental Engineering, 149*(8), 04023061.
Kim, R. H., Lee, N. H., Yoon, S. P., Song, S. H., & Park, J. K. (2023). Considerations on the methane correction factor and fraction of methane parameters in the IPCC first-order decay model for active aeration landfills. *Waste Management, 169*, 232–242.
Kim, Y. (2023). *Quantification of Methane Emissions by Surface Mass Balance Method.*
Kissas, K., Ibrom, A., Kjeldsen, P., & Scheutz, C. (2022). Methane emission dynamics from a Danish landfill: The effect of changes in barometric pressure. *Waste Management, 138*, 234–242.
Kriipsalu, M., Somani, M., Pehme, K., Tamm, O., Truu, J., Truu, M., & Orupold, K. (2023). Performance of biocover in controlling methane emissions from landfill: A decade of full-scale investigation. *Process Safety and Environmental Protection, 172*, 486–495.
Kotze, J. P., Thiel, P. G., & Hattingh, W. H. J.(1969). *Water Research 3*, 459. Pergamon Press, Great
Kumar, A., Bhardwaj, S., & Samadder, S. R. (2023a). Evaluation of methane generation rate and energy recovery potential of municipal solid waste using anaerobic digestion and landfilling: A case study of Dhanbad. *India. Waste Management & Research, 41*(2), 407–417.
Kumar, P., Caldow, C., Broquet, G., Shah, A., Laurent, O., Yver-Kwok, C., & Ciais, P. (2023b). Detection and long-term quantification of methane emissions from an active landfill. *Atmospheric Measurement Techniques Discussions, 2023*, 1–30.
Kwaśny, M., & Bombalska, A. (2023). Optical methods of methane detection. *Sensors, 23*(5), 2834.
Li, S., Liu, Y., & Du, K. (2023). Quantifying emissions from fugitive area sources using a hybrid method of multi-path optical remote sensing and tomographic inverse-dispersion techniques. *Remote Sensing, 15*(4), 1043.
Lin, Y. H. (1966). Ph.D. Thesis, West Virginia University.
Liu, Y., Paris, J. D., Vrekoussis, M., Quéhé, P. Y., Desservettaz, M., Kushta, J., & Sciare, J. (2023). Reconciling a national methane emission inventory with in-situ measurements. *Science of the Total Environment*, 165896.
Chakraborty, M., Sharma, C., Pandey, J., Singh, N., & Gupta, P. K. (2011). Methane emission estimation from landfills in Delhi: A comparative assessment of different methodologies. *Atmospheric Environment, 45*, 7135–7142. https://doi.org/10.1016/j.atmosenv.2011.09.015
Ma, L., Cui, Y., Liu, B., Liao, B., Wei, J., Han, H., & Tian, W. (2023). A GIS-based method for modeling methane emissions from paddy fields by fusing multiple sources of data. *Science of the Total Environment, 859*, 159917.
Manheim, D. C., Newman, S., Yeşiller, N., Hanson, J. L., & Guha, A. (2023). Application of cavity ring-down spectroscopy and a novel near surface Gaussian plume estimation approach to inverse model landfill methane emissions. *MethodsX, 10*, 102048.
Mellqvist, J., Conde, V., Bandodkar, S., Brohede, S., Andersson, P., Offerle, B., & Samuelsson, J. (2023). *Methane emissions from industrial activities using drones.*
Mønster, J., Kjeldsen, P., & Scheutz, C. (2019). Methodologies for measuring fugitive methane emissions from landfills–A review. *Waste Management, 87*, 835–859.
Njoku, P. O., & Edokpayi, J. N. (2023). Estimation of landfill gas production and potential utilization in a South Africa landfill. *Journal of the Air & Waste Management Association, 73*(1), 1–14.
Adhikary, P. K., Sengupta, P. P. (2023). Spatial modeling and assessment of environmental contaminants. In *Environmental challenges and solutions*. Springer. https://doi.org/10.1007/978-3-030-63422-3_4
Papale, L. G., Guerrisi, G., De Santis, D., Schiavon, G., & Del Frate, F. (2023). Satellite data potentialities in solid waste landfill monitoring: Review and case studies. *Sensors, 23*(8), 3917.
Pliszka, H. O. (2023). *Characterizing landfill methane through a low-cost ground-based sensor network: Preliminary studies at a Colorado landfill* (Doctoral dissertation). University of Colorado at Boulder.
Ramprasad, C., Teja, H. C., Gowtham, V., & Vikas, V. (2022). Quantification of landfill gas emissions and energy production potential in Tirupati Municipal solid waste disposal site by LandGEM mathematical model. *MethodsX, 9*, 101869.

Scheutz, C., Duan, Z., Møller, J., & Kjeldsen, P. (2023). Environmental assessment of landfill gas mitigation using biocover and gas collection with energy utilisation at aging landfills. *Waste Management, 165*, 40–50.

Fallahizadeh, S., Rahmatinia, M., Mohammadi, Z., Vaezzadeh, M., Tajamiri, A., & Soleimani, H. (2019). Estimation of methane gas by LandGEM model from Yasuj municipal solid waste landfill. *Iran, MethodsX., 6*, 391–398. https://doi.org/10.1016/j.mex.2019.02.013

Jeong, S., Park, J., Kim, Y. M., Park, M. H., & Kim, J. Y. (2019). Innovation of flux chamber network design for surface methane emission from landfills using spatial interpolation models. *Science of the Total Environment, 688*, 18–25. https://doi.org/10.1016/j.scitotenv.2019.06.142

Kumar, S., Nimchuk, N., Kumar, R., Zietsman, J., Ramani, T., Spiegelman, C., & Kenney, M. (2016). Specific model for the estimation of methane emission from municipal solid waste landfills in India. *Bioresource Technology, 216*, 981–987. https://doi.org/10.1016/j.biortech.2016.06.050

Kumar, S., Gaikwad, S. A., Shekdar, A. V., Kshirsagar, P. S., & Singh, R. N. (2004). Estimation method for national methane emission from solid waste landfills. *Atmospheric Environment, 38*, 3481–3487. https://doi.org/10.1016/j.atmosenv.2004.02.057

Saltykov, M. Y., & Trifonov, S. V. (2023). Analysis of greenhouse gas generation models at household waste landfills for ecosystem applications. In *E3S Web of Conferences* (Vol. 411, p. 02048). EDP Sciences.

Sherwin, E. D., Rutherford, J. S., Chen, Y., Aminfard, S., Kort, E. A., Jackson, R. B., & Brandt, A. R. (2023). Single-blind validation of space-based point-source detection and quantification of onshore methane emissions. *Scientific Reports, 13*(1), 3836.

Kormi, T., Ali, N. B. H., Abichou, T., & Green, R. (2017). Estimation of landfill methane emissions using stochastic search methods. *Atmospheric Pollution Research, 8*, 597–605. https://doi.org/10.1016/j.apr.2016.12.020

Toha, M., & Rahman, M. M. (2023). Estimation and prediction of methane gas generation from landfill sites in Dhaka city, Bangladesh. *Case Studies in Chemical and Environmental Engineering, 7*, 100302.

Troiano, D. T., Hofmann, T., Brethauer, S., & Studer, M. H. (2023). Toward optimal use of biomass as carbon source for chemical bioproduction. *Current Opinion in Biotechnology, 81*, 102942.

Wang, Y. N., Wang, Q., Li, Y., Wang, H., Gao, Y., Sun, Y., & Zhan, M. (2023). Impact of incineration slag co-disposed with municipal solid waste on methane production and methanogens ecology in landfills. *Bioresource Technology, 377*, 128978.

Yu, K., Xiao, S., Zheng, F., Fang, X., Zou, J., & Liu, S. (2023). A greater source of methane from drainage rivers than from rice paddies with drainage practices in southeast China. *Agriculture, Ecosystems & Environment, 345*, 108321.

Zhang, S., Ma, J., Zhang, X., & Guo, C. (2023). Atmospheric remote sensing for anthropogenic methane emissions: Applications and research opportunities. *Science of the Total Environment, 893*, 164701.

Zifarelli, A., Sampaolo, A., Patimisco, P., Giglio, M., Gonzalez, M., Wu, H., & Spagnolo, V. (2023). Methane and ethane detection from natural gas level down to trace concentrations using a compact mid-IR LITES sensor based on univariate calibration. *Photoacoustics, 29*, 100448.

Prospective Energy Production from Incineration: A Brief Overview on Different Types of Energy Produced by Incineration

Sadia Sikder, Mohammad Toha, and Md. Mostafizur Rahman

Abstract Through incineration, the volume of municipal solid garbage, medical waste with its possible infectious properties, and the volume of hazardous chemical and biological waste with its potential toxicity are all reduced. Municipal solid waste (MSW) must be incinerated to be converted into heat and steam. The process produces heat and energy, reduces the amount of municipal solid waste (MSW) that must be managed, and eliminates harmful elements. Waste incineration may reduce greenhouse gas emissions by substituting fossil fuels and lowering methane gas emissions at landfills. Incinerating plants are critical in the local energy market since they produce tremendous energy, especially in minimal areas. The garbage's biogenic (plant-based) makeup, which includes food and other organic waste, cardboard, and wood, contributes to the perception that the energy generated is somewhat renewable. This cremation process produces heat, gas, or fuel. The goal of this chapter was to provide a thorough analysis of the prospects for incineration-based energy production. Later, this chapter briefly discussed the several kinds of energy incineration produces.

1 Introduction

According to the World Bank's projections for 2018, there is an anticipated increase in worldwide waste generation from 2.01 billion tonnes in 2016 to 3.40 billion tonnes over 30 years. This growth is expected to be driven mainly by growing economies in Asia and Africa. This observation highlights the prevailing global inclination towards a societal model characterized by disability and the limited advancements made in

S. Sikder · M. Toha
Department of Environmental Science and Disaster Management , Daffodil International University, Savar, Dhaka 1216, Bangladesh

Md. M. Rahman (✉)
Laboratory of Environmental Health and Ecotoxicology, Department of Environmental Sciences, Jahangirnagar University, Savar, Dhaka 1342, Bangladesh
e-mail: rahmanmm@juniv.edu

A. Anouzla and S. Souabi (eds.), *Generation of Energy from Municipal Solid Waste*,
https://doi.org/10.1007/978-3-031-74334-4_12

mitigating the generation of municipal solid waste (MSW). The incinerating destruction has effectively decreased the amount of waste generated while also harnessing energy (IETC, 2020).

Waste incineration is a treatment process frequently used worldwide to recover the energy component of leftover waste. The recovery of materials and energy, followed by residue disposal, are the primary goals of an integrated waste management system. Waste recycling, composting, waste combustion with energy recovery, and landfilling of leftovers each play a crucial part individually and collectively in a balanced manner (Grosso et al., 2010).

Incineration is the process of treating garbage by igniting any organic materials that are present. Without additional processing, it transforms the waste material into heat, flue gas, and ash released into the atmosphere. The study examines the different waste products created and can be effectively used again with correct treatment, reaction, and analysis. A significant portion of heat can be used to produce electricity. Nitrogen, carbon dioxide, and sulfur dioxide are present in trace amounts in flue gases, each of which has better utilization when used optimally. Carbon dioxide can be utilized as fire extinguishers, nitrogen produced can be used as fertilizer to boost crop productivity, and sulfur, once separated from sulfur dioxide, can be used in dental care. Ash can be utilized in buildings since it can be obtained as solid lumps. The good thing about incineration is that it reduces organic waste's volume and solid mass by 95–96% and 80–85%, respectively. Bottom ash is the most significant byproduct of MSWI, which makes up between 80 and 87% (by weight) of solid waste (Nidoni, 2017).

Incineration is still the most widely used technique for turning trash into energy. The advantages of incinerating solid waste include significantly reducing the amounts and volumes (up to 90%) that need to be dumped in landfills and eradicating diseases (Zhang, 2012). Incineration facilities may also shorten the distance that municipal trash must be transported, depending on their location. However, these benefits are outweighed by the incinerator's emissions of carbon dioxide, sulfur dioxide, particulates, heavy metals, and other pollutants. Particular focus has been placed on the emissions of dioxins and furans, which are more harmful and expensive to manage than other pollutants (Cheng & Hu, 2010; Nadal et al., 2016; Zhao & Wang, 2018). One of the leading causes of the release of dioxins into the environment is thought to be waste incineration. It is feasible to manage the pollution produced by incinerators, as is true for all types of pollutants. By putting rigorous pollution control systems in place, some economically developed nations, like Austria, have been able to incinerate MSW in a very "clean" manner (Abbasi, 2018).

While there may be notable variations in the implementation strategies across different regions and territories, both nationally and internationally, there is a prevailing trend towards adopting waste management systems that involve segregating collected materials and facilitating reusing secondary byproducts. The waste stream consists of the materials discarded during the sorting and collective screening processes and the unsorted wastes that require proper disposal. Furthermore, a substantial proportion of the overall waste generated consists of lesser quantities of commonly utilized materials, like metals, glass, and biodegradable organic matter.

The substance can be further divided into combustible moist organic ingredients and primarily inert/mineral constituents. Considering the valuation of potential energy within the waste flow downstream is of utmost importance, particularly in light of the growing interest in harnessing thermal and thermoelectric power from renewable sources, as highlighted by Stehl'k in 2009 (Panepinto et al., 2015).

The primary waste-to-energy treatment method is incineration. The treatment involves controlled burning solid waste at high temperatures to destroy it. Heat is released concurrently. Burning thermal energy can be converted into usable energy. Incinerating municipal solid waste (MSW) decreases the amount and size of trash that must be disposed of in landfills, making it an effective waste management method in densely populated urban areas. This technology can gather energy in cities. However, a significant financial investment is needed. To ensure environmental safety, incineration ambient conditions must be carefully managed. Due to environmental laws like emissions control, more budget is required. To examine incineration as an option, consider various factors. These include the amount of garbage produced, the waste's heat of combustion, the site's location, the facility's dimensions, the expenses of management and maintenance, and the required investment (Patil et al., 2014).

This paper focuses on the different types of energy production through incineration. It also made an effort to examine its significance.

2 MSW Waste to Energy Incineration

WtE incineration directly controls waste burning at temperatures of 850 °C and above in the presence of oxygen, along with simple mechanisms to recover heat and energy and more complex means to clean flue gas, use wastewater, and assimilate various waste streams. MSW incineration's reliable thermal treatment method has advanced significantly over time, along with controls for air pollution and dioxins (Makarichi et al., 2018). Incineration of MSW is a feasible approach to handling MSW with the significant advantages of volume reduction and disease control.

Waste can be decreased by 80% and reduced in volume by more than 85% after treatment for incineration. Additionally, high-temperature burning can destroy a significant amount of poisonous materials and pathogenic germs, reducing the amount of harmful bacteria contaminating the environment. Ash leftovers from waste treatment can also be used as road paving materials, transformed into building materials, turning trash into gold, the waste's thermal energy. Incineration waste can be gathered and used for power generation, cooling, and heating. This will lower the cost of trash disposal and make it possible to use energy completely. In other words, in a setting where the environmental situation is worsening and the need for environmental preservation is becoming more critical, the future of garbage incineration is bright (Wang & Dong, 2019). Table 1 shows the typical heat values of MSW components found by Islam (2016). This represents that MSW has high heat values.

Waste incineration was developed to address disease outbreaks and mitigate the escalating waste volume resulting from the persistent growth of urban populations

Table 1 Typical heat values of MSW components found by Islam (2016)

MSW components	Heat value (KJ/Kg, dry weight)
Food waste	957.13
Paper	3445.68
Plastics	6699.94
Textiles	3589.25
Glass and ceramic	28.71
Metals	143.57
Grass and straw	1339.99
Others	1435.70

in big and densely populated cities. Another benefit of utilizing waste as a resource for power generation is an additional advantage of waste-to-energy incineration. According to the Intergovernmental Panel on Climate Change (IPCC, 2007), incineration as an alternative to landfilling can effectively decrease carbon emissions by diminishing the reliance on fossil fuel energy sources and mitigating methane production in landfills. Nevertheless, implementing municipal solid waste (MSW) incineration encounters various obstacles, as highlighted in the literature (GAIA, 2021; Karim & Corazzini, 2019). These challenges encompass the substantial expenses associated with constructing and operating incinerators, the absence of adequate revenue generated from energy and waste sales to offset all costs, the potential diversion of waste from the principles of reduce, reuse, and recycle due to the minimal feedstock requirements for incineration operations, and the potential health hazards associated with this waste management method (Liu et al., 2020). Figure 1 illustrates the municipal solid waste to energy incineration precisely.

3 Energy Recovery Through Incineration

Incineration refers to the deliberate and regulated process of subjecting waste materials to high temperatures exceeding 8000C while ensuring the presence of oxygen. This thermal treatment results in the liberation of heat energy, the production of gaseous byproducts, and the formation of inert ash. The density and composition of the trash determine the net energy yield. Various factors can influence heat loss in a combustion system, such as the proportion of inert materials and moisture, the ignition temperature, the size and form of the elements, and the structure of the combustion system. According to Patil et al. (2014), it has been observed that a significant proportion of the energy contained in organic matter, ranging from 65 to 80%, can be effectively harnessed as heat energy. This heat energy can be utilized for several purposes, including direct thermal applications or electricity production through steam turbine generators. Figure 2 illustrates the flow diagram of energy generation through incineration of municipal solid waste.

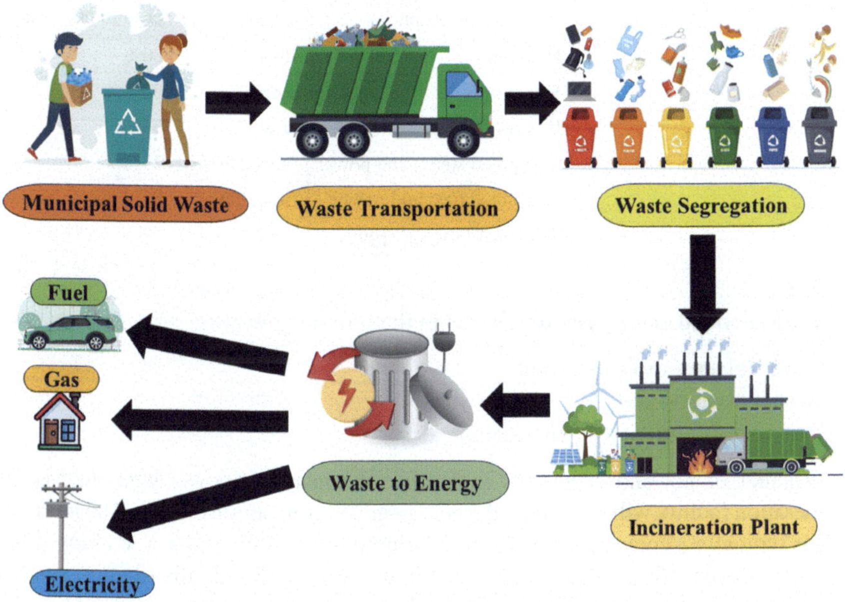

Fig. 1 Municipal solid waste to energy by incineration process

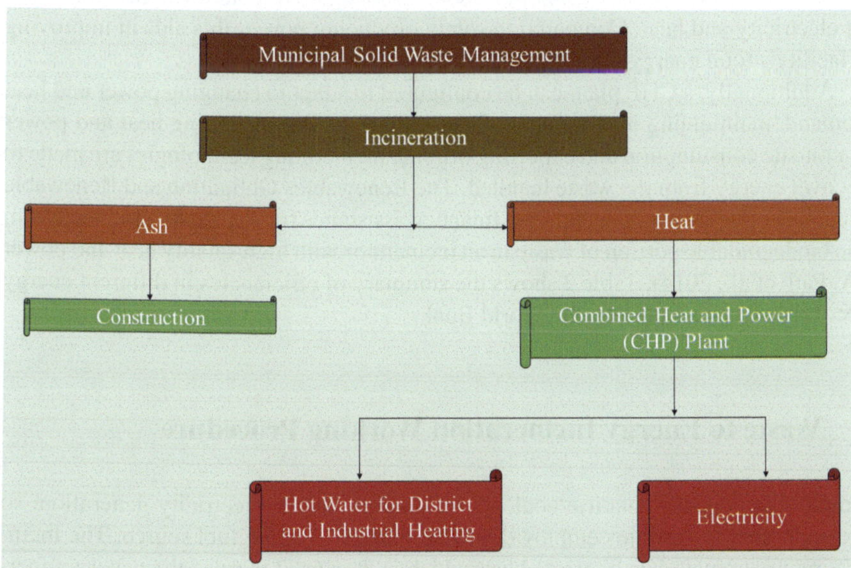

Fig. 2 Flow diagram of energy generation through incineration of municipal solid waste

Table 2 Summary of efficiencies in different energy

Energy utilization	Recovery	Overall efficiency
Heat only	Heat	80%
Steam only	Steam	80%
Power only	Power	35%
Combined steam and power	Steam, Power	35–75%
Combined heat and power	Heat, Power	85%

*Source*World Bank (1999)

Waste-derived energy can be applied in the following ways:

- Power (electricity) generation.
- Heat generation and
- Heat and power (CHP) generation.

Whether or not end customers can utilize the heat and power generated at an incineration facility will determine the energy generation option chosen. The national grid is typically a convenient means of distributing and selling power, making it the type of energy recovery that occurs most frequently. A specific distribution system (network) is needed for heat, and the consumer must be close to the plant providing the heat. The generating facility will only sometimes be working at its most efficient level if all of the available heat cannot be used. Utilizing CHP combines the production of electricity and heat. Compared to solely producing power, this aids in improving a facility's total energy efficiency.

Additionally, a CHP plant can be configured to adapt to changing power and heat demand, maintaining maximum levels of efficiency. By generating heat and power for on-site consumption and exporting off-site, incinerating technologies are made to recover energy from the waste handled. The Renewables Obligation and Renewable Heat Incentive scheme may provide financial assistance for electricity produced from the biodegradable portion of waste in an incinerator with high-quality heat and power (A. Patil et al., 2014). Table 2 shows the summary of efficiencies in different energy recovery systems stated by the World Bank.

4 Waste to Energy Incineration Working Procedure

Traditional power plants use coal, oil, or natural gas for electricity generation, so waste-to-energy facilities employ domestic trash as a viable fuel source. The incinerating waste materials generate heat, which is then used to raise the water temperature. Subsequently, the resulting steam is harnessed to drive a turbine, facilitating electrical energy generation. The waste is introduced and completely blended within an enclosed reception zone with the intention of subsequent incineration. A timed moving grate mechanism introduces a waste mixture into the combustion chamber,

Table 3 Outputs from incineration technology

Outputs	State	Quantity by weight of original waste
Bottom Ash	Solid	20–30%
Metals (Ferrous and Non-Ferrous)	Mostly solid, require separation from residues	2–5%
APC Residues (Fly Ash, Reagents and Wastewater)	Solid and Liquid	2–6%
Gases	Gaseous	70–75%

Source Liu et al. (2020)

which is continuously rotated to ensure consistent exposure and combustion. Table 3 presents the results obtained from the utilization of the incineration method. The filter baghouse is designed to remove minuscule airborne particles, specifically fly ash. Either lime or sodium hydroxide is introduced into the system to counteract the acidic nature of the combustion emissions. In the process of recycling, magnets and eddy current separators are employed to effectively segregate ferrous metals (steel and iron) and non-ferrous metals (such as copper, brass, nickel, and aluminum) from the residual trash generated by combustion, sometimes referred to as "bottom ash." The leftover ash has the potential to be effectively employed as an aggregate material for the construction of rail embankments and roadbeds. The steam turbine's generator is driven by steam that has been heated to a high temperature. The condenser facilitates the conversion of cooled steam into liquid water while alternatively serving as a heat source for various constructions or commercial establishments. The chilled stream undergoes reheating in the economizer and superheater to complete the steam cycle. The mixture is supplemented with activated carbon to effectively capture and eliminate heavy metals such as mercury and cadmium from the hot gases. This activated carbon, essentially charcoal that has undergone oxygen treatment to enhance its porosity, is introduced into the system. To mitigate the presence of nitrogen oxide in the ascending combustion emissions, the injection of either ammonia or urea is employed to neutralize. According to Hockenos (2021), the flue gases undergo a process of being heated to very elevated temperatures to mitigate the harmful effects of dioxins and furans.

5 Effective Utilization of by Products During Incineration

The waste is gathered collectively, and the organic components are burned during incineration. When released into the atmosphere, this incineration process produces waste heat energy, acid rain, and infectious diseases (Nidoni, 2017).

5.1 Ash: The upper inlet introduces waste, while the ash door removes the residue after incineration. Ash substance is solid and can be used in commercial and residential buildings. Nidoni (2017) lists road construction as another usage for the technique.

Incinerators produce bottom and fly ash. Monitoring landfill reclamation is the primary therapy. Incinerators can change bottom ash and APC residue (fly ash), according to Liu et al. (2020). Loss on ignition must be assessed to assure incinerator bottom ash combustion quality. The LOI of Japanese bottom ash is 5% or less. Dioxins and heavy metals must be restricted. The magnetic separation procedure can remove steel from bottom ash. There are numerous ways to remediate bottom ash besides trash dump reclamation. The recycling systems cannot process all bottom ash. Maintaining regulated landfills is essential.

Bottom ash as cement raw material. Bottom ash replaces 0.5–1% cement. The presence of chlorine limits several compounds (Liu et al., 2020). Sintering, chemical solidification and stabilization (S/S), and ageing are used in construction. Safely handling hazardous chemicals for bottom ash treatment is essential.

Slag aggregate is made by melting at high temperatures. Bottom ash melting requires 1,250 °C or more elevated temperatures in burners, coke beds, and electric furnaces. Discharge and cooling form "slag"—a dense, glass-like material. This slag is used in buildings. Despite the high quality of this slag, melting furnaces are expensive to build and operate. Smelting can use secondary fly ash and metal ingots as feedstock. Melting furnaces directly dissolve municipal solid waste (MSW) ash, while gasification melting systems produce slag, according to Liu et al. (2020).

5.2 Flue gases: As a byproduct of incineration, flue gas, a mixture of several gases, is produced. These gases, discharged into the atmosphere, are the primary cause of several illnesses, including heart attacks, lung cancer, and asthma, which indirectly cause early mortality in living things. Nitrogen dioxide, carbon dioxide, sulfur dioxide, water vapor, and traces of oxygen are the primary gases that evolved during burning. Since sulfur dioxide is a potent oxidizer, sulfur can be recovered via the Claus method. Thus, hydrogen sulfide reduces sulfur dioxide to produce elemental sulfur.

Chemical interactions with the elementary sulfur can produce sulfur compounds like sodium thiosulfate, utilized in dentistry. Water, oxygen, and nitrogen dioxide are other byproducts of combustion. Using Ostwald's method, when NO2, oxygen, and water combine at a pressure of 4 to 10 atm at a temperature of 217 degrees, around 98% of nitric acid can be produced. Crop cultivation makes use of nitric acid. This will increase output and assist farmers in growing higher-quality crops (Nidoni, 2017).

5.3 Heat: Waste heat can be used by transforming waste heat into power. Thermoelectric generators employ the Seebeck effect to turn heat into energy. Thermoelectric materials use phenomena that change the temperature differential into an electric potential. Bismuth telluride is a thermoelectric material that is frequently utilized. These days, there are frequent power outages in every city in Bangladesh and India. This power generation technology will use waste materials to generate electricity, easing these countries' electricity crisis (Nidoni, 2017).

5.1 Material Recycling from Incineration

The potential for high-value markets is generally lower for recyclables derived from a separate household recycle collection system than those derived from the front-end preparation stage of an incinerator or metals extracted from the back end of the process (i.e. out of the ash). They can, however, be sold into the secondary metals markets for recycling because of their high value. Most UK incinerators recover and recycle non-ferrous or ferrous metals, which can raise overall recycling rates. Additionally, they will significantly reduce carbon emissions, which will lessen the effects of climate change. The generated Bottom Ash (or IBA) can be recycled and used as a secondary aggregate in several construction projects. In Box 3, a case is given. While the Environment Agency is now working on a Quality Protocol project10 to try and develop a standard where IBA may be categorized as a product rather than trash, recycling IBA would need to be done in conformity with pertinent legislation and guidelines. Bottom ash is typically used as infill or in construction goods under certain restrictions about the placement of the materials (DEFRA, 2013).

6 Different Types of Application of Incineration Procedure in Energy Production

6.1 Hot Water Boiler and Its Application

The hot water boiler is relatively easy to design, finance, run, and maintain. It is also easy to accommodate in building layouts. Technically, special consideration must be given to the corrosive properties of the waste incineration flue gases. Hot water production (between 110 and 160 C) is possible. Depending on the boiler's working pressure, it can be heated to higher temperatures. It is possible to attain boiler efficiencies of up to about 80%. The temperature of the cooling water that is being reclaimed places a limit on the recovery. The Circuit for a Hot Water Boiler The hot flue gas energy is transported to an internal water circuit via a hot water boiler, which transfers the power to the end user circuit (the district heating system). Heat exchangers divide the boiler and end-user circuits (World Bank, 1999).

Hot water has few uses because of its low energy content. It is frequently used in district heating systems to heat homes and public buildings in temperate areas. The most cost-effective technique to cool the flue gases before cleaning is with a hot water boiler connected to a cooling facility when there is no valuable use for the waste energy generated. Several well-known techniques for cooling are cooling towers, fan-mounted air coolers (radiators), and heat exchangers that use rivers or seawater. A cooling tower produces steam while using water as the cooling medium. River or sea water is needed for a heat exchanger, and hot water is led back to the recipient. A fan-mounted air cooler can be used instead of cooling water, albeit less energy-efficient. Aside from that, an air cooler is less suitable for hot conditions because the cooler's size grows more significant as the air temperature rises (World Bank, 1999).

6.2 Steam Boiler and Its Application

Low-pressure steam boiler: The Low-Pressure Steam Boiler is a low-pressure steam generator. A low-pressure boiler may be an alternative to a hot water boiler if a district heating network is not accessible and there is a need for processing steam. In terms of complexity, building accommodations, finance, operation, and maintenance, the low-pressure boiler is comparable to the hot water boiler—although the design needs more care because the flue gases are so corrosive. The temperature of the steam can range from 120 to 250 °C, depending on the operating pressure of the boiler and the degree of superheating. Steam pressures of up to 20 bar may be considered to be modest. This enables saturated steam to be produced at about 210 °C. Depending on the location of the end users, a certain amount of superheating can be required because the uses for low-pressure steam depend on its energy content. Since there might not always be a need for process steam, cooling capacity must always be available.

To prevent significant heat loss and the requirement for costly pipelines, the businesses that require process steam should be situated close to the plant. It is necessary to do a detailed assessment of the hazards involved in providing moisture to one or more industrial plants. It is possible to attain boiler efficiencies of up to about 80%. The temperature of the water returned to the boiler limits recovery (World Bank, 1999).

High-pressure steam boiler: Because the flue gas from a steam boiler is so much more corrosive than that from a hot water-generating or low-pressure steam boiler, more excellent care must be taken in its design. Additionally, it needs more room and care to operate well. Several steam boiler characteristics, including design, arrangement, and steam parameters, require special consideration (World Bank, 1999).

Steam Application: Only for the generation of electrical power using steam. Up to 35% of the energy in the garbage can be recovered and used to create electrical power alone. The excess heat needs to be cooled in a cooling tower or condenser. This alternative may be appealing if the factory is far from customers who desire it or sectors of the economy that need process steam. A condensing turbine is utilized when only power is produced. In this condenser, the excess heat is generated at such a low temperature that it is not conducive to recovery. Air or seawater is typically used as the cooling medium (World Bank, 1999).

6.3 Combined Heat and Power Generation

Up to 85% of the waste's energy can be used to generate heat and electricity simultaneously. A boiler made for trash incineration (with modest steam parameters) can produce between 20 and 25% power and between 65 and 60% heat. A so-called back pressure turbine is utilized to generate both power and district heating. The coolant's temperature and velocity, typically water from a district heating network, affect the back pressure (World Bank, 1999).

6.4 Combined Process Steam and Power Generation

Depending on how much process steam is extracted from the turbine, the electrical output for producing both process steam and power may fall somewhere between the figures for producing only power and having power and heat. This ranges from 35 to 20 percent. During this procedure, minimal steam must flow through the turbine ultimately. This implies that at least 10% of the low-pressure steam must be removed through cooling. An extraction turbine is used to generate electricity and process steam, combining the previously mentioned ideas. Steam can be recovered from a bleed in the turbine at pertinent conditions (pressure and temperature) when necessary. In contrast, the turbine operates as a fully condensed turbine that is cooled by seawater or air. The companies that need to process steam should be close to the plant to prevent significant heat loss and save money on costly pipelines (World Bank, 1999).

6.5 Thermal Energy Generation

Research to boost the energy efficiency of trash incineration power generation has also switched to combining the system with other thermal systems with better thermal properties. The combined operation of massive coal-fired units and waste incinerator power-producing technology saves energy. This technique combines the technological properties of two power generation modes (Song et al., 2021). The main steam from the garbage incinerator's output is fed to the coal-fired unit's thermal system, and some of the low-grade incinerator's heat is transferred to the high-grade unit for power generation to maximize garbage heat utilization. The coal-fired unit's flue gas treatment system receives rubbish incinerator tail flue gas. It will reduce the waste heat boiler outlet flue gas purification system equipment investment. Traditional waste incineration power generation uses waste to generate heat, which the flue gas absorbs through the boiler's heating surfaces to power the steam turbine and generator. Mechanical grate incinerators and liquidized bed boilers are widely used in waste incineration. Kozioi (2020) compares these technologies' features, pros, and cons. Due to improved development and experimentation, mechanical grate incinerators are frequently used. The power generation efficiency of typical trash incineration generating facilities is 20% to 32% lower than massive coal-fired systems. Small unit capacity, low steam characteristics, an inadequate regenerative cycle, and other issues cause low efficiency. The coupling technology can boost waste incineration power production thermal efficiency to 32% and efficiency by 10% (Bandarra & Quina, 2021).

The quality of garbage fed into the incinerator affects its combustion qualities. It will be very damp if waste is not allowed to dry out, especially in the lower waste storage pit. The waste moisture will evaporate from the middle and lower storage pits and collect in the high part. Thus, burning rubbish in the top half is more complex than in the bottom. The fuel factory separates debris from sawdust, plastics, and slag. After crushing, drying, pressing, and solidifying, they become safe RDF fuel (refuse-derived fuel) (Dadario et al., 2021). After delivery, the power plant produces RDF power from RDF fuel. After treatment, RDF fuel has a significant calorific value. Power generation, boiler efficiency, and steady combustion can be improved (Khandelwal et al., 2019).

6.6 Electricity Generation

Utilizing the waste-to-energy (WTE) or energy recovery process, incinerating garbage can produce power. In this method, waste is burned in an incinerator to produce heat that creates steam, which drives turbines attached to generators to make electricity. Here is a detailed explanation of how incineration produces electricity (EIA, 2022):

Waste Combustion: Wastes, such as municipal solid waste, are fed into an incinerator that has been specially made. The garbage is burned under regulated conditions within the incinerator at temperatures typically between 800–1000 °C (1472–1832°F). Organic materials are broken down and turned into heat energy during combustion.

Heat Recovery: During combustion, heat is created and transported to a heat recovery system. The boiler or heat exchanger is the typical component of this system. The heat recovery system absorbs the combustion gases' thermal energy, which then uses it to heat water and produce high-pressure steam.

Steam Generation: A steam turbine receives the high-pressure steam generated by the heat recovery system. The high pressure and temperature of the steam accelerate the rotation of the turbine's blades.

Turbine operation: The steam turbine's rotating blades turn a shaft attached to a generator. Kinetic energy is produced by mechanical rotation, and utilizing the principles of electromagnetic induction, this kinetic energy is subsequently transformed into electrical power.

Electricity Production: The generator transforms mechanical energy from the turning turbine into electrical energy. The power grid then receives this electrical energy and distributes it to end users, including households, companies, and industries.

Environmental Controls: Before combustion gases are released into the atmosphere in modern waste-to-energy facilities, they are subjected to environmental controls. These measures ensure that the process complies with air quality regulations while reducing the emissions of pollutants (EIA, 2022).

It is significant to highlight that waste-to-energy facilities are outfitted with cutting-edge pollution control systems to reduce their adverse environmental effects. To catch and treat pollutants, including particulate matter, sulfur dioxide, nitrogen oxides, and heavy metals, before they are released into the air, these technologies include scrubbers, baghouses, and other machinery.

While waste-to-energy technologies can aid in producing electricity and reducing garbage, it is crucial to manage the waste stream properly, monitor emissions, and abide by legal requirements to minimise any adverse environmental and human health effects. Waste composition, incinerator design, and the effectiveness of pollution control measures are only a few examples of the variables that affect how adequate and environmentally responsible waste-to-energy facilities are (Thomson, 2023).

7 Environmental Considerations

7.1 Potential Environmental Impacts

Potential challenges associated with garbage incineration techniques include opposition from the local community, fly ash and bottom ash management, and the significant release of flue gas. In practical applications, lime and activated carbon are commonly employed to eliminate dioxins and acidic gases such as sulfur dioxide (SO2) and hydrogen chloride (HCl). Due to the potential unreliability of baghouses, it is imperative to replace them when broken promptly. Failure to do so may result in releasing contaminated lime and activated carbon, compromising the quality of the emitted flue gases. Constantly eliminating pollutants poses challenges, as lime and activated carbon infusion need easier control and modification capabilities. The procedure for incineration of municipal solid waste and the subsequent generation of bottom and fly ash.

Furthermore, most incineration facilities fail to effectively eliminate nitrogen oxides (NOx) in the flue gases, leading to their subsequent discharge into the atmosphere. The potential for recycling and utilizing bottom ash in brick production is considered throughout the planning phase. However, due to their high production costs and substandard quality, the marketability of bottom ash bricks could be much better. Stabilizing fly ash is necessary before its disposal due to its classification as hazardous waste. The execution and upkeep of stabilizing and landfilling processes are challenging and costly, as highlighted by Youcai (2017).

7.2 Emission Control and Pollution Reduction Measures

The IED/WID and Environmental Permitting Regulations limit MSW incinerator flue gas emissions for specific pollutants. An environmental permit is required to operate an MSW incinerator in the UK, which includes operating and monitoring requirements and emission limits. The combustion process must be regulated, and flue gases must be cleaned before release to meet emission limits. The incinerator plant's technology provider will deliver exact emissions clean-up methods to meet criteria using Best Available Techniques. The typical emissions control technique is the reduction of NOx emissions by injecting ammonia into hot flue gases and lime or sodium bicarbonate injection to reduce SO2 and HCl emissions. The system comprises a filtration mechanism designed for separating fly ash and solids, alongside implementing carbon injection to effectively capture heavy metals.

Ensuring appropriate combustion conditions is crucial in effectively managing carbon monoxide (CO), volatile organic compounds (VOCs), and dioxins. Once the flue gases have undergone the cleaning process, the resulting solid residues consist of fly ash, lime/bicarbonate, and carbon. Air pollution control wastes, classified as hazardous waste, are typically consolidated, although specific systems may opt to segregate fly ash and other constituents. Consequently, waste disposal must adhere to the appropriate regulations and guidelines outlined by DEFRA (2013).

Before putting in place a Waste-to-Energy (WtE) incineration facility, it is necessary to conduct a thorough review of the environmental implications connected to air pollution, with a focus on public acceptance and knowledge. New air pollution control (APC) methods have proven to reduce pollution from various sources significantly. It is crucial to enact regulations on air pollutants such as dust, acidic gases, NOx, dioxins, and mercury, as suggested by the Takuma Environmental Technology Research Group (2017) and Liu et al. (2020).

Multiple establishments have switched to using closed wastewater systems. Wastewater is treated using flue gas treatment systems before being pumped into furnaces for coolant use. Leakage from trash pits is possible if they hold wet municipal solid waste (MSW). Therefore, it is essential to employ therapeutic strategies. Using bag filters to clean flue gas of contaminants is a regular practice. Alkali agents, such as lime powder or powdered activated carbon, are added to the bag filter before the flue gas is passed through. Besides reducing NOx levels, the following actions have been shown to reduce air pollution.

The filtration process is used to get rid of the dust particles. Dioxins, mercury, and acidic gases like hydrogen chloride (HCl) and sulfur dioxide (SO2) can all be absorbed by powdered activated carbon, which can then be removed.

Continuous flue gas monitoring requires ongoing monitoring of particulate matter, hydrogen chloride, sulfur dioxide, and nitrogen oxides. As a result of the Minamata

Convention in 2013, there is a pressing need for improved technology to monitor mercury levels continuously. Since these metrics are so crucial, competent analytical institutions should be measuring them regularly. Regular assessments are necessary due to the limits of continuous dioxin monitoring. Bag filters release a slurry of fly ash, alkali, and activated carbon pumped into the system as part of the Air Pollution Control (APC) process. Fly ash results from air pollution control (APC) residue, and its safe disposal is of the utmost importance, according to Liu et al. (2020).

8 Case Studies

8.1 Clean Plaza (Yokote City, Japan)

The Clean Plaza, located in Yokote City, Japan, is notable for its cleanliness and maintenance.

Japan has substantial municipal solid waste (MSW) incineration facilities. Many incinerators that handle municipal solid waste (MSW) are often distinguished by their modest size. However, they need help in achieving adequate levels of energy recovery. Nevertheless, notwithstanding the limited capacities of such facilities, there has been a noteworthy surge in the proliferation of municipal solid waste (MSW) incineration facilities. The current scenario can be mainly attributed to the widespread use of technological breakthroughs in thermal treatment procedures for various waste materials. The Clean Plaza MSW incinerator, situated in Yokote City, with an estimated population of 90,000, is a pertinent example of a compact-scale incineration facility. The plant was established in March 2016 and has a limited capacity of 95 tonnes per day, attributed to its tiny dimensions measuring 47.5×2. In addition, the incinerator was explicitly engineered to attain a power generation efficiency of approximately 20%. The plant reaches a high-efficiency level through boiler conditions characterized by elevated temperatures and pressures.

Despite its limited capacity, the primary objective behind developing the incinerator was to maximize energy recovery. In addition, the power generation system must operate at specific conditions of heightened pressure and temperature, specifically at 400 degrees Celsius and four megapascals, to maximize the efficiency of the recovery processes.

Moreover, it is necessary to augment the vacuum level of condensers to optimize the operational efficiency of boilers. The improvement of materials used in pipelines and equipment also plays a significant role in increasing operational efficiency. After the boiler's operation, the flue gas undergoes a water injection cooling procedure. However, in recent years, there has been a growing inclination towards adopting economizers for cooling purposes, primarily focusing on reducing energy consumption. The injection method of introducing urea into the furnace is employed to minimize the emissions of nitrogen oxide (NOx). The utilization of this non-catalytic reduction strategy possesses the capacity to make a valuable contribution towards endeavours aimed at conserving energy. Activated carbon injection is employed as a method to eradicate dioxins and mercury. Yokote City, located in a northern region of Japan known for its frequent snowfall, uses waste incinerator energy as a heat source to facilitate snow removal from highways during winter (Liu et al., 2020).

8.2 *Joetsu Clean Center (Joetsu City, Japan)*

The Joetsu Clean Center, located in Joetsu City, Japan, is a facility that warrants attention due to its significance in waste management and environmental conservation efforts.

The characteristics of municipal solid waste (MSW) are determined by the person's behaviours and the strategies implemented for its collection in a particular geographic area. The moisture content of municipal solid waste (MSW) in Japan is estimated to be around 40% (w/w). However, if the amount of kitchen waste with higher moisture content could be reduced during waste collection, the collected garbage's lower calorific value (LCV) would increase compared to ordinary waste. The municipal solid waste (MSW) designated for incineration at the Joetsu City plant, which has a population of 190,000, exhibits a lower calorific value (LCV) range of 8,100 to 15,900 kJ/kg. The value given above is obtained from the unique collection of organic waste emanating from households' kitchens within the municipality. This particular plant exhibits noteworthy characteristics regarding its steam conditions, surpassing those of other reactors. It operates under a pressure of 5.0 MPa and a temperature of 420 oC. A power output of 6,290 kilowatts is achieved by utilising generators, leading to a generation efficiency that surpasses 20%.

Furthermore, the facility uses non-catalytic NOx reduction technology to optimize energy recovery efficiency. Before being finally disposed of, bottom ash and fly ash undergo extensive processing to ensure their safe and effective management. After using appropriate chemical treatment methods to mitigate the leaching of heavy metals, fly ash is directed towards disposal, whereas bottom ash is promptly deposited without undergoing additional treatment. Furthermore, it has been observed that bottom ash possesses the capacity to serve as the principal constituent in the production of cement. Liu et al. (2020) assert that Japan is home to many contemporary incineration facilities that demonstrate adeptness in energy production.

To optimize the effectiveness of incineration procedures, it is essential to ascertain the suitable composition of solid waste. In addition, it is crucial to emphasize the regular production and effective gathering of municipal solid waste (MSW) (Liu et al., 2020).

8.3 *Incinerator Operation (Phuket, Thailand)*

Phuket Province, situated in the southern region of Thailand, has two incinerators, one constructed in 1996 and the other in 2012. In 2018, the waste disposal complex, comprising a landfill, incineration plant, wastewater treatment facility, and buffer zone, successfully managed a daily processing capacity of 928 metric tons of municipal solid waste (MSW). In 2017, the project yielded a total revenue of THB 545 million and incurred expenses of THB 275 million. It is worth noting that the project had a daily incineration capacity of 700 tons and operated for 320 days within that year. The Waste Management Executive Council (SWM) was founded in 1996, with the Waste-to-Energy (WtE) incineration facility commencing energy production of 12 MW in 2012. The proposal put forth by the Pollution Control Department (PCD) in Thailand suggests the use of community clustering as a means to govern and oversee waste policy and technology effectively. The 2014 Phuket Office of Natural Resources and Environment Solid Waste Management (SWM) Master Plan has elicited varied perspectives, and the effectiveness of these institutions and policies hinges on the comprehension of concerns by the general public and competent authorities, as well as their collaborative efforts to implement reforms (Liu et al., 2020).

9 Conclusion

Incineration plays a pivotal role in implementing ecologically conscious waste management practices and producing electricity. The proposed solution presents a comprehensive approach to addressing several pressing challenges of contemporary society. Incineration is a highly efficient method of converting solid waste into energy through controlled combustion. This process effectively mitigates the demand for additional landfill space and minimizes the release of detrimental greenhouse gases associated with fires in landfills and conventional waste disposal techniques. This strategy not only aids in mitigating the adverse effects of trash accumulation on the surrounding ecosystem but also contributes to producing renewable energy, reducing reliance on fossil fuels. The utilization of advanced technological solutions, such as waste-to-energy incineration facilities integrated with efficient pollution abatement systems, represents a viable approach for achieving the dual objectives of emission reduction and maintenance of air quality. Using incineration for generating electricity and heat means two forms of renewable energy that offer significant advantages to the local communities where they are implemented. Moreover, these methods contribute to advancing a power system characterized by enhanced security and efficiency. Overall, incineration is widely recognized as a notable and comprehensive approach that effectively combines waste reduction, environmental conservation, and energy production, highlighting its crucial role in advancing a more environmentally friendly and sustainable trajectory.

References

Abbasi, S. A. (2018). The myth and the reality of energy recovery from municipal solid waste. *Energy, Sustainability and Society, 8*(1). https://doi.org/10.1186/s13705-018-0175-y

Bandarra, B. S., & Quina, M. J. (2021). Municipal solid waste incineration and sustainable development. In *Advances in sustainable energy*, 653–680. Springer.

Cheng, H., & Hu, Y. (2010). Curbing dioxin emissions from municipal solid waste incineration in China: Re-thinking about management policies and practices. *Environmental Pollution, 158*(9), 2809–2814. https://doi.org/10.1016/j.envpol.2010.06.014

Dadario, N., Neto, M. M., Bernardo, C. H. C5, Bernardo, R., Almeida Gabriel Filho, L. R., Cremasco, C. P., et al. (2021). Waste-to-energy technology for the Brazilian context: A review article. *International Journal for Innovation Education, 9*(9), 427–448. https://doi.org/10.31686/ijier.vol9.iss9.3367

DEFRA. (2013). *Incineration of municipal solid waste.* https://assets.publishing.service.gov.uk/government/uploads/system/uploads/attachment_data/file/221036/pb13889-incineration-municipal-waste.pdf

EIA. (2022). Waste-to-Energy (MSW) in Depth—U.S. Energy Information Administration (EIA). www.eia.gov. https://www.eia.gov/energyexplained/biomass/waste-to-energy-in-depth.php#:~:text=The%20waste%20(fuel)%20is%20burned

GAIA. (2021, November 11). *Waste-to-energy has no place in Africa—GAIA.* GAIA. https://www.no-burn.org/waste-to-energy-has-no-place-in-africa/

Grosso, M., Motta, A., & Rigamonti, L. (2010). Efficiency of energy recovery from waste incineration, in the light of the new Waste framework directive. *Waste Management, 30*(7), 1238–1243. https://doi.org/10.1016/j.wasman.2010.02.036

Hockenos, P. (2021, May 26). *Waste to energy—Controversial power generation by incineration.* Clean Energy Wire. https://www.cleanenergywire.org/factsheets/waste-energy-controversial-power-generation-incineration

IETC. (2020, June 23). *Waste-to-energy incineration.* International Environmental Technology Centre. https://www.unep.org/ietc/resources/toolkits-manuals-and-guides/waste-energy-incineration

IPCC. (2007). Climate change 2007: Synthesis report. *Contribution of Working Groups I, II and III to the Fourth Assessment Report of the Intergovernmental Panel on Climate Change* (Core Writing Team, Pachauri, R.K and Reisinger, A., Eds., 104 pp.). IPCC, Geneva, Switzerland.

Islam, K. M. N. (2016). Municipal solid waste to energy generation in Bangladesh: Possible scenarios to generate renewable electricity in Dhaka and Chittagong City. *Journal of Renewable Energy, 2016*, 1–16. https://doi.org/10.1155/2016/1712370

Karim, M., & Corazzini, B. (2019). The current status of MSW disposal and energy production: a brief review of waste incineration. *MOJ Ecology & Environmental Sciences, 4*(1). https://doi.org/10.15406/mojes.2019.04.00129

Khandelwal, H., Dhar, H., ThaliaK, A., & Kumar, S. (2019). Application of life cycle assessment in municipal solid waste management: A worldwide critical review. *Journal of Cleaner Production, 209*, 630–654. https://doi.org/10.1016/j.jclepro.2018.10.233

Kozioi, M. (2020). The effect of the ratio of excess air and the process time on CO and NOx emissions from the combustion of alternative fuel in a grate furnace. *E3S Web Conference, 154*, 03003. EDP Sciences. https://doi.org/10.1051/e3sconf/202015403003

Liu, C., Nishiyama, T., Kawamoto, K., & Sasaki, S. (2020). *CCET guideline series on intermediate municipal solid waste treatment technologies: Waste-to-Energy Incineration.* https://www.iges.or.jp/en/publication_documents/pub/policysubmission/en/10877/WtEI_guideline_web_200615.pdf

Makarichi, L., Jutidamrongphan, W., & Techato, K. (2018). The evolution of waste-to-energy incineration: A review. *Renewable and Sustainable Energy Reviews, 91*, 812–821. https://doi.org/10.1016/j.rser.2018.04.088

Nadal, M., Rovira, J., Díaz-Ferrero, J., Schuhmacher, M., & Domingo, J. L. (2016). Human exposure to environmental pollutants after a tire landfill fire in Spain: Health risks. *Environment International, 97*, 37–44. https://doi.org/10.1016/j.envint.2016.10.016

Nidoni, P. (2017). Incineration process for solid waste management and effective utilization of by products. *International Research Journal of Engineering and Technology (IRJET)* e-ISSN, pp. 2395–0056. https://www.irjet.net/archives/V4/i12/IRJET-V4I1270.pdf

Panepinto, D., Senor, A., & Genon, G. (2015). Energy recovery from waste incineration: Economic aspects. *Clean Technologies and Environmental Policy, 18*(2), 517–527. https://doi.org/10.1007/s10098-015-1033-7

Patil, A., Kulkarni, A., & Patil, B. (2014). Waste to energy by incineration. *Journal of Computing Technologies, 3*(6), 2278–3814.

Song, W., Gao, Y., Bashir, S., & Liu, L. J. (2021). Postface: Conclusion on renewable energy strategies for a sustainable future: Part A: Role of energy storage. In *Advances in sustainable energy*, 839–846. Springer.

Stehl'k, P. (2009). Efficient waste processing and waste to energy: Challenge for the future. *Clean Technologies and Environmental Policy, 11*, 7–9

Takuma Environmental Technology Research Group. (2017). *Waste burning technology* (268 pp.). Ohmsha.

Thomson, E. (2023). *The incinerator giving waste-to-energy plants a green future*. Spectra. https://spectra.mhi.com/the-incinerator-giving-waste-to-energy-plants-a-green-future

Wang, A., & Dong, W. (2019). Discussion on waste incineration power generation and its process calculation. *IOP Conference Series: Materials Science and Engineering, 490*, 072047. https://doi.org/10.1088/1757-899x/490/7/072047

World Bank. (1999). *Municipal solid waste incineration*. https://web.mit.edu/urbanupgrading/urbanenvironment/resources/references/pdfs/MunicipalSWIncin.pdf

Youcai, Z. (2017). Municipal solid waste incineration process and generation of bottom ash and fly ash. *Pollution Control and Resource Recovery: Municipal Solid Wastes Incineration*, 1–59. https://doi.org/10.1016/b978-0-12-812165-8.00001-9

Zhang, J. H. (2012). Characterization of a Rhizobium larrymoorei FJ Exhibiting High Level Cr(VI) Reduction Potential. *Advanced Materials Research, 356–360*, 1009–1014. https://doi.org/10.4028/www.scientific.net/AMR.356-360.1009

Zhao, H., & Wang, J. (2018). Chemical-looping combustion of plastic wastes for in situ inhibition of dioxins. *Combustion and Flame, 191*, 9–18. https://doi.org/10.1016/j.combustflame.2017.12.026

Biogas Purification, Upgrading and Utilization: Focusing on Biological Systems

J. J. González-Cortés, F. Almenglo, and M. Ramírez

Abstract Biogas, a renewable energy source generated by organic matter degradation without oxygen, consists primarily of CH_4 (45–75%) and CO_2 (20–50%), along with minor components like O_2, H_2S, N_2, NH_3, H_2, volatile methyl siloxanes (VMSs), and volatile organic compounds (VOCs). While conventional physicochemical methods have been used for biogas purification, biological technologies have emerged due to their cost-effectiveness and eco-friendliness. The chapter provides an overview of biological systems for biogas purification and upgrading. It covers in-situ desulfurization through microaeration and aerobic and anoxic ex-situ desulfurization techniques within different bioreactor configurations such as fixed-bed (biofilters, biotrickling filters) and suspended biomass (CSTR, gas-lift, bubble column, bioscrubber). The chapter also addresses the removal of poorly biodegradable volatile methyl siloxanes (VMSs) and volatile organic compounds (VOCs), known for causing combustion-related deposit formation and emission of toxic compounds during combustion, respectively. Biogas upgrading has gained momentum, producing high CH_4 concentration effluent (biomethane) comparable to natural gas. Various upgrading technologies exist, including physical and chemical methods and biological approaches. Among biological options, microalgae and hydrogenotrophic archaea are prominent for biogas upgrading, utilizing photobioreactors for CO_2 assimilation and hydrogenotrophic methanogens for CO_2 and H_2 conversion. The implementation of biological techniques can reduce investment and operational costs compared to physicochemical methods, promoting sustainable integration of upgraded biogas into the energy mix. The chapter concludes by outlining future research directions for biological biogas purification and upgrading.

J. J. González-Cortés · F. Almenglo (✉) · M. Ramírez
Department of Chemical Engineering and Food Technologies, Wine and Agrifood Research Institute (IVAGRO), Faculty of Sciences, University of Cadiz, Puerto Real, 11510 Cádiz, Spain
e-mail: fernando.almenglo@uca.es

A. Anouzla and S. Souabi (eds.), *Generation of Energy from Municipal Solid Waste*,
https://doi.org/10.1007/978-3-031-74334-4_13

1 Introduction

Biogas is a renewable energy source obtained from the biodegradation of organic matter in the absence of oxygen. Biogas is mainly composed of CH_4 (45–75%) and CO_2 (20–50%), although several minority compounds can be found, such as O_2 (<1%–5%), H_2S (0.5–80,000 ppm_V), N_2 (<1%–25%), NH_3 (traces), H_2 (0–5%), volatile methyl siloxanes (VMSs, 0–50 mg m^{-3}), volatile organic compounds (VOCs, 0–330 mg m^{-3}), etc. (Ramírez et al., 2015). The composition of biogas depends mainly on the type of organic matter used as substrate. The main types of biogases are: biogas from landfills, from sludge digestion in wastewater treatment plants (WWTP) and from agricultural and/or livestock manure.

The lower calorific value (LCV) of pure CH_4 is approximately 34,300 kJ m^{-3} (standard pressure and temperature) (Hosseini & Wahid, 2013). Then, the LCV of biogas is approximately 15,435–25,725 kJ m^{-3}, for CH_4 concentrations of 45 and 75%, respectively. Generally, biogas needs to be purified depending on its composition of minority compounds, mainly H_2S and VMSs. The main use of biogas is combined heat and power (CHP) cogeneration. However, if its concentration in CH_4 is increased (upgrading processes) by CO_2 separation or the conversion of CO_2 into CH_4 (catalytic or biological processes), the main applications are its use as fuel in vehicles or injection into natural gas grids.

Biogas has traditionally been purified using physicochemical methods. However, in recent years, biological technologies have emerged because of their cost-effectiveness and low environmental impact compared with conventional physicochemical methods (Almenglo et al., 2023). This chapter provides an overview of biological systems used for biogas purification and upgrading. The main ex-situ desulfurization technologies, aerobic and anoxic, used in various types of bioreactors are detailed: fixed-bed bioreactors (biofilters and biotrickling filters) and suspended biomass bioreactors (CSTR, gas-lift, bubble column, bioscrubber, etc.). In situ desulfurization by microaeration is also an attractive option, depending on the H_2S concentration and digester configuration (Azizi et al., 2022). The book chapter also addresses the removal of VMSs, which are poorly biodegradable compounds that cause problems of deposit formation (silicon dioxide) during combustion, leading to problems of abrasion and clogging of pipes (Rivera-Montenegro et al., 2023). Finally, in relation to purification processes, the elimination of VOCs (monoaromatic hydrocarbons, terpenes, halogenated compounds, etc.), which cause the emission of toxic compounds during combustion or catalyst poisoning in catalytic biogas conversion applications, is considered. As shown in Fig. 1, most published studies have focused on H_2S removal (76% in 2023), with a total of 919 in 2023 on the removal of these three compounds.

Biogas upgrading has been increasing in recent years, as it produces an effluent with a high CH_4 concentration (biomethane) and similar composition and LCV to natural gas, but renewable. There are many existing technologies to carry out the biogas upgrading, including physical absorption processes (water scrubbing and organic physical scrubbing), chemical absorption (amines, alkaline salts, etc.),

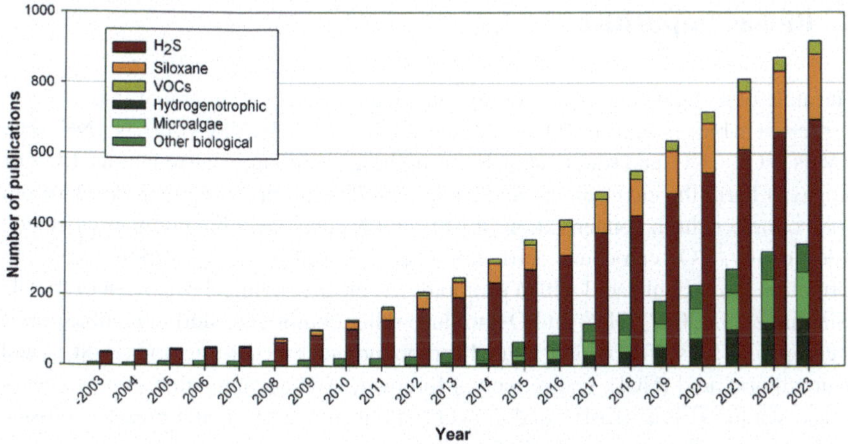

Fig. 1 The cumulative number of publications containing either the concept: biogas upgrading (first column) and biogas purification (second column). Search filters (TITLE-ABS-KEY): "biogas AND hydrogen AND sulfide AND removal"; "biogas AND siloxane AND removal"; "biogas AND VOCs AND removal"; "biogas AND upgrading AND hydrogenotrophic"; "biogas AND upgrading AND microalgae"; "biogas AND upgrading AND biological". Search filters: TITLE-ABS-KEY (biogas AND upgrading AND biological) AND (hydrogenotrophic); TITLE-ABS-KEY (biogas AND upgrading AND biological) AND (microalgae). [Scopus, accessed 05 July 2023]

adsorption processes (pressure swing adsorption and vacuum swing adsorption), membrane separation, cryogenic separation, hybrid technologies and biological technologies (Aghel et al., 2022). Microalgae are the most commonly used microorganisms in biogas upgrading (total publications in 2023 of 133) followed by hydrogenotrophic archaea (total publications in 2023 of 133) (Fig. 1) (Cattaneo et al., 2023).

Photobioreactors are of great interest because of the high CO_2 assimilation capacity of microalgae, and the use of photobioreactors in open and closed systems is described. Hydrogenotrophic methanogens are microorganisms that are capable of converting CO_2 and H_2 into CH_4. These processes consider the use of green H_2, the production of which is a reality given the expansion of renewable energy coupled with a reduction in production costs. The main types of bioprocesses (in situ, ex situ, and hybrid systems) were described.

The implementation of the biological techniques discussed in the present chapter can reduce the investment and operating costs of currently used physicochemical techniques, thereby promoting the sustainable integration of upgraded biogas into the current energy mix. Finally, future research directions for biological biogas purification and upgrading are discussed.

2 Biogas Impurities

The anaerobic digestion process of organic matter causes certain minor impurities to be present in biogas such as CO_2, volatile sulfur compounds (VSCs), O_2, N_2, NH_3, H_2, VMSs, VOCs, CO, and water vapor. Some of these compounds can reduce the LCV of biogas or harm the environment and equipment directly or during their combustion. For example, a high concentration of CO_2 and N_2 decreases biogas energy density. Additionally, CO_2 can cause corrosion when exposed to wet condition due to the formation of carbonic acid which promotes the electrochemical corrosion of metals (Singhal et al., 2017). H_2O and O_2 produce other compounds during combustion if accompanied by sulfur or halogenated compounds such as sulfuric acid (H_2SO_4) and hydrochloric acid (HCl), respectively. Moreover, H_2O accumulation in pipes poses a risk, while O_2 is corrosive and concentrations above 6% could create explosive environments (Bragança et al., 2020).

Among the VSCs present in biogas, hydrogen sulfide (H_2S), carbonyl sulfide (COS) and organic sulfur such as mercaptans (RSH), are the most commonly found. They are originated due to the decomposition of sulfur-rich organic matter such as sulfolipids or amino acids (methionine, cysteine, and their derivatives) and the reduction of inorganic sulfur, mainly sulfate. Therefore, the final concentrations of VSC will depend on the feedstock and process conditions of the anaerobic digestion. The presence of VSCs, especially H_2S, can cause corrosion in pipes and equipment, catalytic converter poisoning, health hazards and formation of SO_2 and SO_3. Typical H_2S concentrations range from 50–5,000 ppm_V, while other VSCs are usually present in lower amounts (Chen et al., 2008). For example, mercaptans are among the most abundant VSC after H_2S. Methyl mercaptan (CH_3SH) is the most common representative of this group, which can be present in biogas in a concentration range of 1–20 ppm_V with a reported maximum of 100 ppm_V (Montebello et al., 2012). The concentration of COS is low compared to H_2S, with usual values below 1 ppm_V (Andersson et al., 2004). The intended use of biogas will restrict the maximum allowed concentration of VSCs. Maximum limits for VSC vary depending on its application: turbines (<10,000 ppm_V); heating systems like boilers or stirling engines (<1.000 ppm_V); internal combustion engines (<50–500 ppm_V depending on engine type); injection into natural gas grid (<4 ppm_V); or fuel cells (<1–20 ppm_V depending on the type of fuel cell) (Awe, 2017).

VMSs are linear (L or M nomenclature) o cyclic (D) organosilicon compounds in which a silicon atom is bonded to at least one carbon and one oxygen atom. While they are typically present in low proportions (0–50 mg m^{-3}) (Muñoz et al., 2015), their presence can be extremely problematic due to generated silicon causing abrasion on the equipment and components. Their presence leads to operational problems for current biogas utilization technologies, such as turbines, engines or catalysts for biogas steam reforming or fuel cells. The most common siloxane compounds found in biogas include hexamethylcyclotrisiloxane (D3), octamethylcyclotetrasiloxane (D4), decamethylcyclopentasiloxane (D5), hexamethyldisiloxane (L2), and octamethyltrisiloxane (L3) (8). The concentration of VMSs in biogas depends on the location and origin of the waste used as a substrate for biogas production. The required concentration of VMSs for biogas use depends on both country-specific legislative frameworks and intended application. Currently, it is recommended that the maximum concentration does not exceed 5 mg m^{-3} (Mendiara et al., 2021) although some industries may require levels below 0.1 mg m^{-3} (Mendiara et al., 2021).

The degradation of nitrogenous matter, primarily proteins and urea, produces both ammonium (NH_4^+) and free ammonia (NH_3). The concentration of the latter is dependent on three factors: total nitrogen concentration, temperature, and pH. Free ammonia is toxic to anaerobic digestion processes (Chen et al., 2008) and highly soluble, so it is essential to closely monitor and control the ammonia concentration in anaerobic digestion plants. Typical ammonia concentration in landfill gas is 5 ppm_V and in anaerobic digestion plants below 100 ppm_V (Rasi et al., 2011). The presence of ammonia in biogas poses significant drawbacks because it is highly corrosive. Upon combustion, it forms NO_X, which has notable environmental and health effects. Depending on the country and utilization application ammonia content is recommended to be under 20 mg m^{-3} (Rasi et al., 2011).

The digestion of halogen-rich waste is responsible for the presence of halogenated compounds in biogas (e.g., carbon tetrachloride, chlorobenzene, and chloroform), which mainly originate from halogenated polymers such as polytetrafluoroethylene (PTFE) and salts in edibles (Abdelkareem et al., 2019). During combustion, these compounds break down and cause highly corrosive compounds to form in the aqueous phase (HCl, HBr, and HF) which can affect the quality of engine oil by reducing its lifespan. The presence of halogenated compounds is especially significant in landfill biogas where concentrations can reach 225 mg m^{-3} (Werkneh, 2022). However, this concentration may vary depending on their location within the landfill site and age. When using biogas for electricity production, it is necessary to limit the amount of halogenated compounds. This requirement becomes even more stringent when solid oxide fuel cells (SOFCs) are used as they should not exceed 5 ppm_V (Turco et al., 2016).

3 Biogas Purification Through Biological Techniques

3.1 Biological Removal of H_2S

As stated above, H_2S is the main VSC present in biogas. Its concentration in biogas can be reduced by biological methods using two distinct strategies: (i) decreasing H_2S production within the anaerobic reactor by altering the reducing conditions (in-situ), and (ii) downstream desulfurization using sulfur-oxidizing microorganisms in independent bioreactors (ex-situ).

In-situ biological desulfurization (microaeration). A strategy for minimizing the concentration of H_2S in biogas is to inject air (or oxygen) into the anaerobic digester, creating microaerobic conditions. This method is reliable, simple, and cost-effective. However, there are some drawbacks such as partial oxidation of the soluble substrate and potential clogging of pipes and equipment with elemental sulfur (Krayzelova et al., 2015). Traditionally, aerobic conditions have been avoided inside anaerobic digesters to protect obligate anaerobic microorganisms. However, in the past two decades, injecting controlled amounts of air has shown certain benefits, improving biogas quality by removing sulfide, enhancing hydrolysis processes, and improving stability and robustness of the microbiome (Azizi et al., 2022). These microaerobic conditions allow for biological oxidation of dissolved sulfide to elemental sulfur or sulfate whose concentration can be effectively regulated by adjusting the oxygen supply. The closer this concentration is to the stoichiometric value for elemental sulfur production (Eq. 4), the lower proportion of sulfate will be generated. Some authors have reported elemental sulfur yield ranging from 73 to 85% when employing oxygen quantities within the range of 0.5 to 0.7 mol O_2 mol^{-1} S^{2-} (29).

The main operational options include choosing between air or pure oxygen, determining the injection point in either the liquid phase or gas phase (such as in the headspace or through biogas recirculation), and implementing intermittent or continuous dosing (Azizi et al., 2022; Krayzelova et al., 2015). Studies indicate that regardless of the injection point, the majority of H_2S removal occurs within the headspace, facilitated by a population of sulfide-oxidizing bacteria that thrive on the walls of the digester's headspace (Azizi et al., 2022). When injecting air into the liquid phase, it is important to note that a portion of the supplied oxygen may be utilized for organic matter oxidation. This can result in a decrease in available oxygen for sulfide-oxidizing bacteria and reduce recoverable substrate as methane. In such cases, up to approximately 60% of oxygen loss has been observed (Azizi et al., 2022). It is crucial to exercise thorough control when injecting into the headspace to avoid potentially hazardous concentrations that could lead to explosive atmospheres and excessive biogas dilution.

Ex-situ biological desulfurization. Ex-situ biological desulfurization processes involve passing the biogas stream generated during anaerobic digestion through an external biological reactor.

In the biological removal of contaminants, it is common to employ concepts widely used in gas treatment technologies to describe the performance biological reactors. These include the removal efficiency (RE), inlet load (IL), and elimination capacity (EC). They are related to the inlet concentration (C_{in}), outlet concentration (C_{out}), volume (V), and volumetric flow rate of biogas (Q) as follows (Eqs. 1–3):

$$\mathrm{RE} = \frac{C_{\mathrm{in}} - C_{\mathrm{out}}}{C_{\mathrm{in}}} \bullet 100(\%) \tag{1}$$

$$\mathrm{IL} = \frac{C_{\mathrm{in}}}{V} \bullet Q(gm^{-3}h^{-1}) \tag{2}$$

$$\mathrm{EC} = \frac{C_{\mathrm{in}} - C_{\mathrm{out}}}{V} \bullet Q = L \bullet \frac{\mathrm{RE}}{100}(gm^{-3}h^{-1}) \tag{3}$$

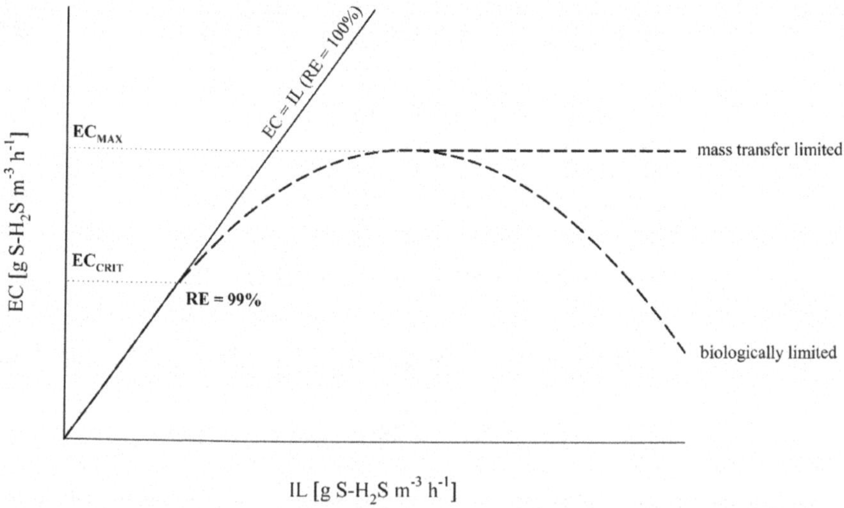

Fig. 2 Schematic representation of H_2S elimination performance: elimination capacity as a function of inlet load

The behavior of the biological reactor can be characterized using the critical elimination capacity (EC_{CRIT}), which represents the EC when the RE exceeds 95–99%. For IL lower than EC_{CRIT}, the relationship between IL and EC is linear. However, when IL surpasses EC_{CRIT}, EC increases up to EC_{MAX} while RE decreases. At this point, two behaviors may occur if IL continues to increase: either EC remains constant due to mass transfer limitation or it decreases because the bioreactor is biologically limited by pollutant accumulation and inhibition (Fig. 2).

The most extensively studied group of microorganisms responsible for sulfur oxidation comprises chemotrophic bacterial species that utilize reduced sulfur compounds as an energy source.

The biological reaction mechanism for desulfurization through chemolithotrophic bacteria depends on the electron acceptor (Eqs. 4–11). There are two main groups: one where the electron acceptor is O_2 (aerobic biodesulfurization), and another where the reaction occurs in its absence, with nitrate (NO_3^-) and/or nitrite (NO_2^-) serving as the electron acceptors.(27). The oxidation products consist of elemental sulfur (S^0) and sulfate (SO_4^{2-}). The molar ratio between the electron acceptor and sulfide determines which reaction product is more favored. Experimental evidence, supported by a significant number of authors, demonstrates that a restriction in O_2, NO_3^-, or NO_2^- leads to an increased production of S^0 as the primary outcome (Almenglo et al., 2023).

$$H_2S + 0.5O_2 \rightarrow S^0 + H_2O \quad (4)$$

$$H_2S + 2O_2 \rightarrow SO_4^{2-} + 2H^+ \quad (5)$$

$$5H_2S + 2NO_3^- \rightarrow 5S^0 + N_2 + 4H_2O + 2OH^- \quad (6)$$

$$H_2S + NO_3^- \rightarrow S^0 + NO_2^- + H_2O \quad (7)$$

$$3H_2S + 2NO_2^- + 2H^+ \rightarrow 3S^0 + N_2 + 4H_2O \quad (8)$$

$$5H_2S + 8NO_3^- \rightarrow 5SO_4^{2-} + 4N_2 + 4H_2O + 2H^+ \quad (9)$$

$$H_2S + 4NO_3^- \rightarrow SO_4^{2-} + 4NO_2^- + 2H^+ \quad (10)$$

$$3H_2S + 8NO_2^- + 2H^+ \rightarrow 3SO_4^{2-} + 4N_2 + 4H_2O \quad (11)$$

Several bioreactor configurations have been reported to be capable of desulfurizing biogas. They are divided mainly depending on how biomass is present in the bioreactor. Fixed bed bioreactors, where the biomass forms a biofilm on a support material (Fig. 3a, b), or suspended growth bioreactors, where the biomass is suspended within the culture medium. Fixed-bed bioreactors typically take the form of biofilters (BF) or biotrickling filters (BTF), with biogas comprising the majority phase and a small liquid layer present in BTF or no liquid layer at all in BF. On the other hand, there are two main approaches to use suspended growth bioreactors: direct dispersion of biogas into the liquid phase (utilizing continuous stirred tank reactors (CSTR), gas-lift systems, or bubble columns) or absorption of biogas in an absorption column prior to biological oxidation within the reactor (known as bioscrubber).

Biofilters (BFs). BFs are the simplest type of fixed-bed bioreactors (Fig. 3a). They are used to treat contaminants with low solubility in water ($H > 0.10$ at 25 °C) and low concentrations (below 1 g m^{-3}). These reactors consist of one or multiple beds with high porosity and low pressure drop. A biofilm develops on these beds, which can sometimes utilize the bed material itself as a source of nutrients or require periodic supplementation. Typically, gas passes through the bed in an upward flow, facilitating the transfer of contaminants from the gas phase to the biofilm. To ensure moisture in the bed, it may be necessary to humidify the gas stream or intermittently introduce liquid media through percolation. Percolation should provide sufficient wetting of the bed and nutrient supply while avoiding continuous formation of a liquid layer that could interfere with contaminant transport. For biogas treatment, closed system designs are utilized compared to open designs commonly employed for odor treatment using BFs. Generally, BFs offer advantages such as low construction, operation, and maintenance costs; ease of startup and operation; and a significant surface area between gas and biofilm. However, controlling reaction conditions can be challenging; periodic replacement of support materials is necessary, and BF systems have slow adaptability to fluctuations in gas inlet conditions. In addition, the use of aerobic biofilters results in biogas dilution due to the utilization of air as an O_2

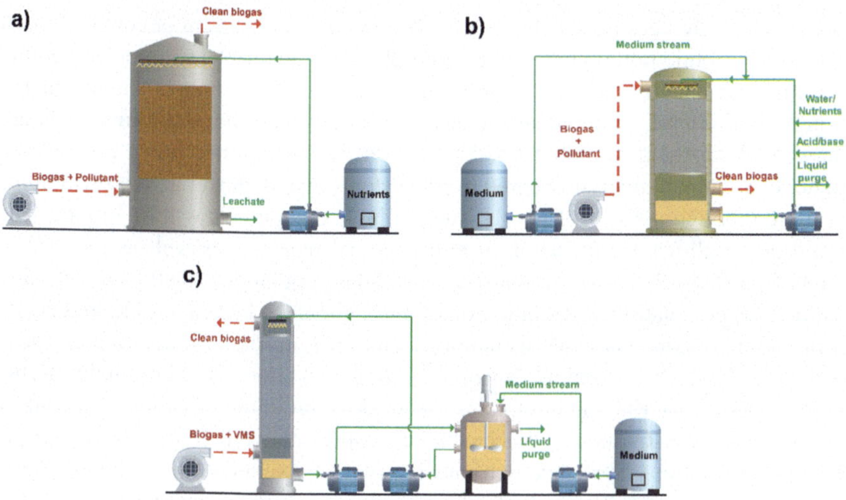

Fig. 3 Compilation of the most used bioreactors to perform the removal of pollutants from biogas. Adapted from Almenglo et al. (2023)

source, with nitrogen (N_2) present in air remaining unconsumed (Cattaneo et al., 2023).

The use of BFs has been extensively investigated for the mitigation of odors caused by H_2S in air, where concentrations typically range from 0.3 to less than 15 ppm_V (Bu et al., 2021). However, as previously mentioned, certain characteristics specific to biofilters suggest that their application for higher concentrations found in biogas may be limited. This limitation primarily stems from the difficulty in removing the oxidation products. The aerobic oxidation to sulfate (Eq. 5) can induce significant acidification within the biofilm and consequently lead to microbial inhibition. In contrast, under limited oxygen conditions that promote the formation of elemental sulfur (Eq. 4); the internal pH of the biofilm must remain unchanged. In the case of using anoxic BFs, whether using NO_3^- or NO_2^- (Eqs. 6–11), it can cause a pH imbalance that may significantly affect microbial activity. Nevertheless, reports have documented aerobic BFs capable of treating H_2S loading rates up to 64 g S-H_2S m^{-3} h^{-1} with an RE of 99% (Cattaneo et al., 2023), and anoxic BFs achieving elimination capacities as high as 30 g S-H_2S m^{-3} h^{-1} (Almenglo et al., 2023). In addition to EC_{CRIT}, two important parameters in the design of BFs are the empty bed residence time (EBRT), what refers to the amount of time that the air or gas remains in contact with the filter media or bed in a biofilter system, and operating pH. For aerobic BFs, an EBRT range between 16 s and 1 min has been reported, with a pH range between 6.1 and 7.5 (Cattaneo et al., 2023). In the case of anoxic BFs, the EBRT ranges from 1 to 5 min, while the operating pH ranges between 7 and 9 (Almenglo et al., 2023).

Biotrickling filters (BTFs). A BTF consists of a fixed-bed column with high-surface-area packed with synthetic material on which the biofilm develops (Fig. 3b).

The contaminated gas passes through the bed in either concurrent or countercurrent flow to the mobile liquid phase. The liquid phase is an aqueous solution containing essential nutrients for the growth and/or maintenance of microorganisms within the biofilm. In this case, contaminants must be absorbed and diffused through the liquid phase, making them available for degradation by the biofilm. BTFs can tolerate higher contaminant loads compared to BF systems due to greater activity within the biofilm, although transfer of poorly water-soluble compounds will be limited. The liquid phase allows for diffusion of nutrients and reactants as well as removal of reaction products that may become toxic or inhibit population growth in the biofilm. Continuous percolation of the liquid phase throughout the bed ensures homogeneity of important parameters along its height, such as pH, nutrient concentrations, intermediate or final toxic compounds from reactions. Additionally, it eliminates drying out of support material and eliminates the need for pre-humidification of incoming gas streams. Less frequent replacement of support material (typically organic in BFs) is required in BTFs since nutrients are typically supplied in the liquid phase. However, compared to BF systems, these units incur higher construction costs and complexity while demanding greater investment in process control measures.

The low solubility of oxygen in water plays a crucial role in the design of aerobic BTFs, where particular attention must be given to its transfer into the liquid phase. Typically, the oxygen stream (air) is directly supplied to the biogas stream at the inlet of the BTF. However, this approach necessitates a significant amount of air for oxidation, presenting challenges to process viability due to the associated operational costs related to blower operation and biogas dilution. To reduce these costs, the enhancement of oxygen transfer has been investigated through the utilization of jet-venturi systems, which offer a greater capacity for oxygenation than conventional diffuser-based devices (Rodriguez et al., 2014). Another crucial parameter is the liquid percolation velocity (TLV), which relates to the flow rate of liquid dripping through the bed in relation to the perpendicular area to the flow (Eq. 14). This parameter has a significant impact on two important aspects: the supply of soluble nutrients and oxygen transfer to the biofilm. Higher TLV values can result in thickening of the liquid layer, leading to a reduction in interfacial area and impaired mass transfer. In extreme cases with excessively high TLVs, column flooding may occur, causing detrimental effects such as increased pressure drop and a drastic decrease in surface area The range of TLVs used in aerobic and anoxic BTFs is similar, typically varying from 5 to 20 m h^{-1} (Almenglo et al., 2023). However, it holds greater importance in aerobic BTFs as it is a more effective variable for improving the H_2S RE and sulfate production compared to air supply (López et al., 2016). Aerobic BTFs can operate within a wide pH range, covering conditions close to neutrality (5.5–7.5) as well as acidic conditions (0.5–5.5).

$$\mathrm{TLV} = \frac{Q_L}{A} \tag{12}$$

In anoxic BTFs, the operating pH is near neutrality (pH 6.8–7.8), and the optimal temperature is 30 °C. The dosing of NO_3^- and/or NO_2^- is done by dissolving them in water, either at the injection point within the recirculation volume or directly into the recirculation stream prior to dispersing it at the top of the bed. There are three distinct strategies for automating the dosing of NO_3^- and/or NO_2^-:

1. Using the value of the redox potential in the liquid phase: An increase in sulfide concentration correlates with a decrease in ORP. Consequently, the ORP value can be employed as a feedback control variable within a closed-loop control system, with dosing of NO_3^-/NO_2^- serving as the manipulated variable.
2. Using the value of H_2S concentration in the outlet biogas: A closed-loop feedback control system can be implemented using this H_2S concentration as a control variable and utilizing both PI and PID controls to dose NO_3^-/NO_2^- (manipulated variable).
3. Using the value of IL: measuring the concentration and flow rate at the inlet of the BTF enables the incorporation of a feedforward closed-loop control system. This control system is based on maintaining a constant N/S feeding ratio by modifying the feed flow rate of medium with NO_3^- or NO_2^-.

One of the main advantages of aerobic BTFs is their low operating cost, as they utilize air as the source of oxygen. Therefore, the costs primarily involve electricity consumption and blower installation. Furthermore, the utilization of air results in a reduction in methane concentration, with a smaller impact when external systems like jet-venturi are employed to enhance transfer efficiency and decrease the required air flow rate (Rodriguez et al., 2014). However, the main disadvantage is related to safety concerns. In an event of a failure in the aeration control system, there is a risk of reaching explosive conditions (Rodriguez et al., 2014). In contrast, anoxic BTFs offer a reduced risk of operating in explosive conditions, they do not decrease methane concentration, and the soluble nature of nitrates and nitrites allows higher electron acceptor concentrations to be achieved. However, one significant drawback of using anoxic BTFs is a higher operational cost and environmental impact associated with the use of commercial nitrate or nitrite compounds (Almenglo et al., 2023). An alternative to the use of commercial NO_3^-/NO_2^- is to utilize the effluent from a nitrification process (either total or partial) of ammonia-rich sources. This integration allows for simultaneous treatment of two contaminated effluents. Consequently, implementing this integrated system would result in a significant reduction in environmental impact compared to other treatment alternatives (Cano et al., 2018).

One of the major challenges in using fixed-bed bioreactors is the accumulation of elemental sulfur on the packing material in the bed. This issue becomes more significant at low molar ratios between electron acceptors (O_2, NO_3^-, NO_2^-) and sulfide. Sulfur gradually deposits on the bed, leading to a gradual decrease in its effective surface area. Additionally, it reduces the gas passage volume and can eventually cause clogging under extreme accumulation conditions. Several options have been proposed to solve or minimize these negative effects and extend the lifespan of the packing materials. One approach is to increase the concentration of electron acceptors, which promotes sulfate production as the main oxidation product. This solution requires improved oxygen transfer and enhanced air supply in aerobic bioreactors or higher feed flow rates for NO_3^-/NO_2^- in anoxic bioreactors, albeit with increased operational costs.

Another proposed option is to alternate between concurrent and countercurrent flow modes. In countercurrent operation, biogas is introduced from the bottom, resulting in minimal molar ratio of sulfide and electron acceptor in the lowest part of the BTF due to the higher concentration of sulfide. On the other hand, in concurrent operation, this condition occurs at the top section. The alternation between these modes allows for a more homogeneous distribution of elemental sulfur along the height of the bed (Almenglo et al., 2016). Once sulfur has accumulated on the bed, there are several methods to remove it being the best approach to stop biogas feeding while maintaining electron acceptor supply or through partial or total replacement of the bed. Another potential solution to this problem is using suspended growth bioreactors since elemental sulfur formed in suspension can be easily removed during a subsequent separation stage. Additionally, in these bioreactors, the incomplete oxidation of H_2S can be promoted, maximizing elemental sulfur production and reducing reagent consumption costs.

Suspended growth bioreactors. Suspended growth bioreactors present specific challenges when applied to gas contaminant removal, particularly regarding the mass transfer of these contaminants into the liquid phase. Achieving high mass transfer rates necessitates a large specific surface area, which can be accomplished by dispersing small bubbles from the bottom of the bioreactor and operating at a high liquid velocity to reduce the boundary layer through agitation. However, these bioreactors have been relatively underutilized for H_2S removal in biogas treatment, with the exception of bioscrubbers as they address mass transfer issues by employing an absorption unit connected to the stirred tank bioreactor (STBR). Bioscrubbers (Fig. 3c) consist of two units: the first unit is a scrubbing tower where H_2S absorption from biogas takes place, and the second unit involves the biological oxidation of H_2S in the liquid stream that exits the absorber. The most prominent example is THIOPAQ®, a patented technology widely marketed and extensively used for industrial biodesulfurization. THIOPAQ® operates under slightly alkaline conditions (pH 8–9), which promotes effective H_2S absorption. The biological reactor employed is a SBTR that utilizes air for oxygen supply and generates elemental sulfur as the primary oxidation product.

Apart from bio-scrubbers, the main equipment employed includes STRs, gas-lift systems, and bubble columns. These bioreactors have been extensively described in the literature for various biological processes. When it comes to biogas desulfurization, particular attention should be paid to STRs. In these reactors, agitation is achieved through mechanical stirring using blades. Increasing the agitation speed enhances material transfer; however, it should be noted that high-speed contact between the blades and water can generate shear forces that may adversely affect the integrity of bacteria present, leading to their disruption and death (González-Cortés et al., 2021). Agitation in gas-lift systems and bubble columns is achieved through the pneumatic action of ascending gas bubbles, thereby addressing the aforementioned issue. Additionally, gas-lift systems exhibit good mass transfer and energy efficiency, they do not have movable mechanical parts and are easy to clean (Almenglo et al.,

2023). Both aerobic and anoxic suspended growth bioreactors can yield comparable ECs to those obtained in fixed bed bioreactors. However, it is important to consider in terms of capital and expenditure costs that the absence of an absorption column in these bioreactors makes it mandatory to install a blower to inject gas from the bottom of the bioreactor to overcome the water column pressure.

3.2 *Biological Removal of Volatile Methyl Siloxanes*

The treatment of gas pollutants using biological processes is often the most cost-effective and environmentally friendly option compared to physical–chemical technologies. Biological technologies can operate at ambient pressure and temperatures, eliminating the generation of potentially toxic waste such as spent adsorbents. However, these technologies face limitations in removing hydrophobic gas pollutants due to poor gas/liquid mass transfer. Pollutants such as VMSs and other hydrophobic VOCs such as alkanes have high dimensionless Henry's law constants ($H > 0.10$ at 25–30 °C and 1 atm). This makes it difficult for these compounds to reach the aqueous phase, where microorganisms can degrade them. Specifically for VMS, their dimensionless Henry's law constants are significantly high ($72 < H < 4637$) what makes its removal through biological techniques difficult (Max-Planck Institute, 2023).

As stated above, in the biological removal of hydrophobic compounds like VMS, the efficiency of mass transfer between the gas and liquid phases is a crucial factor to consider. Numerous studies have highlighted the significance of mass transfer in bioreactors when dealing with these compounds (Pascual et al., 2021a; Popat & Deshusses, 2008). Typically, the mass transfer in bioreactors is evaluated using an overall mass transfer coefficient (K_La), which is derived from a mass balance equation represented by Eq. 15. This equation relates the gas–liquid mass transfer rate ($F_{G/L}$) to K_La and the concentration gradient. The solubility of VMS is characterized by CG in and the dimensionless Henry's law constant (H).

$$F_{G/L} = K_La \cdot (\frac{C_{G,\text{in}}}{H} - C_L) \quad (13)$$

As reported by Rivera-Montenegro et al. (2023), if a typical VMS concentration in biogas is chosen (200 mg m^{-3}) and an H value of a model VMS such as D4 is considered (3.1), the maximum gas liquid concentration gradient to be achieved when the liquid VMS concentration (C_L) equals zero is 64.5 mg m^{-3}. From their example it can be concluded that the low concentration of VMS in the biogas together with the high H values cause mass transfer to be the limiting factor for the degradation of these compounds by biological techniques.

To address this challenge, researchers have been investigating various bioreactor configurations to enhance mass transfer and improve the removal of VMS from biogas. BFs and BTFs have been studied for the removal of these biogas pollutants alone and with other biogas pollutants (Boada et al., 2020; Li et al., 2014; Pascual et al., 2020; Yang & Corsolini, 2019; Zhang et al., 2020). BTFs has been the most studied bioreactor configuration in literature for VMS removal. In this bioreactor, the VMS-laden biogas pass through a continuously irrigated packing material to which biofilm is attached. This configuration allows for good control of operational conditions and are commonly used for full-scale gas pollutant removal. On the other hand, BFs have been scarcely studied. BFs are composed of a packed bed where biofilm develops that is periodically irrigated by nutrients. The polluted gas stream passes through this relatively dry packing material that allows hydrophobic compounds to be more easily adsorbed onto the media for subsequent biodegradation compared to BTFs where the presence of a polar (trickling) liquid can hinder the adsorption of pollutants onto the media.

Among all the operating parameters, the contact time between the gas and the microorganisms has been highlighted as the most critical parameter in these bioreactors (Boada et al., 2021; Pascual et al., 2020, 2021b; Popat & Deshusses, 2008). Packed bed bioreactors measure this through the EBRT. Additionally, the EC and RE are key performance indicators for biological VMS removal technologies. Considering that fair comparisons of VMS removal performance should consider both EC and RE, a table with the best data obtained from all the studies reporting VMS removal from gas streams is shown in Table 1.

In view of Table 1, studies have shown that longer EBRT values are associated with better VMS mass transfer and higher removal performance in BTFs.

Pascual et al. (2021b) investigated the relationship between EBRT and RE in a BTF that treated D4, D5, L2, and L3. According to them, when the EBRT was reduced from 60 to 15 min, the total VMS RE dropped from 76 to 49%. Similar results were observed by Boada et al. (2021) who treated D4 and D5 in a BTF and reported a drop in RE from 40 to 15% when the EBRT was shortened from 14.5 to 4 min. In fact, according to Zhang et al. (2020), the EBRT required to achieve a high level of D4 and D5 elimination in aqueous conditions is between 2 and 4 h.

The research of VMS removal has been mainly studied in aerobic conditions (Table 1). The complete metabolic pathway for the degradation of VMS has not yet been confirmed. However, a number of reactions involved in the biological oxidation of VMS have been proposed, including the hydroxylation of D4 to dimethylsilanediol (DMSD) by bacterial enzymes and the generation of intermediates, such as silicic acid and methanol identified in the aerobic metabolism of D4 (Li et al., 2014). In any case, the degradation of VMS from biogas using aerobic conditions is dangerous because the introduction of air or oxygen into the biogas significantly increases the risk of explosion and dilutes the biogas, reducing its energy content. A possible solution to these problems would be to use anoxic conditions for VMS degradation. In this way, the microorganisms would use nitrate or nitrite as a source of electron acceptor instead of oxygen, which would not only eliminate the previously mentioned disadvantages of aerobic processes but would also simultaneously eliminate a NO_3^-/NO_2^- contaminated effluent. Although the use of NO_3^-/NO_2^- may increase operating costs, the ubiquity of nitrogen-rich effluents in biogas plants makes the anoxic degradation of VMS in biogas plants economically viable. Biological degradation of VMS under anoxic conditions has been reported in BTFs (Pascual et al., 2022;

Table 1 Comparison of biological VMS removal from gas streams. IC stands for inlet concentration, EBRT for empty bed residence time, EC_{max} for maximum reported elimination capacity and RE for removal efficiency

Type of bioreactor	Types of VMS[1]	Inlet concentration ($mg\ m^{-3}$)	EBRT (min)	VMS EC_{max} ($g\ m^{-3}\ h^{-1}$)	RE (%)	Dominant microorganisms or isolates	Reference
BTF (aerobic, pall rings)	D3	45–77	2.1–3.6	~0.20	10–20	*Pseudomonas*	Accettola et al. (2008)
BTF (porous ceramic)	D4					n.r	Popat and Deshusses (2008)
aerobic		~45	~5–19.5	0.03–0.06	~10–43		
anaerobic[2]		n.r	4	n.r	15		
BTF (aerobic, lava rock)	D4	20–110	3.3	0.44	48	*Pseudomonas aeruginosa*	Li et al. (2014)
		50	13.2	0.17	74		
		150	480		~15		
BF (compost) + activated carbon filter	D3	330–1190	n.r	n.r	98.1	n.r	Yang and Corsolini (2019)
	D4	73–170			96.6		
	D5	13–38			78.6		
BTF (anoxic, lava rock)	D4	54	870	0.03–0.06	< 13	n.r	Santos-Clotas et al. (2019)
	D5	102	14.5–4	0.14–0.23	37–16		
BTF (anoxic, lava rock)	D4	62	15	~0.03	12–14	*Pseudomonas aeruginosa* *Methylibium sp.*	Boada et al. (2020)
	D5			~0.15	~37		
BTF[2] (aerobic)	L2	~625	60	0.02–0.15	20–60	*Pseudoxanthomonas* *KMBC-112* *Reynarella* *Chitinophaga*	Pascual et al. (2020)
	L3			0.11–0.12	78–70		
	D4			0.17–0.14	81–91		
	D5			0.17–0.12	88–80		

(continued)

Table 1 (continued)

Type of bioreactor	Types of VMS[1]	Inlet concentration (mg m^{-3})	EBRT (min)	VMS EC_{max} (g m^{-3} h^{-1})	RE (%)	Dominant microorganisms or isolates	Reference
BTF[2] (aerobic, Kaldnes K1 Micro rings)	L2	~700–900	60–15	0.07–0.14	45–19	*KCM-B-112*	Pascual et al. (2021b)
	L3			0.13–0.24	81–31		
	D4			0.14–0.48	84–53		
	D5			0.21–0.95	87–69		
BTF[2] (anoxic, Kaldnes K1 Micro rings)	L2	138–239	60	~0.00–0.01	12–15	*Parvibaculum, Zarvaz inia KCM-B-11, Arenimonas Moraxellaceae Luteimonas*	Pascual et al. (2022)
	L3			~0.01–0.02	22–38		
	D4			~0.01–0.03	41–54		
	D5			~0.02–0.04	48–73		
BF (aerobic, woodchips + compost)	L3	500–1361	37.5	0.26	26.5	*Mycobacterium Rhodococcus Planctomicrobium Other VMS degraders such as Bacillus, Pseudomonas* and *Dokdonella*	González-Cortés et al. (2023)
	D4			0.37	30.4		
BF (aerobic, perlite)	L3			0.28	31.2	n.r	
	D4			0.23	32.8		
BF (aerobic, woodchips + compost)	L3	250–1354	37.5	0.17	21.7	*Pseudomonas, Sphingobium, Rhodococcus* and *Bacillus*	(González-Cortés et al., 2022)
	D4			0.41	28.4		
BF (aerobic, perlite)	L3			0.16	18.7	n.r	
	D4			0.41	28.8		

n.r.—Non reported
1—Hexamethyldisiloxane (L2), Octamethyltrisiloxane (L3), Hexamethylcyclotrisiloxane (D3), Octamethylcyclotetrasiloxane (D4) and Decamethylcyclopentasiloxane (D5)
2—Two phase partitioning
3—Electron acceptors (NO_3^-/SO_4^{2-})

Santos-Clotas et al., 2019). Although the reported ECs have generally been low, efforts should be made to optimize this process as it offers major advantages from an operational and environmental point of view.

Aiming to enhance VMS removal, researchers have explored other configurations with less amount of polar phase such as BFs in an effort to improve the mass transfer of these compounds to the biofilm. Yang and Corsolini (Yang & Corsolini, 2019) coupled a BF with an activated carbon filter, achieving high REs (78.6–98.1%) for D3, D4 and D5 compounds. The BF not only removed VMS but also partially removed water vapor from the biogas, enhancing subsequent VMS removal in the activated carbon filter. Other study operated two BFS packed with different packing materials (wood chips + compost (WC) and perlite) removing L3 and D4 (González-Cortés et al., 2022). They concluded that there was not a significant difference between both packing materials and the supplementation of a more biodegradable VOC in the IL such as toluene enhanced the EC of both BFs up to 0.58 g VMS m^{-3} h^{-1} (D4 EC = 0.41 g D4 m^{-3} h^{-1}, D4 RE = 28.4%; L3 EC = 0.17 g L3 m^{-3} h^{-1}, L3 RE = 21.7%).

While longer EBRT values are desirable for better removal performance, they result in larger working volumes, making them impractical for industrial-scale applications. Bioreactors used for biogas purification typically operate at shorter EBRT values, such as a few minutes. Efforts are still needed to reduce EBRT while achieving high VMS removal.

For this reason, recent efforts are being made to improve the mass transfer of VMS from the gas phase to the aqueous phase where the microorganisms are located. González-Cortés et al. (2023) studied the effect of adding surfactants (Tween 80 (1–3 CMC) and Saponin (1 CMC) to two biofilters with different packing materials (Woodchips + compost and perlite) removing D4 and L3. They demonstrated how the addition of 1 CMC of Tween 80 and 1 CMC of Saponin enhanced the removal of both VMS. On the other hand, Pascual et al. (2020) studied the effect of adding an organic phase such as silicone oil to the trickling liquid of a BTF fed by L2, L3, D4 and D5. They concluded that the addition of silicone oil (30%) improved the RE the BTF from < 30% to ~ 70% obtaining the highest EC (1.80 g VMS m^{-3} h^{-1}; RE = 49%) of VMS at the date of writing of this book chapter.

In short, the biological removal of VMS is promising but efforts must be made to improve the mass transfer of VMS from the biogas to the liquid phase. In addition, the study of the process under anoxic conditions is little studied and the use of bioscrubbers has not yet been reported.

3.3 *Biological Removal of VOCs (Monoaromatic and Linear Hydrocarbons, Terpenes, Halogenated Compounds and Organosulfur Compounds*

Apart from VMS, various types of VOCs have been identified in raw biogas from municipal and sewage sludge digestion. These include monoaromatic hydrocarbons like toluene and p-cymene, linear hydrocarbons (generally C9-C12) such as undecane, terpenes like α-pinene, halogenated compounds including iodomethane, and organosulfur compounds like carbon disulfide (Paolini et al., 2018).

Although the presence of some VOCs may be relatively harmless, the presence of others can significantly limit the uses to which biogas can be applied. For example, the organosulfur compounds, such as carbonyl sulfide, carbon disulfide, dimethyl sulfide, and must be stringently removed prior to emerging applications of biogas, such as its use as fuel in SOFCs (Gaj, 2017) or in catalytic methanation reactors (Dannesboe et al., 2021), to avoid sulfur poisoning of catalysts. On the other hand, the presence of halogenated compounds (solvents) in biogas can lead to the generation of corrosive acids during combustion (Scheutz et al., 2004), restriction in combustion reactions and the emission of toxic by-products during combustion, such as polyhalogenated dioxins and furans (Rey et al., 2013).

Different bioreactor configurations, such as BFs, BTFs, BSs have been utilized for VOCs removal. Most studies have focused on waste air streams, and very few investigations reports the biological removal of VOCs from raw biogas (Franco-Morgado et al., 2018; Lakhouit et al., 2016).

Although BFs may appear better from a mass transfer point of view because they lack trickling liquid continuously irrigating the packing material, their operation under anaerobic or anoxic conditions is not simple and the oxygen content of the biogas is usually low (0.5–3%). In addition, the addition of air to the BF would not be desirable (increased risks of explosion and biogas dilution). Therefore, the removal of these compounds using BTFs or BSs would be the most suitable option as these configurations allow to operate anoxically or aerobically in the case of BS, in which an air stream may be supplied to the stirred tank without affecting the treated biogas.

Despite the biological removal of VOCs depends on several factors (i.e. presence of a suitable microbial community, type of VOC mixture, etc.) the removal performance of a bioreactor removing VOCs can be estimated by the VOC bioavailability in the biofilm phase. This bioavailability can be determined mainly by their Henry's law coefficient (solubility in water). Therefore, VOCs can be classified based on their dimensionless Henry's law coefficient, with VOCs having a large coefficient ($H > 0.10$ at 25 °C) being limited by poor solubility in water (<500 g m^{-3}), while VOCs with low coefficients ($H < 0.10$ at 25 °C) are often kinetically limited due to their high miscibility in water and ease of transfer into the biofilm phase (Lamprea Pineda et al., 2021).

Among the bacterial strains that have been reported to play a significant role in the biodegradation of VOCs in BTFs some can be highlighted such as *Pseudomonas putida*, *Rhodococcus erythropolis* and *Cladosporium sphaerospermum* for benzene, toluene, ethylbenzene and xylene (BTEX); *Burkholderia sp.* strain T3 and *Bacillus cereus S1* and *S2* for toluene; *Pseudomonas oleovorans* DT4 for ethylbenzene; *Ralstonia pickettii L2* for chlorobenzene; *Ralstonia eutropha* for styrene; and *Bacillus cereus GIGAN2*, *Pseudomonas putida* and *Pseudomonas acidovorans* for dimethyl disulfide (Das et al., 2022; Wu et al., 2018). Fungi, particularly those with a filamentous structure like *Fusarium sp.*, are also capable of degrading VOCs and exhibit efficient mass transfer of hydrophobic VOCs even at low humidity (Mezgebe et al., 2019).

4 Biogas Upgrading

4.1 *Chemoautotrophic Upgrading Using Hydrogenotrophic Methanogens (CO_2 to CH_4 Using H_2)*

Chemoautotrophic upgrading using hydrogenotrophic methanogens is an innovative technology that shows great promise in the field of biogas upgrading. This process involves the conversion of carbon dioxide (CO_2) into methane (CH_4) by utilizing hydrogen (H_2) as an energy source (Eq. 16) (Zabranska & Pokorna, 2018).

$$4H_2 + CO_2 \rightarrow CH_4 + 2H_2O \ \Delta G^{\circ'} = -130.7 \text{ KJ mol}^{-1} \quad (14)$$

In the context of chemoautotrophic biogas upgrading, H_2 is typically obtained through water electrolysis using renewable energy sources, such as wind or solar, what convert these intermittent energy sources into more stable energy sources (biomethane) that can be easily stored (Farghali et al., 2022). In this section, we will explore the three distinct types of upgrading configurations for chemoautotrophic biogas upgrading technology: ex-situ, in situ, and hybrid-situ systems (Fig. 4).

In-situ technologies. In situ biogas upgrading systems involve the direct injection of hydrogen into the anaerobic digester. This method stimulates the activity of hydrogenotrophic methanogens, such as *Methanobacterium*, *Methanosarcina*, and *Methanosaeta*, present in the digester (Kim et al., 2021). The delivered hydrogen enhances the utilization and conversion of endogenous CO_2 into biomethane. However, in situ hydrogen delivery systems face several challenges due to the conditions in which anaerobic digesters are operated, including low hydrogen solubility (Rafrafi et al., 2021), low hydrogen utilization rates (Zhu et al., 2019) and increased system pH (Sarker et al., 2018). These challenges need to be addressed to optimize the performance of in situ systems. Despite these challenges, in situ systems offer the advantage of simplicity and direct integration with the anaerobic digestion process, eliminating the need for separate reactors.

Ex-situ technologies. In contrast, ex-situ biogas upgrading systems involve the extraction of biogas from the anaerobic digestion process and its subsequent upgrading in a separate reactor in which H_2 is injected. The hydrogenotrophic methanogens present in the anaerobic digestate play a crucial role in this process consuming the injected hydrogen gas and converting CO_2 into CH_4. Various reactor configurations have been employed in ex-situ systems, including two-stage continuous stirred reactors (Bassani et al., 2015), up-flow reactors (Bassani et al., 2017; Kougias et al., 2017), and BTFs (Dupnock & Deshusses, 2019; Rachbauer et al., 2016). These configurations have shown high biomethane production rates, reaching

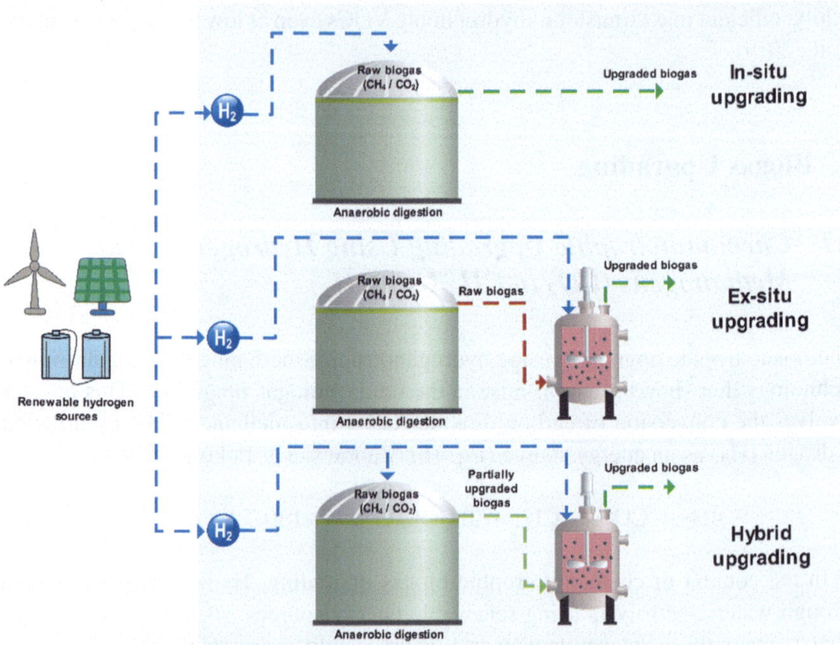

Fig. 4 Configurations studied to perform the chemoautotrophic biogas upgrading

up to 98% and hydrogen utilization rate ranging from 93 to 100%, resulting in a significant increase in biomethane content. These good results are derived from the fact that ex-situ systems offer greater flexibility and control over the upgrading process, which may have long-term operational benefits. This makes ex-situ systems a promising means of biogas upgrading, offering a potential substitute for conventional upgrading technologies. However, ex-situ upgrading require additional infrastructure and equipment what result in an increase of the capital and operational costs compared to in situ systems.

Hybrid technologies. Hybrid-situ systems, bring together the best features of ex-situ and in situ biogas upgrading technologies. By combining these two approaches, hybrid systems aim to overcome the limitations of each method and maximize the overall benefits (Corbellini et al., 2018) by merging the in-situ conversion of CO_2 to biomethane with the biomethanation of residual CO_2 in a separate unit. This integration offers several advantages, including the potential to reduce the low mass transfer of unused hydrogen. Although the hybrid approach is still largely conceptual and lacks extensive experimental data, it holds significant promise for future research and development.

4.2 Photosynthetic Biogas Upgrading

These processes are based on the ability of microalgae to utilize CO_2 as a carbon source. Generally, microalgae can utilize approximately between 1.8 and 2.5 g of CO_2 per gram of biomass produced. However, this value is subject to variation owing to different factors, such as the specific microalgal strain, nutrient availability, and environmental cultivation conditions. Factors such as nitrogen abundance and the carbon-to-nitrogen ratio can also influence this relationship. Nonetheless, the significant capacity of microalgae to assimilate substantial amounts of CO_2 while producing biomass with versatile applications makes them candidates for biogas upgrading (Pulz & Gross, 2004).

The optimal growth temperature range for microalgae is typically between 15 and 25 °C. Seasonal changes can lead to a decrease in their activity or may require seasonal operation, depending on the location of the plant. Microalgae require significant amounts of nutrients, and the use of synthetic or fertilizer-based media can severely affect process profitability. Therefore, utilization of nutrients from wastewater, landfill leachate, or reject water from the digestate drying process is preferred. However, it should be noted that the origin of nutrient sources could greatly limit the potential uses of the biomass obtained from microalgae growth owing to possible contamination with heavy metals, pathogens, or other pollutants. Most microalgae thrive within a growth pH range of 6.5–10. The use of alkaliphilic microalgal species with an optimum pH of approximately 10 is preferable. Operating at elevated pH improves the CO_2 solubility in the medium from biogas, minimize the risk of contamination and provides a higher buffering capacity. For this purpose, it is highly desirable to use water with high alkalinity to buffer cost-effectively the rapid decrease in pH caused by the absorption of CO_2 in the medium. However, basic media may lead to a loss of nitrogen as NH_3 and the precipitation of phosphorus as calcium and magnesium salts (Bose et al., 2019).

Biogas upgrading with microalgae can be carried out in either a single- or two-stage process. In the single-stage approach, biogas is passed through a photobioreactor, allowing for simultaneous absorption and assimilation of CO_2 by microalgae within the same equipment. Alternatively, in the two-stage process, there is an initial absorption stage, followed by the subsequent assimilation of inorganic carbon in the photobioreactor. It is crucial to consider the generation of O_2 during photosynthetic reactions (Eq. 17). For every mole of CO_2 assimilated, one mole of O_2 is produced. To avoid inhibitory concentrations on microalgal activity and to prevent the replacement of CO_2 with O_2 in the outlet biogas stream, it is necessary to remove part of the dissolved oxygen. In single-stage systems, O_2 can be stripped by feeding biogas, air, or N_2 into the photobioreactor. However, the implementation of this method poses several challenges for industrial application. Biogas upgrading occurs in batches rather than continuously, no complete desorption of O_2 from the photobioreactor occurs, and mixing between biogas and air during transition periods leads to biogas dilution and CH_4 emission with air. An alternative option involves supplying biogas on demand, based on the target pH values. This approach has shown potential CO_2

REs ranging from 30 to 50% (Acién Fernández et al., 2012).

$$6H_2O + 6CO_2 \overset{\text{light}}{\rightarrow} C_6H_{12}O_6 + 6O_2 \tag{15}$$

The use of two-stage systems does not completely solve the problem of biogas stream enrichment with O_2, as dissolved oxygen concentrations typically reach saturation levels when air is introduced into the photobioreactor. Therefore, recirculating a portion of the culture medium to the absorption unit results in the desorption of some of the contained O_2 in the biogas outlet stream. However, this issue can be minimized by operating at minimal gas/liquid ratios in the absorber without compromising the efficiency. A viable alternative is employing symbiotic cultivation of microalgae and aerobic bacteria. This approach allows the generated O_2 to be consumed through aerobic processes by bacteria. For instance, Bahr et al. (2014) conducted research utilizing a consortium of microalgae and sulfur-oxidizing bacteria. Their study demonstrated simultaneous CO_2 assimilation and removal of H_2S from biogas, achieving 100% RE for H_2S and 90% for CO_2, while maintaining a final concentration of O_2 in biogas of 0.2%.

Average biomass productivity ranges from 15 to 22 g m^{-2} d^{-1} (Bose et al., 2019). It is important to consider this in terms of energy efficiency per unit area because incident light serves as the primary source of energy for biomass formation and CO_2 assimilation. By performing a simple calculation, it is possible to estimate the required surface area based on the biogas flow rate and microalgae composition. For instance, if we suppose a biogas flow rate of 1 m^3 h^{-1} with a CO_2 composition of 40% v/v, it would be necessary to have a surface area of approximately 314 m^2 under the best-case scenario operating at a GRT of 94 h.

5 Biogas and Biomethane Applications

Biogas is a renewable energy source generated via the anaerobic digestion of organic waste materials. It offers a convenient and easily combustible energy option due to its significant methane (CH_4) content. However, the presence of carbon dioxide (CO_2), which does not readily burn, reduces the calorific value of biogas and imposes limitations on its potential applications. Therefore, the suitability of biogas for various applications may necessitate pretreatment steps to address the specific requirements and optimize its use, depending on the intended application. The different uses of biogas, biomethane, and the purified CO_2 (bio-CO_2) are depicted in Fig. 5

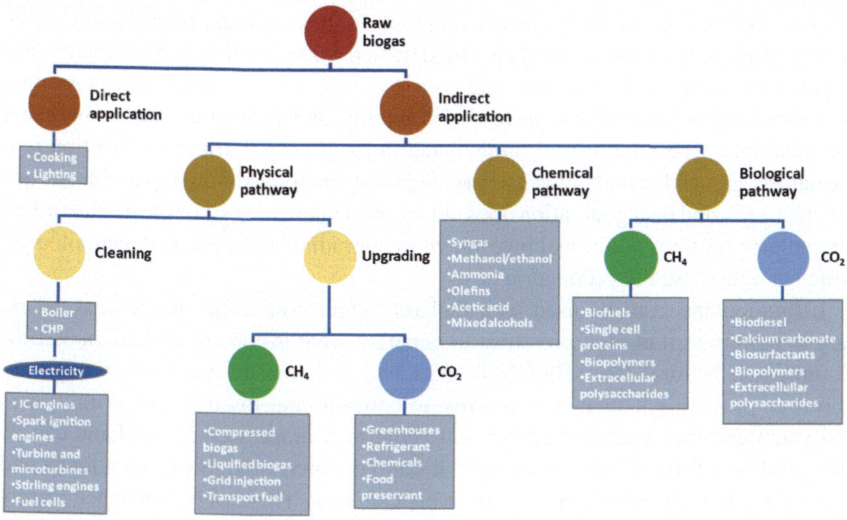

Fig. 5 Diagrammatic overview of the utilization pathways available for biogas, biomethane, and bio-CO_2. Adapted from Kapoor et al. (2020)

5.1 Valorization of Biogas as a Renewable Energy Vector

The direct use of biogas in its crude form is a simple and cost-effective way to utilize the energy produced from organic waste. It can be used for cooking, lighting, heating and power generation, providing a sustainable alternative to traditional fossil fuels. This direct use of biogas is particularly relevant in developing countries, where access to modern energy sources is limited, and the use of traditional biomass, charcoal or kerosene for cooking, heating or lighting is prevalent. In these contexts, the use of biogas in its crude form can have significant social and environmental benefits (Firmo et al., 2022).

One of the most common ways of using biogas in developed countries is its direct use in combined heat and power (CHP) systems where the heat generated during electricity production is captured and utilized for heating purposes. This maximizes the energy efficiency of biogas utilization (up to 90%) compared to the lower efficiency of separate electricity and heat generation systems saving costs, reducing the environmental impact of the energy production, and providing a level of energy independence by generating on-site electricity and heat. This independence can be particularly beneficial for industries, commercial buildings, and agricultural operations that require a reliable and continuous energy supply. Furthermore, CHP systems can provide support to the electrical grid during peak demand periods alleviating strain on the grid and contributing to grid stability. In addition, these units can operate with biogas laden with H_2S up to 500 ppm_v which, despite its removal being desired to extend the working life of the CHP systems, makes the previous purification stage requirements less exhaustive (Firmo et al., 2022).

However, aiming to offer a wider range of applications and higher energy efficiency, biogas can be purified and upgraded to increase the methane content removing impurities such as carbon dioxide and corrosive constituents. In this way, a gas that is comparable to natural gas in terms of quality and properties can be obtained (biomethane). The main use of biomethane in its gaseous form is its injection into the natural gas grid, which allows a very wide distribution. Although possible, its use for electricity and heat generation does not seem so interesting since the production of biomethane requires additional investment in upgrading and treatment technologies, which can increase the production cost.

If biomethane is to be used as a fuel for vehicles or to be stored, it is necessary to compress it at high pressures to generate what is commonly known as Bio-Compressed Natural Gas (BioCNG). BioCNG is primarily used as a transportation fuel, providing a cleaner alternative to conventional fossil fuels for light (cars and vans) and heavy (trucks and buses) vehicles. Compressed biomethane can be also used as a form of energy storage. It can be stored in high-pressure tanks and used as needed to generate electricity or provide heat. This can help balance energy supply and demand, particularly in situations where renewable energy sources have intermittent generation.

The use of processed forms of biogas requires additional investment in upgrading and treatment technologies, which can increase the production cost. However, the higher energy efficiency and wider range of applications can provide significant economic and environmental benefits in the long run.

5.2 Biological Valorization of Biogas.

Biogas can be also used as substrate for microorganisms to convert their CH_4 and CO_2 into valuable chemicals and products. This section will provide an overview of the main uses of the CH_4 and CO_2 contained in biogas for its biotransformation. The biovalorization of methane derived from biogas involves the conversion of methane into high-value products such as biopolymers, biofuels, ectoine, and single-cell proteins (SCPs) through biological processes. This process utilizes specific microorganisms known as methanotrophs, which have the ability to oxidize methane and transform it into various valuable compounds. Methanotrophs are gram-negative bacteria that act as biological sinks for methane. They utilize methane as a carbon and energy source through two main pathways: the ribulose monophosphate (RUMP) pathway and the serine pathway. These pathways involve the enzymatic oxidation of methane, resulting in the production of CO_2, biomass, and various bioproducts.

One of the key products obtained through the biovalorization of methane is biopolymers. Methanotrophs can produce biopolymers such as polyhydroxyalkanoates (PHA) using methane as a carbon source. PHA is a biodegradable polymer that has applications in various industries, including packaging, agriculture, and biomedical fields. Biofuels are another valuable product derived from the biovalorization of methane. Methanotrophs can produce biofuels such as methane-derived methanol and methane-derived biodiesel which also serve as a sustainable alternative to fossil fuels. Unlike biomethane, these biofuels do not require compression for its storage, making it a viable option for energy storage and distribution, particularly in areas with limited access to traditional energy infrastructure.

In addition to biopolymers and biofuels, methanotrophs can also produce other high-value compounds. For example, they can synthesize ectoine, a cyclic imino acid synthesized by halotolerant bacteria to maintain osmotic balance in high-salt environments, which has applications in pharmaceutical and cosmetic industries. Methanotrophs can also produce SCP through the conversion of methane into microbial biomass. SCPs are rich in essential amino acids and can be used as nutritional supplements or animal feed with the potential to reduce reliance on traditional protein sources such as meat and soy.

Generally, the biotransformation of methane derived from biogas into high-value bioproducts offers several advantages, such as its contribution to the reduction of methane emissions, while offering economic opportunities and diversification in various industries. However, there are challenges associated with the biovalorization of methane. Methanotrophs have specific growth requirements and can be sensitive to environmental conditions so the optimization of cultivation conditions, such as temperature, pH, and nutrient availability, is necessary to enhance their productivity. Additionally, the scalability and cost-effectiveness of the biovalorization process and the subsequent downstream steps need to be further explored and improved for large-scale implementation.

On the other hand, CO_2 valorization through biological methods also offers an opportunity to convert CO_2 into biomass and various value-added products what promotes eco-friendly and sustainable development. CO_2 fixing microbes, including bacteria, cyanobacteria, and algae, are equipped with enzymatic machinery such as Ribulose-1,5-bisphosphate carboxylase/oxygenase (RuBisCO) and carbonic anhydrase (CA). These microbes play a vital role in sequestering CO_2 and transforming it into various products including biodiesel, calcium carbonate ($CaCO_3$) through biomineralization, biosurfactants, polyhydroxyalkanoates (PHA), and extracellular polymeric substances (EPS) (Kapoor et al., 2020). These biomaterials can act as substitutes for conventional materials, reducing reliance on fossil fuels, synthetic plastics, and chemically synthesized surfactants, and flocculants.

Biodiesel production and research have predominantly centered around the utilization of microalgae as a promising feedstock due to its high oil content and potential for sustainable cultivation. Nonetheless, despite extensive investigations, the commercial viability and profitability of microalgae-based biodiesel production still pose significant hurdles such as high production costs, low lipid extraction efficiency and scale-up challenges. To address these challenges, researchers have explored alternative approaches that involve the use of CO_2-sequestering microorganisms like *Serratia* sp. ISTD04 and *Bacillus* sp. SS105, which have been reported as environmentally friendly alternatives to conventional diesel (Bharti et al., 2014; Maheshwari et al., 2018).

In addition to biodiesel, CO_2-sequestering bacteria have the potential to produce other bio-products. Biosurfactants, generated by microorganisms, have diverse applications and, while traditionally produced using substrates such as fats, vegetable oils, carbohydrates, and glycerol, their production by CO_2 sequestering bacteria, specifically *Bacillus* sp., has been recently reported (Maheshwari et al., 2017). Microbial strains capable of calcification, including cyanobacteria, microalgae, and various bacteria, contribute to $CaCO_3$ precipitation from the biomineralization of CO_2 (Bharti et al., 2014). PHA accumulation, a survival strategy for microorganisms, and EPS production, influenced by nutrient conditions, temperature, pH, and microbial type, also offer potential avenues for CO_2 sequestration and utilization (Kumar et al., 2019).

Overall, the biovalorization of CH_4 and CO_2 provides a promising approach to not only mitigate their emissions but also generate valuable products, contributing to a more sustainable and environmentally friendly future. However, continued research, technological advancements, and supportive policies are essential to further unlock the benefits of this approach and accelerate its adoption in various sectors.

6 Conclusions and Future Directions for Research on Biological Biogas Purification and Upgrading

As outlined in this chapter, biological processes have numerous advantages, such as lower operational expenditures (OPEX) and environmental impact. For these reasons, in the medium term, biological desulfurization processes will undoubtedly be more widespread than physicochemical processes. Many laboratory-scale studies have been carried out, but pilot-scale studies are still scarce, and there is a need to increase the technology readiness level (TRL) of biological technologies. It is essential to operate pilot-and/or demonstration-scale plants to determine bottlenecks that are impossible to identify at the laboratory scale as well as to obtain a better estimation of environmental impacts and OPEX. At the laboratory scale, there are gaps mainly related to the avoidance of elemental sulfur accumulation in fixed bed bioreactors and the separation of elemental sulfur from the liquid medium in suspended growth bioreactors. Improved mathematical models are also required to implement feedforward control systems to anticipate responses to external perturbations. The development of microsensors to measure sulfide, nitrate, and nitrite in biofilm bioreactors is necessary to improve these mathematical models. In relation to microbial characterization, this has already been carried out in most of the studies using NGS, but it is necessary to deepen the understanding of the role of microorganisms and the effects of operational variables and/or operation failures on these communities. It is also essential to define the application ranges of each technology according to the biogas composition, the flowrate to be treated, the maximum outlet H_2S concentration, and, in the case of anoxic systems, to determine the characteristics of the water to be nitrified (ammonium concentration, conductivity, pH, etc.) or the characteristics of the nitrified water (NO_3^-/NO_2^- concentration, pH, etc.). Another important factor is temperature control in places with high temperatures during summer or very low temperatures in winter.

Regarding the removal of VMS, the REs at the laboratory scale are still low, and the EBRTs are very high, which is necessary to improve this aspect before increasing the TRL. It is necessary to improve the gas–liquid mass transfer and determine populations with high degradation capacity. In this sense, it may be difficult to compete in the medium-to long-term with activated carbon filters. The elimination of VOCs is not a priority for biogas stakeholders; however, regulations will likely become more restrictive in the future, which will require the availability of technologies for their elimination.

In relation to biogas upgrading, the use of green H_2 will encourage an increase in studies utilizing hydrogenotrophic microorganisms. It is essential to improve the gas–liquid transfer of H_2 and to increase the number of studies at the laboratory and pilot scales. Biogas upgrading with microalgae also requires decreasing the gas residence times and low-cost alternatives to buffer the pH drop due to CO_2 absorption.

Biogas upgrading will be predominant in the medium-to short-term, pushing the biogas industry towards biomethane production. Thus, we are at the beginning of the Golden Age of biomethane production, and new uses of biomethane are continually emerging. Therefore, it is necessary to increase the TRL of bioreactors that use methane as a carbon source to obtain added-value products. In the long term, biogas producers with large volumetric production capacities can become producers of biocommodities under the concepts of biorefinery and circular economy.

References

Abdelkareem, M. A., Tanveer, W. H., Sayed, E. T., Assad, M. E. H., Allagui, A., & Cha, S. W. (2019). On the technical challenges affecting the performance of direct internal reforming biogas solid oxide fuel cells. *Renewable and Sustainable Energy Reviews, 101*, 361–375. https://doi.org/10.1016/j.rser.2018.10.025

Accettola, F., Guebitz, G. M., & Schoeftner, R. (2008). Siloxane removal from biogas by biofiltration: Biodegradation studies. *Clean Technologies and Environmental Policy, 10*, 211–218. https://doi.org/10.1007/s10098-007-0141-4

Acién Fernández, F. G., González-López, C. V., Fernández Sevilla, J. M., & Molina Grima, E. (2012). Conversion of CO_2 into biomass by microalgae: How realistic a contribution may it be to significant CO_2 removal? *Applied Microbiology and Biotechnology, 96*, 577–586. https://doi.org/10.1007/s00253-012-4362-z

Aghel, B., Behaein, S., Wongwises, S., & Shadloo, M. S. (2022). A review of recent progress in biogas upgrading: With emphasis on carbon capture. *Biomass and Bioenergy, 160*, 106422. https://doi.org/10.1016/j.biombioe.2022.106422

Almenglo, F., Bezerra, T., Lafuente, J., Gabriel, D., Ramírez, M., & Cantero, D. (2016). Effect of gas-liquid flow pattern and microbial diversity analysis of a pilot-scale biotrickling filter for anoxic biogas desulfurization. *Chemosphere, 157*, 215–223. https://doi.org/10.1016/j.chemosphere.2016.05.016

Almenglo, F., González-Cortés, J. J., Ramírez, M., & Cantero, D. (2023). Recent advances in biological technologies for anoxic biogas desulfurization. *Chemosphere*, 138084.https://doi.org/10.1016/j.chemosphere.2023.138084

Andersson, F. A. T., Karlsson, A., Svensson, B. H., & Ejlertsson, J. (2004). Occurrence and abatement of volatile sulfur compounds during biogas production. *Journal of Air Waste Manage Association, 54*, 855–861. https://doi.org/10.1080/10473289.2004.10470953

Awe, O. W. (2017). A review of biogas utilisation, purification and upgrading technologies. *Waste and Biomass Valorization, 8*, 267–283. https://doi.org/10.1007/s12649-016-9826-4

Azizi, S. M. M., Zakaria, B. S., Haffiez, N., Niknejad, P., & Dhar, B. R. (2022). A critical review of prospects and operational challenges of microaeration and iron dosing for in-situ biogas desulfurization. *Bioresource Technology Reports, 20*, 101265. https://doi.org/10.1016/j.biteb.2022.101265

Bahr, M., Díaz, I., Dominguez, A., González Sánchez, A., & Muñoz, R. (2014). Microalgal-biotechnology as a platform for an integral biogas upgrading and nutrient removal from anaerobic effluents. *Environmental Science and Technology, 48*, 573–581. https://doi.org/10.1021/es403596m

Bassani, I., Kougias, P. G., Treu, L., & Angelidaki, I. (2015). Biogas upgrading via hydrogenotrophic methanogenesis in two-stage continuous stirred tank reactors at mesophilic and thermophilic conditions. *Environmental Science and Technology, 49*, 12585–12593. https://doi.org/10.1021/acs.est.5b03451

Bassani, I., Kougias, P. G., Treu, L., Porté, H., Campanaro, S., & Angelidaki, I. (2017). Optimization of hydrogen dispersion in thermophilic up-flow reactors for ex situ biogas upgrading. *Bioresource Technology, 234*, 310–319. https://doi.org/10.1016/j.biortech.2017.03.055

Bharti, R. K., Srivastava, S., & Thakur, I. S. (2014). Proteomic analysis of carbon concentrating chemolithotrophic bacteria *Serratia* sp. for sequestration of carbon dioxide. *PLoS One, 9*, e91300. https://doi.org/10.1371/journal.pone.0091300

Boada, E., Santos-Clotas, E., Bertran, S., Cabrera-Codony, A., Martín, M. J., Bañeras, L., & Gich, F. (2020). Potential use of Methylibium sp. as a biodegradation tool in organosilicon and volatile compounds removal for biogas upgrading. *Chemosphere*, 240. https://doi.org/10.1016/j.chemosphere.2019.124908

Boada, E., Santos-Clotas, E., Cabrera-Codony, A., Martín, M. J., Bañeras, L., & Gich, F. (2021). The core microbiome is responsible for volatile silicon and organic compounds degradation during anoxic lab scale biotrickling filter performance. *Science of the Total Environment, 798*, 149162. https://doi.org/10.1016/j.scitotenv.2021.149162

Bose, A., Lin, R., Rajendran, K., O'Shea, R., Xia, A., & Murphy, J. D. (2019). How to optimise photosynthetic biogas upgrading: A perspective on system design and microalgae selection. *Biotechnology Advances, 37*, 107444. https://doi.org/10.1016/j.biotechadv.2019.107444

Bragança, I., Sánchez-Soberón, F., Pantuzza, G. F., Alves, A., & Ratola, N. (2020). Impurities in biogas: Analytical strategies, occurrence, effects and removal technologies. *Biomass and Bioenergy, 143*, 105878. https://doi.org/10.1016/j.biombioe.2020.105878

Bu, H., Carvalho, G., Yuan, Z., Bond, P., & Jiang, G. (2021). Biotrickling filter for the removal of volatile sulfur compounds from sewers: A review. *Chemosphere, 277*, 130333. https://doi.org/10.1016/j.chemosphere.2021.130333

Cano, P. I., Colón, J., Ramírez, M., Lafuente, J., Gabriel, D., & Cantero, D. (2018). Life cycle assessment of different physical-chemical and biological technologies for biogas desulfurization in sewage treatment plants. *Journal of Cleaner Production, 181*, 663–674. https://doi.org/10.1016/j.jclepro.2018.02.018

Cattaneo, C. R., Muñoz, R., Korshin, G. V., Naddeo, V., Belgiorno, V., & Zarra, T. (2023). Biological desulfurization of biogas: A comprehensive review on sulfide microbial metabolism and treatment biotechnologies. *Science of the Total Environment, 893*, 164689. https://doi.org/10.1016/j.scitotenv.2023.164689

Chen, Y., Cheng, J. J., & Creamer, K. S. (2008). Inhibition of anaerobic digestion process: A review. *Bioresource Technology, 99*, 4044–4064. https://doi.org/10.1016/j.biortech.2007.01.057

Corbellini, V., Kougias, P. G., Treu, L., Bassani, I., Malpei, F., & Angelidaki, I. (2018). Hybrid biogas upgrading in a two-stage thermophilic reactor. *Energy Convers Manag, 168*, 1–10. https://doi.org/10.1016/j.enconman.2018.04.074

Dannesboe, C., Hansen, J. B., & Johannsen, I. (2021). Removal of sulfur contaminants from biogas to enable direct catalytic methanation. *Biomass Convers Biorefinery, 11*, 1823–1834. https://doi.org/10.1007/s13399-019-00570-7

Das, J., Ravishankar, H., & Lens, P. N. L. (2022). Biological biogas purification: Recent developments, challenges and future prospects. *Journal of Environmental Management, 304*, 114198. https://doi.org/10.1016/j.jenvman.2021.114198

Dupnock, T. L., & Deshusses, M. A. (2019). Detailed investigations of dissolved hydrogen and hydrogen mass transfer in a biotrickling filter for upgrading biogas. *Bioresource Technology, 290*, 121780. https://doi.org/10.1016/j.biortech.2019.121780

Farghali, M., Osman, A. I., Umetsu, K., & Rooney, D. W. (2022). Integration of biogas systems into a carbon zero and hydrogen economy: A review. *Environmental Chemistry Letters, 20*, 2853–2927. https://doi.org/10.1007/s10311-022-01468-z

Firmo, A. L. B., Silva, F. M. S., de F. S. Alves, I. R., de Brito, E. P. L., & da Silva, L. C. S. (2022). Biogas, biomethane and BioCNG: Definitions, technologies and solutions. In *Production technologies for gaseous and solid biofuels* (pp. 1–31). Wiley.

Franco-Morgado, M., Toledo-Cervantes, A., González-Sánchez, A., Lebrero, R., & Muñoz, R. (2018). Integral (VOCs, CO_2, mercaptans and H_2S) photosynthetic biogas upgrading using innovative biogas and digestate supply strategies. *Chemical Engineering Journal, 354*, 363–369. https://doi.org/10.1016/j.cej.2018.08.026

Gaj, K. (2017). Applicability of selected methods and sorbents to simultaneous removal of siloxanes and other impurities from biogas. *Clean Technologies and Environmental Policy, 19*, 2181–2189. https://doi.org/10.1007/s10098-017-1422-1

González-Cortés, J. J., Lamprea-Pineda, P. A., Ramírez, M., Demeestere, K., & Walgraeve, C. (2022). Enhancement of siloxanes biofiltration from waste air by toluene addition. In *1st International Conference on Sustainable Chemical and Environmental Engineering*. Rethymno, Crete, Greece.

González-Cortés, J. J., Lamprea-Pineda, P. A., Ramírez, M., Langenhove, H. Van, Demeestere, K., & Walgraeve, C. (2023). Impact of surfactant addition on biofiltration of siloxane loaded waste gas. *In 5th Euro-Mediterranenan Conference For Environmental Integration*. Rende, Italy.

González-Cortés, J. J., Torres-Herrera, S., Almenglo, F., Ramírez, M., Cantero, D. (2021). Anoxic biogas biodesulfurization promoting elemental sulfur production in a continuous stirred tank bioreactor. *Journal of Hazardous Materials, 401*.https://doi.org/10.1016/j.jhazmat.2020.123785

Hosseini, S. E., & Wahid, M. A. (2013). Biogas utilization: Experimental investigation on biogas flameless combustion in lab-scale furnace. *Energy Convers Management, 74*, 426–432. https://doi.org/10.1016/j.enconman.2013.06.026

Kapoor, R., Ghosh, P., Tyagi, B., Vijay, V. K., Vijay, V., Thakur, I. S., Kamyab, H., Nguyen, D. D., & Kumar, A. (2020). Advances in biogas valorization and utilization systems: A comprehensive review. *Journal of Cleaner Production, 273*, 123052. https://doi.org/10.1016/j.jclepro.2020.123052

Kim, S., Mostafa, A., Im, S., Lee, M.-K., Kang, S., Na, J.-G., & Kim, D.-H. (2021). Production of high-calorific biogas from food waste by integrating two approaches: Autogenerative high-pressure and hydrogen injection. *Water Research, 194*, 116920. https://doi.org/10.1016/j.watres.2021.116920

Kougias, P. G., Treu, L., Benavente, D. P., Boe, K., Campanaro, S., & Angelidaki, I. (2017). Ex-situ biogas upgrading and enhancement in different reactor systems. *Bioresource Technology, 225*, 429–437. https://doi.org/10.1016/j.biortech.2016.11.124

Krayzelova, L., Bartacek, J., Díaz, I., Jeison, D., Volcke, E. I. P., & Jenicek, P. (2015). Microaeration for hydrogen sulfide removal during anaerobic treatment: A review. *Reviews in Environmental Science and Bio/Technology, 14*, 703–725. https://doi.org/10.1007/s11157-015-9386-2

Kumar, M., Kumar, M., Pandey, A., & Thakur, I. S. (2019). Genomic analysis of carbon dioxide sequestering bacterium for exopolysaccharides production. *Science and Reports, 9*, 4270. https://doi.org/10.1038/s41598-019-41052-0

Lakhouit, A., Cabral, A. R., & Cabana, H. (2016). Two Novel Biofilters to remove volatile organic compounds emitted by landfill sites. *Water, Air, and Soil Pollution, 227*, 113. https://doi.org/10.1007/s11270-016-2813-3

Lamprea Pineda, P. A., Demeestere, K., Toledo, M., Van Langenhove, H., & Walgraeve, C. (2021). Enhanced removal of hydrophobic volatile organic compounds in biofilters and biotrickling filters: A review on the use of surfactants and the addition of hydrophilic compounds. *Chemosphere, 279*, 130757. https://doi.org/10.1016/j.chemosphere.2021.130757

Li, Y., Zhang, W., & Xu, J. (2014). Siloxanes removal from biogas by a lab-scale biotrickling filter inoculated with *Pseudomonas aeruginosa* S240. *Journal of Hazardous Materials, 275*, 175–184. https://doi.org/10.1016/j.jhazmat.2014.05.008

López, L. R., Dorado, A. D., Mora, M., Gamisans, X., Lafuente, J., & Gabriel, D. (2016). Modeling an aerobic biotrickling filter for biogas desulfurization through a multi-step oxidation mechanism. *Chemical Engineering Journal, 294*, 447–457. https://doi.org/10.1016/j.cej.2016.03.013

Maheshwari, N., Kumar, M., Thakur, I. S., & Srivastava, S. (2018). Carbon dioxide biofixation by free air CO_2 enriched (FACE) bacterium for biodiesel production. *Journal of CO_2 Utilization, 27*, 423–432. https://doi.org/10.1016/j.jcou.2018.08.010

Maheshwari, N., Kumar, M., Thakur, I. S., & Srivastava, S. (2017). Recycling of carbon dioxide by free air CO_2 enriched (FACE) *Bacillus* sp. SS105 for enhanced production and optimization of biosurfactant. *Bioresource Technology, 242*, 2–6. https://doi.org/10.1016/j.biortech.2017.03.124

Max-Planck Institute. (2023). *Henry's Law Constants*. www.henrys-law.org. Accessed 22 August 2023.

Mendiara, T., Cabello, A., Izquierdo, M. T., Abad, A., Mattisson, T., & Adánez, J. (2021). Effect of the presence of siloxanes in biogas chemical looping combustion. *Energy & Fuels, 35*, 14984–14994. https://doi.org/10.1021/acs.energyfuels.1c02031

Mezgebe, B., Sorial, G., Wendell, D., & Sahle-Demessie, E. (2019). Effectiveness of biosurfactant for the removal of trihalomethanes by biotrickling filter. *Engineering Reports, 1*, e12031. https://doi.org/10.1002/eng2.12031

Montebello, A. M., Fernández, M., Almenglo, F., Ramírez, M., Cantero, D., Baeza, M., & Gabriel, D. (2012). Simultaneous methylmercaptan and hydrogen sulfide removal in the desulfurization of biogas in aerobic and anoxic biotrickling filters. *Chemical Engineering Journal, 200–202*, 237–246. https://doi.org/10.1016/j.cej.2012.06.043

Muñoz, R., Meier, L., Diaz, I., & Jeison, D. (2015). A review on the state-of-the-art of physical/chemical and biological technologies for biogas upgrading. *Reviews in Environmental Science & Biotechnology, 14*, 727–759. https://doi.org/10.1007/s11157-015-9379-1

Paolini, V., Petracchini, F., Segreto, M., Tomassetti, L., Naja, N., & Cecinato, A. (2018). Environmental impact of biogas: A short review of current knowledge. Journal of Environmental Science and Health—Part A Toxic/Hazardous Substances and Environmental Engineering 53, 899–906. https://doi.org/10.1080/10934529.2018.1459076

Pascual, C., Cantera, S., & Lebrero, R. (2021a). Volatile siloxanes emissions: Impact and sustainable abatement perspectives. *Trends in Biotechnology, 39*, 1245–1248. https://doi.org/10.1016/j.tibtech.2021.05.003

Pascual, C., Cantera, S., Muñoz, R., & Lebrero, R. (2020). Comparative assessment of two biotrickling filters for siloxanes removal: Effect of the addition of an organic phase. *Chemosphere, 251*, 7. https://doi.org/10.1016/j.chemosphere.2020.126359

Pascual, C., Cantera, S., Muñoz, R., & Lebrero, R. (2021b). Siloxanes removal in a two-phase partitioning biotrickling filter: Influence of the EBRT and the organic phase. *Renewable Energy, 177*, 52–60. https://doi.org/10.1016/j.renene.2021.05.144

Pascual, C., Cantera, S., Muñoz, R., & Lebrero, R. (2022). Innovative polishing stage in biogas upgrading: Siloxanes abatement in an anoxic two-phase partitioning biotrickling filter. *Journal of Cleaner Production, 371*, 133427. https://doi.org/10.1016/j.jclepro.2022.133427

Popat, S. C., & Deshusses, M. A. (2008). Biological removal of siloxanes from landfill and digester gases: Opportunities and challenges. *Environmental Science and Technology, 42*, 8510–8515. https://doi.org/10.1021/es801320w

Pulz, O., & Gross, W. (2004). Valuable products from biotechnology of microalgae. *Applied Microbiology and Biotechnology, 65*, 635–648. https://doi.org/10.1007/s00253-004-1647-x

Rachbauer, L., Voitl, G., Bochmann, G., & Fuchs, W. (2016). Biological biogas upgrading capacity of a hydrogenotrophic community in a trickle-bed reactor. *Applied Energy, 180*, 483–490. https://doi.org/10.1016/j.apenergy.2016.07.109

Rafrafi, Y., Laguillaumie, L., & Dumas, C. (2021). Biological methanation of H2 and CO2 with mixed cultures: Current advances, hurdles and challenges. *Waste and Biomass Valorization, 12*, 5259–5282. https://doi.org/10.1007/s12649-020-01283-z

Ramírez, M., Gómez, J. M., Cantero, D. (2015). Biogas: Sources, purification and uses. In U. C. Sharma (Ed.), *Hydrogen and other technologies* (pp. 296–323)

Rasi, S., Läntelä, J., & Rintala, J. (2011). Trace compounds affecting biogas energy utilization—A review. *Energy Conversion Management, 52*, 3369–3375. https://doi.org/10.1016/j.enconman.2011.07.005

Rey, M. D., Font, R., & Aracil, I. (2013). Biogas from MSW landfill: Composition and determination of chlorine content with the AOX (adsorbable organically bound halogens) technique. *Energy, 63*, 161–167. https://doi.org/10.1016/j.energy.2013.09.017

Rivera-Montenegro, L., Valenzuela, E. I., González-Sánchez, A., Muñoz, R., & Quijano, G. (2023). Volatile methyl siloxanes as key biogas pollutants: occurrence, impacts and treatment technologies. *BioEnergy Res, 16*, 801–816. https://doi.org/10.1007/s12155-022-10525-y

Rodriguez, G., Dorado, A. D., Fortuny, M., Gabriel, D., & Gamisans, X. (2014). Biotrickling filters for biogas sweetening: Oxygen transfer improvement for a reliable operation. *Process Safety and Environment Protection, 92*, 261–268. https://doi.org/10.1016/j.psep.2013.02.002

Santos-Clotas, E., Cabrera-Codony, A., Boada, E., Gich, F., Muñoz, R., & Martín, M. J. (2019). Efficient removal of siloxanes and volatile organic compounds from sewage biogas by an anoxic biotrickling filter supplemented with activated carbon. *Bioresource Technology, 294*, 122136. https://doi.org/10.1016/j.biortech.2019.122136

Sarker, S., Lamb, J. J., Hjelme, D. R., & Lien, K. M. (2018). Overview of recent progress towards in-situ biogas upgradation techniques. *Fuel, 226*, 686–697. https://doi.org/10.1016/j.fuel.2018.04.021

Scheutz, C., Mosbaek, H., & Kjeldsen, P. (2004). Attenuation of methane and volatile organic compounds in landfill soil covers. *Journal of Environmental Quality, 33*, 61–71. https://doi.org/10.2134/jeq2004.6100

Singhal, S., Agarwal, S., Arora, S., Sharma, P., & Singhal, N. (2017). Upgrading techniques for transformation of biogas to bio-CNG: A review. *International Journal of Energy Research, 41*, 1657–1669. https://doi.org/10.1002/er.3719

Turco, M., Ausiello, A., & Micoli, L. (2016). The effect of biogas impurities on SOFC. In *Treatment of biogas for feeding high temperature fuel cells. Green energy and technology* (pp. 137–149). Springer

Werkneh, A. A. (2022). Biogas impurities: Environmental and health implications, removal technologies and future perspectives. *Heliyon, 8*, e10929. https://doi.org/10.1016/j.heliyon.2022.e10929

Wu, H., Yan, H., Quan, Y., Zhao, H., Jiang, N., & Yin, C. (2018). Recent progress and perspectives in biotrickling filters for VOCs and odorous gases treatment. *Journal of Environmental Management, 222*, 409–419. https://doi.org/10.1016/j.jenvman.2018.06.001

Yang, L., & Corsolini, S. I. (2019). Online removal of volatile siloxanes in solid-state anaerobic digester biogas using a biofilter and an activated carbon filter. *Journal of Environmental Chemical Engineering, 7*, 103284. https://doi.org/10.1016/j.jece.2019.103284

Zabranska, J., & Pokorna, D. (2018). Bioconversion of carbon dioxide to methane using hydrogen and hydrogenotrophic methanogens. *Biotechnology Advances, 36*, 707–720. https://doi.org/10.1016/j.biotechadv.2017.12.003

Zhang, Y., Oshita, K., Kusakabe, T., Takaoka, M., Kawasaki, Y., Minami, D., & Tanaka, T. (2020). Simultaneous removal of siloxanes and H_2S from biogas using an aerobic biotrickling filter. *Journal of Hazardous Materials, 391*, 122187. https://doi.org/10.1016/j.jhazmat.2020.122187

Zhu, X., Chen, L., Chen, Y., Cao, Q., Liu, X., & Li, D. (2019). Differences of methanogenesis between mesophilic and thermophilic in situ biogas-upgrading systems by hydrogen addition. *Journal of Industrial Microbiology and Biotechnology, 46*, 1569–1581. https://doi.org/10.1007/s10295-019-02219-w

Biogas Production from Municipal Solid Waste (MSW) Insights into Feedstock Preparation and Upgrading Techniques

Mohammed Abdillah Ahmad Farid, Mohd Nor Faiz Norrahim, Jacqueline Lease, and Yoshito Andou

Abstract This chapter comprehensively explores biogas production from municipal solid waste (MSW), revealing its intricate process. Covering crucial aspects, it enhances our grasp of renewable energy extraction from waste. Examining feedstock preparation emphasizes its role in optimizing biogas production. Identifying suitable MSW types targets efficient waste utilization, redefining waste as a resource. The various stages involved in biogas production from MSW are outlined, and process optimization strategies to enhance biogas production are discussed in detail, including the influence of types of waste, feedstock preparation, and operating conditions on biogas yield. Strategies like co-digestion and innovative reactor design boost biogas production from MSW, enhancing efficiency. Technological advancements, including physiochemical and adsorption techniques, maximize biogas yield. Biogas applications in energy and heat underscore its positive impact. The article also highlights the opportunities and potential for biogas production from MSW, explaining its significance for the wider biogas industry. Yet, challenges, from finance

M. A. A. Farid · J. Lease · Y. Andou (✉)
Department of Biological Functions Engineering, Graduate School of Life Science and Systems Engineering, Kyushu Institute of Technology, 2-4 Hibikino, Wakamatsu, Kitakyushu, Fukuoka 808-0196, Japan
e-mail: yando@life.kyutech.ac.jp

M. A. A. Farid
e-mail: ahmad-farid.abdillah-mohammed335@mail.kyutech.jp

J. Lease
e-mail: lease.jacqueline708@mail.kyutech.jp

M. A. A. Farid
Institute of Tropical Forestry and Forest Products (INTROP), Universiti Putra Malaysia (UPM), Serdang, 43400 Selangor, Malaysia

M. N. F. Norrahim
Research Center for Chemical Defence, Universiti Pertahanan Nasional Malaysia, Kem Sungai Besi, 57000 Kuala Lumpur, Malaysia
e-mail: faiz@upnm.edu.m

Y. Andou
Collaborative Research Centre for Green Materials on Environmental Technology, Kyushu Institute of Technology, 2-4 Hibikino, Wakamatsu, Kitakyushu, Fukuoka 808-0196, Japan

A. Anouzla and S. Souabi (eds.), *Generation of Energy from Municipal Solid Waste*,
https://doi.org/10.1007/978-3-031-74334-4_14

to tech complexities, must be recognized for effective solutions. Acknowledging hurdles paves the way for realizing biogas potential. Considering opportunities and concerns offers a comprehensive view of MSW-based biogas production's future. This analytical work is a valuable resource for advanced researchers, policymakers, and practitioners interested in sustainable waste management and renewable energy production.

1 Introduction

Renewable energy has become increasingly important in the global effort to mitigate climate change and reduce greenhouse gas emissions. The burning of fossil fuels for energy production is a major contributor to greenhouse gas emissions, which have been linked to climate change and its associated impacts, such as sea-level rise, droughts, and extreme weather events (UN, 2018). Therefore, the development and adoption of renewable energy sources, such as biogas, have become crucial for achieving a sustainable energy future. Municipal Solid Waste (MSW) biogas production is a promising technology for sustainable energy generation and waste management. MSW is a significant contributor to greenhouse gas emissions and can cause environmental problems if not properly managed. By converting MSW into biogas, we can not only generate renewable energy but also reduce greenhouse gas emissions and mitigate environmental impacts (Magazzino et al., 2020). Moreover, MSW biogas production can provide a local source of renewable energy, reduce dependence on fossil fuels, and create job opportunities in the waste management and energy sectors.

In 2021, worldwide CO_2 emissions were 36.4 billion metric tonnes, a significant contributor to climate change. Biogas, produced by the anaerobic digestion of organic waste materials like MSW, is a renewable energy source that can replace fossil fuels in various applications (Abbasi et al., 2022). According to the International Energy Agency, using biogas could reduce up to 10 to 13% of global greenhouse gas (GHG) emissions (EBA, 2019). Biogas can reduce life-cycle GHG emissions by up to 90% compared with fossil fuel use (GASUM, 2021), and production from MSW can reduce methane emissions from landfills. To put it into perspective, according to the World Nuclear Association, the global production of electricity from biomass was 637 TWh in 2018. Municipal and industrial waste, which refers to the waste produced by households and businesses, accounted for 19% (121 TWh) of total biopower (WNA, 2021). This waste can be used to generate electricity through incineration or by producing biogas through anaerobic digestion. The environmental benefits of biogas as a source of energy have been recognized due to its low carbon footprint. Studies have shown that the carbon footprint of biogas energy is significantly lower than other conventional fuels. The carbon emissions from biogas energy can range between −81 and 251 g of CO_2 equivalent per kilowatt-hour (gCO_2eq per kWh), depending on the feedstock used, the efficiency of the biogas plant, and the method of biogas utilization (Valli et al., 2017).

Apart from leachate generation, the Municipal Solid Waste (MSW) disposal sites are also prone to landfill gas (LFG) emissions comprising CH_4 and CO_2, which poses health hazards to waste managers and residents living nearby (Singh et al., 2023). Hence, the growing need for renewable energy sources and the mounting problem of waste management have made biogas production from MSW an attractive solution for sustainable energy generation. Biogas production from MSW involves the conversion of organic waste materials into a renewable energy source through the process of anaerobic digestion (Tshemese et al., 2023). MSW is a heterogeneous mixture of organic and inorganic waste materials generated from households, commercial establishments, and industries (Abdel-Shafy & Mansour, 2018). The organic fraction of MSW contains biodegradable materials such as food waste, yard waste, and paper that can be converted into biogas through the process of anaerobic digestion.

Anaerobic digestion is a biological process that occurs in the absence of oxygen, and it involves the breakdown of organic matter by microorganisms to produce biogas (USA EPA, 2022). The anaerobic digestion process is typically carried out in a closed system called a digester, which is designed to optimize the conditions for the growth and metabolism of the microorganisms involved in the process. The organic waste materials are introduced into the digester, where they are mixed with water and subjected to a series of conditions such as temperature, pH, and hydraulic retention time that facilitate the anaerobic digestion process. The biogas produced from the anaerobic digestion of MSW is composed mainly of CH_4 and CO_2, with small amounts of other gases such as H_2S and N_2. When burned for energy, biogas emits mainly carbon dioxide, water, and small amounts of other gases, including NO_x and SO_2 (Paolini et al., 2018). The composition of biogas can vary depending on the type of feedstock used, the process conditions, and the upgrading techniques applied to the biogas. The biogas produced from MSW can be used as a renewable energy source in various applications such as heating, electricity generation, and vehicle fuel.

The success of MSW biogas production depends on various factors such as feedstock preparation, process optimization, and upgrading techniques. Feedstock preparation involves the separation and preparation of the organic fraction of MSW to optimize the feedstock for anaerobic digestion. The first step is to sort and separate the organic fraction of MSW from non-organic materials. Feedstock preparation involves the separation and sorting of the organic fraction of MSW from other non-organic materials, such as plastics, glass, and metals, to optimize the feedstock for anaerobic digestion (Carlsson et al., 2015). This may involve shredding or grinding the organic material to a uniform size to ensure efficient digestion. After that, pre-treatment can be applied to the organic fraction to enhance its digestibility. Various methods such as thermal hydrolysis, acid/alkaline pre-treatment, mechanical pre-treatment, and enzymatic pre-treatment can be used to increase the organic content of the feedstock, reduce the presence of inorganic materials that can hinder the anaerobic digestion process, and enhance its digestibility and improve biogas yield (Carrere et al., 2016).

The anaerobic digestion process is sensitive to a variety of parameters, including temperature from (35 to 55 C), pH (6.5–7.5), hydraulic retention time (HRT), and organic loading rate (OLR). Any changes in these parameters can affect the microbial community and the rate of biogas production. Therefore, process optimization aims to maintain optimal conditions for the microorganisms responsible for the anaerobic digestion process. Upgrading techniques such as CO_2 removal, H_2S removal, and biogas purification can be used to improve the quality of the biogas produced and make it suitable for use in various applications. CO_2 removal can increase the energy content of the biogas, while H_2S removal can prevent corrosion and fouling of the equipment used to transport and use the biogas (Tippayawong & Thanompongchart, 2010). Biogas purification involves the removal of impurities such as moisture, siloxanes, and volatile organic compounds (VOCs) to improve the quality of the biogas and make it suitable for use in high-value applications such as natural gas replacement (Giraudet et al., 2014).

Renewable energy is a critical component of efforts to mitigate climate change and transition towards a sustainable energy future. Biogas production from MSW has the potential to contribute significantly to the development of renewable energy sources, while also addressing the challenges of waste management and reducing greenhouse gas emissions. Therefore, this chapter aims to provide insights into feedstock preparation, process optimization, and existing techniques that can be applied to MSW to enhance biogas production, as well as highlight the advantages and limitations of each approach.

2 Feedstock Preparation for Biogas Production from MSW

Feedstock preparation is a critical and foundational stage in the biogas production process from MSW. The success of biogas generation largely hinges on the quality, composition, and suitability of the feedstock. In this section, we will delve into the various aspects of feedstock preparation for biogas production from MSW, including an in-depth discussion on the types of MSW suitable for biogas generation, the necessary steps for feedstock preparation, and the profound impact of feedstock preparation on biogas yield.

2.1 Types of MSW Suitable for Biogas Production

MSW is a diverse mix of organic and non-organic materials generated by households, businesses, and institutions. Figure 1 shows the suitable sources of substrates for biogas production (Tshemese et al., 2023). The organic fraction, comprising food waste, yard waste, and other biodegradable materials, is particularly well-suited for biogas production. This fraction is rich in carbon and serves as a valuable source of energy for anaerobic microorganisms during the biogas generation process. As such, segregating the organic waste from the non-organic materials and recyclables at the

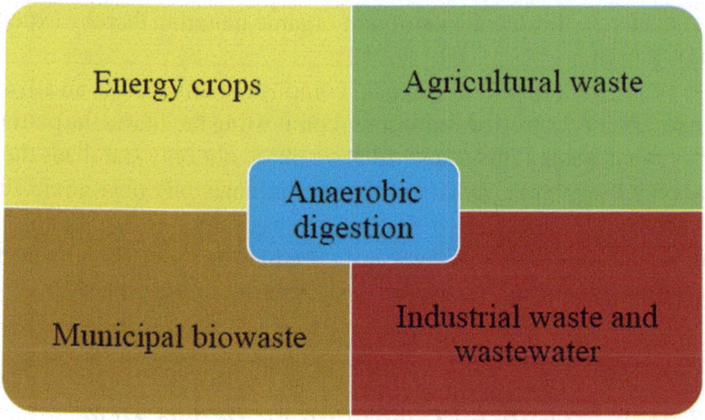

Fig. 1 Sources of suitable substrates for anaerobic digestion Reproduced from Tshemese et al. (2023)

source is of paramount importance to ensure that only biodegradable waste enters the biogas production facility (Kasinath et al., 2021).

2.2 *Feedstock Preparation*

In the domain of sustainable waste management, the effective preparation of MSW feedstock assumes a critical role in maximizing the potential of generating renewable energy through the production of biogas (Kasinath et al., 2021). This complex procedure involves a series of interconnected stages that together contribute to the efficiency and effectiveness of biogas creation (Korbag et al., 2020).

The process initiates with source segregation, a foundational and primary phase that serves as the cornerstone of successful feedstock preparation. To accomplish this, a multifaceted strategy is employed, incorporating campaigns to raise public awareness, active involvement of the community, and well-structured waste collection systems. It is imperative to engage households, businesses, and institutions, educating them about the importance of separating organic waste from non-organic components. This meticulous sorting ensures that only the biodegradable portion is directed to the biogas production facility, thereby optimizing resource utilization, and minimizing waste.

Upon collection of the segregated waste, the challenge posed by its bulkiness and heterogeneity emerges. This is where the steps of shredding and size reduction are started. Through mechanical processes, the organic waste is methodically broken down into smaller, more uniform particles. This physical transformation significantly increases the surface area accessible for microbial action during the subsequent anaerobic digestion phase. By enhancing microbial access and interaction, these procedures accelerate the decomposition of organic material, thereby expediting the generation of biogas.

In the quest for high-quality feedstock, composting emerges as an advantageous optional stage. Under controlled conditions, composting facilitates the partial decomposition of organic waste. This transformative phase not only stabilizes the material but also reduces its volume. Additionally, it eliminates any phytotoxic substances that might hinder the anaerobic digestion process later. An additional benefit of composting is the production of heat, a byproduct that can be utilized for other purposes, thereby enhancing the overall sustainability of the process.

2.3 Impact of Feedstock Preparation on Biogas Yield

The quality and composition of the feedstock have a direct impact on the biogas yield and the stability of the anaerobic digestion process (Meegoda et al., 2018). Effective feedstock preparation positively influences biogas production in several ways as listed below.

(a) *Improved degradation*: Shredding and size reduction of the feedstock provide more surface area for microbial activity, enabling faster and more thorough degradation of organic materials. This results in increased biogas production rates and shorter retention times, making the overall process more efficient.
(b) *Enhanced digestibility*: Composting the feedstock before anaerobic digestion can lead to increased digestibility, as partially decomposed materials are more accessible to the microorganisms, leading to higher biogas yields. The composting process also reduces the C/N ratio, making the feedstock more nutritionally balanced for microbial growth.
(c) *Reduced inhibition*: Proper pre-treatment, such as composting, can reduce the presence of inhibitory substances in the feedstock, such as heavy metals and certain organic compounds. This mitigation of inhibitory substances helps to maintain a stable microbial ecosystem, mitigating the negative effects on the anaerobic digestion process and promoting consistent and robust biogas production.
(d) *Minimized process disruptions*: Well-prepared feedstock ensures a consistent and uniform substrate supply to the anaerobic digester. This helps reduce the risk of process disruptions caused by clogging, floating layers, or stratification, which can negatively impact the overall stability and efficiency of the biogas production process.
(e) *Reduction in residual waste*: Effective feedstock preparation also leads to a reduction in the volume of residual waste after anaerobic digestion. This residual waste, known as digestate, can be further processed through composting or used as a nutrient-rich fertilizer for agricultural purposes, closing the loop on waste management.

3 Biogas Production Process

Biogas production from MSW is a dynamic and intricate process that can be significantly improved through careful optimization (Lisowyj & Wright, 2020). Process optimization aims to maximize biogas yields, enhance the stability of the anaerobic digestion system, and improve the overall energy conversion efficiency.

The biogas production process comprises several interlinked stages, each of which plays a crucial role in converting organic waste into renewable energy. Figure 2 illustrates a simplified representation of a standard anaerobic digestion process, encompassing four distinct stages, which are hydrolysis, acidogenesis, acetogenesis, and methanogenesis.

In the hydrolysis stage, complex organic compounds, such as proteins, lipids, and carbohydrates, are broken down into simpler soluble compounds like amino acids, fatty acids, and sugars. This initial step prepares the substrates for further degradation during subsequent stages. During acidogenesis, the hydrolyzed compounds are fermented by acidogenic bacteria into volatile fatty acids (VFAs), alcohols, and other intermediate products. This process generates acidic conditions and plays a pivotal role in solubilizing organic matter. Next, the acetogenesis stage involves the conversion of VFAs and other intermediates into acetic acid, hydrogen (H_2), carbon dioxide (CO_2), and traces of other compounds (Kadam & Panwar, 2017). The final and most critical stage is methanogenesis, where methanogenic archaea utilize the acetic acid, H_2, and CO_2 produced in the previous stages to generate methane (CH_4), the main component of biogas.

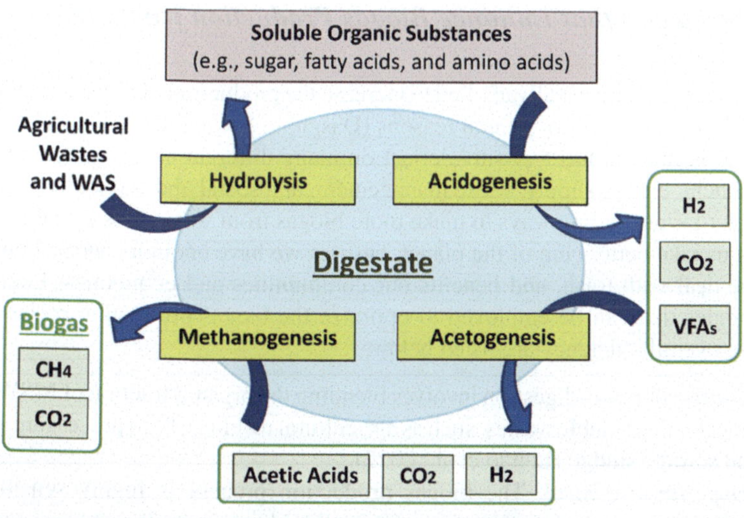

Fig. 2 Process-wise stages of biogas production, including hydrolysis, acidogenesis, acetogenesis, and methanogenesis Reproduced from Pan et al. (2021)

Table 1 Operational conditions that influence the biogas yield from MSW

Operating condition	Influence on biogas yield from MSW
Feedstock characteristics	Composition, organic content, particle size, and C/N ratio impact biogas yields. A balanced feedstock with an optimal C/N ratio supports microbial growth and methane production
Organic loading Rate (OLR)	The amount of organic material introduced per unit volume per day. High OLR can lead to overload and instability, while low OLR may underutilize capacity. Finding the optimal OLR is crucial for maximizing biogas yield
Substrate mixing	Proper mixing promotes uniform conditions, reducing inhibitory substances' accumulation. Effective mixing ensures microorganisms access organic matter, enhancing biogas production
Reactor design	Digester type (e.g., continuous stirred tank, plug-flow, fixed-bed) and configuration impact efficiency. Different designs have advantages and disadvantages, influencing biogas yield and stability
Process monitoring	Regular monitoring of temperature, pH, VFAs, and biogas composition enables timely adjustments and ensures optimal process performance

The biogas yield from MSW is heavily influenced by various operational conditions that impact the efficiency of the anaerobic digestion process. These conditions play a pivotal role in determining the quantity and quality of biogas produced. Table 1 tabulated several operational conditions that influence the biogas yield from MSW.

3.1 Strategies that Enhance Biogas Production from MSW

Various strategies are usually applied to increase the production of biogas from MSW is incredibly important for several reasons (Dasgupta & Chandel, 2020). This effort is crucial because it has a positive effect on many different aspects, including the environment, energy supply, waste management, and social and economic progress. In simpler terms, finding ways to make more biogas from waste is essential because it helps us take better care of the planet, ensures we have enough energy, improves how we deal with trash, and benefits our communities and economies. Examples of strategies that can be employed to optimize the biogas production process and improve overall efficiency are listed below.

(a) *Co-digestion*: Co-digestion involves blending the organic fraction of MSW with other biodegradable wastes, such as agricultural residues, food processing waste, and sewage sludge (Leitão et al., 2022).
(b) *Temperature control*: The biogas production process is highly sensitive to temperature variations (Wang et al., 2019). Maintaining the optimal temperature range for mesophilic or thermophilic digestion can significantly improve the metabolic activity of microorganisms and boost biogas production rates.

(c) *pH adjustment*: The pH level of the anaerobic digester plays a vital role in creating an optimal environment for microbial growth (Singkhala et al., 2021). Monitoring and adjusting the pH within the range from 6.5 to 7.5 can foster the growth of methane-producing microorganisms and prevent the inhibition of biogas production.
(d) *Retention time optimization*: The retention time, or hydraulic retention time, refers to the duration that the substrate remains in the digester (Meegoda et al., 2018). Properly optimizing the retention time ensures sufficient contact between microorganisms and feedstock, allowing for complete digestion and higher biogas production.
(e) *Inoculum selection*: The selection of a suitable inoculum, which contains the desired microorganisms for efficient anaerobic digestion, is crucial (Korbag et al., 2020). Inoculum from well-established digesters or specialized microbial cultures can improve the start-up and stability of the biogas production process.
(f) *Mixing and Homogenization*: Proper mixing and homogenization of the substrate within the digester ensure uniform distribution of nutrients and prevent the formation of zones with low microbial activity. This helps maintain stable digestion conditions and enhances biogas production (Korbag et al., 2020).
(g) *Reactor Design*: The choice of digester types such as continuous stirred-tank reactor, plug-flow reactor, or fixed-bed reactor, and reactor configuration impacts process efficiency. Different digester designs have distinct advantages and disadvantages, influencing biogas yield and system stability (Korbag et al., 2020).

4 Technological Advances in Biogas Production from MSW

Technological advancements have a significant impact on the enhancement and refinement of biogas in the process of widely implementing it on a commercial scale. Several methods for cleaning and improving the quality of raw biogas exist, falling into two categories: Physiochemical and biological technologies. The range of techniques includes conventional approaches like physical absorption, chemical absorption, membrane infiltration, and biological processes, as well as emerging methods like cryogenic enhancement, membrane enrichment, multistage processes, and high-pressure anaerobic digestion. This range of technologies is instrumental in ensuring that biogas meets high standards for various applications (Bharathiraja et al., 2018; Chozhavendhan et al., 2020).

4.1 Physiochemical Technologies

There are various physiochemical technologies utilized for the purification and enhancement of biogas, each offering distinct mechanisms to refine its composition. These technologies can be categorized as follows.

4.1.1 Physical Absorption Method

The physical absorption method employs a water scrubbing system, which is widely used for biogas cleaning and upgrading (Korbag et al., 2020). In this approach, the solubility difference between H_2S, CO_2, and CH_4 in water is exploited. Additionally, the method employs organic solvents to achieve the same principle, enhancing the absorption of CO_2 and H_2S (Benefits et al., 2014).

4.1.2 Chemical Absorption Method

Chemical absorption methods utilize various techniques to bind CO_2 molecules within the biogas. Examples include chemical scrubbers with aqueous amine solutions. This approach offers the advantage of effectively absorbing H_2S in amine scrubbers. These systems often comprise an absorber unit and a stripper (Eze & Agbo, 2010).

4.1.3 Pressure Swing Adsorption

Pressure swing adsorption (PSA) leverages molecular properties and adsorbent compatibility to separate gases from biogas. Adsorbents like zeolites, carbon molecular sieves, and activated carbon are employed. PSA relies on the pressurized gas's property to adhere to solid surfaces; adsorption occurs under high pressure, while a pressure reduction leads to gas release. The process includes four stages: Pressurization, adsorption, blow-down, and purge (Augelletti et al., 2017).

4.1.4 Membrane Technologys

Membrane technology serves as an alternative to traditional absorption-based biogas upgrading. It exploits the selective permeability of membranes to separate biogas components (Benefits et al., 2014).

4.1.5 Cryogenic Technique

The cryogenic technique capitalizes on different liquefaction temperatures for biogas compounds (Yousef et al., 2018). By gradually lowering the biogas temperature, CH_4 is selectively separated from CO_2 and other components. This results in high-purity biomethane meeting Liquefied Natural Gas standards. Cryogenic methods are conducted at around 10 bar pressure (Grande & Blom, 2014; Munoz et al., 2015; Song et al., 2019; Tan, 2017). Different temperature steps are employed to remove pollutants sequentially, yielding high-purity products ranging from 95 to 99% (Baena-Moreno et al., 2019; Walsh et al., 1989).

4.1.6 Chemical Hydrogenation Method

The chemical hydrogenation method involves reducing CO_2 with H_2 using the Sabatier reaction. Various catalysts like Nickel and Ruthenium are employed under elevated temperatures and pressures (Jürgensen et al., 2015; Xia et al., 2016). While efficient, challenges include catalyst degradation due to trace gases, energy consumption, and the need for pure gases (Angelidaki et al., 2018).

4.2 *Biological Approaches*

Biological methods for upgrading biogas fall into two categories: Chemoautotrophic and photosynthetic. Many of these configurations have been practically demonstrated and are in early pilot or full-scale implementation. The significant advantage of these techniques lies in their ability to convert CO_2 into energy-rich or value-added products under mild operational conditions (moderate temperatures, atmospheric pressure). This characteristic contributes significantly to circular economy principles and sustainable bio-based processes.

4.2.1 Chemoautotrophic Methods

Chemoautotrophic biogas upgrading relies on hydrogenotrophic methanogens that utilize H_2 to convert CO_2 into CH_4, as depicted by the equation:

$$4H_2 + CO_2 \rightarrow CH_4 + H_2O(\Delta G^\circ = -130.7^\circ kJ/mol)$$

To ensure the renewable nature of this biological upgrading, the necessary H_2 must originate from a renewable source. Using renewable electricity to generate H_2 by water hydrolysis, especially surplus energy from solar panels or wind turbines, has gained significant attention. Another concept involves injecting H_2 into a biogas digester for in-situ biological biogas upgrading. Here, H_2 is connected with endogenous CO_2 produced within the anaerobic reactor, transformed into CH_4 by native methanogenic archaea (Angelidaki et al., 2018; Kougias et al., 2017).

4.2.2 Photoautotrophic Approaches

Photosynthetic biogas upgrading provides an alternative method to extract CO_2 and produce CH_4-rich gas. These techniques not only achieve H_2S elimination but also consume over 54% of CO_2. Depending on reactor types and algal species selection, methane recovery through photoautotrophic methods can reach approximately 97% (Khan et al., 2021).

5 Applications of Biogas

The applications of biogas span various sectors, offering a range of benefits that contribute to a more sustainable and resource-efficient future. In this section, we will delve into the diverse applications of biogas, ranging from power generation to heat supply and beyond. Figure 3 summarizes the potential uses of biogas, which can be applied in MSW landfill management.

5.1 Power Generation

One of the primary applications of biogas is power generation. Biogas can be combusted in gas engines or turbines to produce electricity (Beschkov, 2021). This electricity can either be used on-site to power the biogas plants themselves or fed into the grid, contributing to the local power supply (see Fig. 4). Biogas power generation helps reduce reliance on fossil fuels and curbs greenhouse gas emissions, making it an important component of the transition to cleaner energy sources. Biogas, derived from organic materials, is a renewable source used for electricity generation. Unlike solar and wind power, which rely on specific conditions, biogas is more dependable due to its connection to consistent crop cultivation. It holds inherent chemical energy that can be transformed into electricity through mechanical processes.

This complex transformation relies on transducers, such as generators and turbines, that convert energy from one form to another. The resulting electricity has diverse applications in both domestic and commercial settings, suitable for various scales of use. To begin the process, establish a link between the biogas source and the gas engine's inlet. The source could be pressurized gas cylinders or a digester designed for organic matter decomposition. The gas engine's operation resembles that of an automobile engine, utilizing pistons to ignite the gas and initiate shaft rotation. This rotation converts chemical energy in the biogas into mechanical energy. Connect

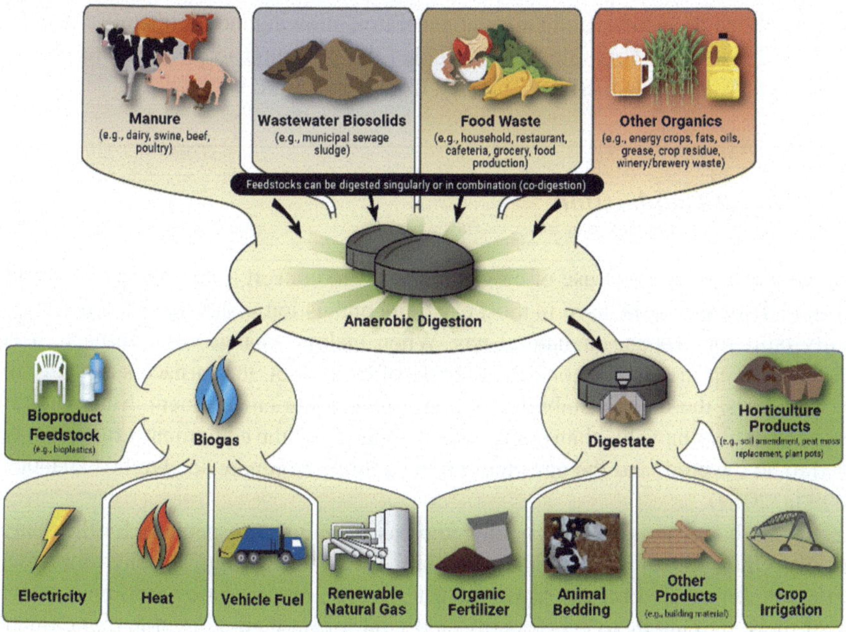

Fig. 3 Potential uses of biogas Adopted from USA EPA (2022)

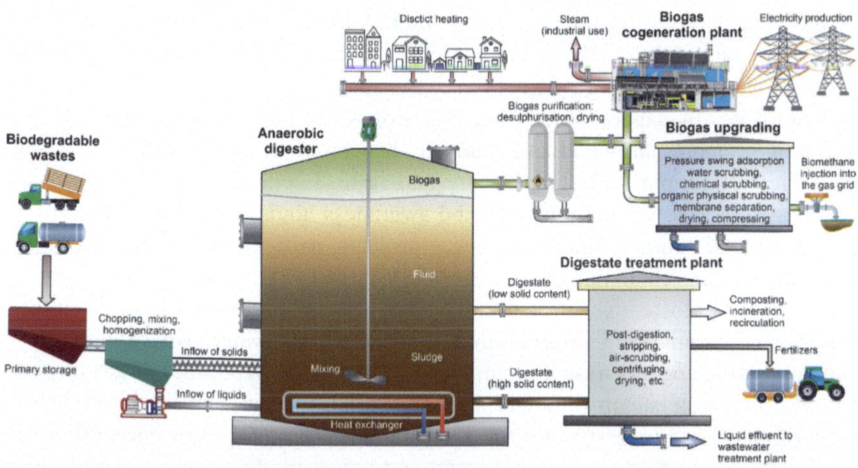

Fig. 4 Schematic representation of biogas production and utilization for electricity generation Adopted from Zupančič et al. (2022)

the gas engine to an alternating current generator, allowing the shaft's movement to drive the generator. This interaction enables the generator to produce electricity, primarily driven by magnetism generated from the motion.

5.2 Direct Combustion

The easiest way to make use of biogas on-site is to convert it into energy by direct burning. Biogas may be used in most existing combustion systems, even those originally built for propane or natural gas. When making changes to a furnace, it is crucial to keep in mind factors like the rate of heat input, the furnace's capacity to handle fluids, the flame's stability, and the surrounding environment. Since biogas has a lower heating value than conventional natural gas, the equipment's ratings may need to be lowered, and the gas input piping's dimension may need to be expanded (Wilkie, 2019).

Biogas shows promise as a cooking and baking solution, especially for the many heavily reliant on inefficient biomass burning. This leads to health and environmental issues. Well-operated biogas stoves emit fewer greenhouse gases than biomass or fossil fuels, making them eco-friendly, and vital for rural cooking affecting climate change, as shown in Fig. 6. They better marginalized communities by replacing polluting methods and managing organic waste, breaking the poverty cycle. However, success hinges on user-generated biogas fuel. Adoption requires assessing biogas technology's suitability for household needs. Nevertheless, the biogas can be directed to the residential zone using a methodology akin to the prevailing technique employed for landfill gas management (see Fig. 4). Biogas mitigates deforestation and soil depletion, replacing firewood and charcoal. Transitioning faces hurdles, as households prefer multiple fuels for security and taste. Economic growth might shift practices to biogas, yet low incomes and awareness hinder adoption, especially in least-developed nations. Financing biogas projects, particularly in underdeveloped countries, remains challenging (Fig. 5).

It is possible to recover heat that would otherwise be forfeited via the biogas conversion system using cogeneration to provide hot water. An example of a combined cycle cogeneration system is one in which the byproduct of gas turbines is used to generate steam to power a second steam turbine, and the byproduct of that turbine is used to heat water for domestic use. The total conversion efficiency of using biogas to produce both hot water and power might range from 65 to 85% (Wilkie, 2019). There are several other potential applications of biogas beyond the ones mentioned earlier. Here are a few additional applications as tabulated in Table 2.

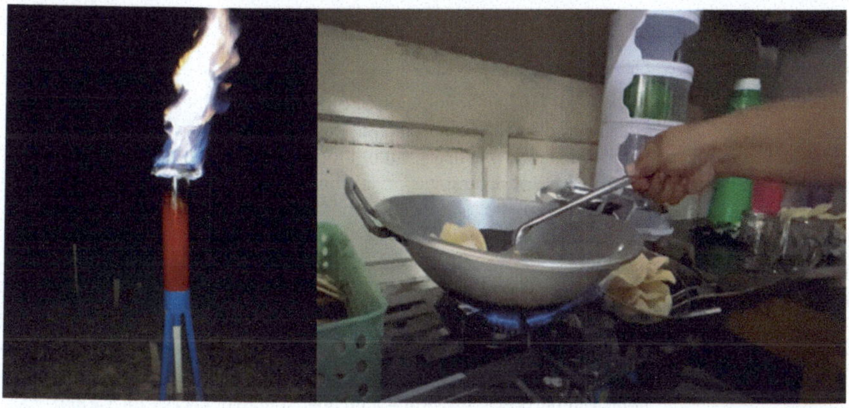

Fig. 5 Flaring of landfill gas in the vicinity (left) and a household that benefits from using landfill gas (right). This initiative has been highly advantageous, as it eliminates the need for LPG Adopted from ICLEI—SEAS (2021)

Fig. 6 Biogas recovery from landfill for household usage Adopted from Mohanty and Das (2019)

6 Challenges Facing Biogas Production from MSW

Biogas production from MSW also faces several challenges that can impact its widespread adoption and efficient implementation. Addressing these challenges is crucial to maximizing the potential of biogas as a renewable energy source and sustainable waste management solution. Several challenges related to technical, economic, regulatory, and social aspects are related to biogas production from MSW.

One of the primary challenges in biogas production from MSW is the variable and heterogeneous nature of the feedstock. MSW consists of a diverse mixture of organic and inorganic materials with varying degrees of biodegradability. Different types of

Table 2 Other applications of biogas

Application	Description
Fuel for vehicles	Another significant application of biogas is as a vehicle fuel, typically after upgrading to compressed natural gas or biomethane standards. Biomethane, the purified form of biogas, can be used as a direct replacement for natural gas or gasoline in vehicles. Biogas-powered vehicles contribute to reduced tailpipe emissions, lower carbon footprint, and decreased dependence on fossil fuels
Injection into the natural gas grid	Biomethane can be injected into the natural gas grid, blending with conventional natural gas to create a cleaner and more sustainable energy mix. This approach extends the use of biogas beyond on-site applications, enabling its distribution to a broader range of consumers and industries. Injected biomethane can be used for residential, commercial, and industrial purposes, providing a greener source of energy for various sectors
Rural electrification	In remote or rural areas with limited access to traditional energy sources, biogas can play a crucial role in providing electricity for lighting and basic appliances. Small-scale decentralized biogas systems can serve as energy hubs for off-grid communities, contributing to improved living conditions and economic development
Fuel cells	It is possible to directly convert biogas into electricity using a fuel cell. However, this approach demands exceptionally clean gas and costly fuel cells, making it a subject of ongoing research rather than a viable solution at present. On the other hand, converting biogas to electric power through a generator set is a far more practical and feasible option

waste, such as food scraps, yard waste, and paper, have different carbon-to-nitrogen (C/N) ratios, affecting the stability and efficiency of the anaerobic digestion process. Managing and optimizing the digestion of such a diverse feedstock require careful feedstock preparation, mixing, and balancing.

Besides that, anaerobic digestion is a complex biological process that is sensitive to changes in environmental conditions. Factors such as pH fluctuations, temperature variations, and the accumulation of certain compounds, such as ammonia and volatile fatty acids, can disrupt the microbial community and lead to process instability or inhibition. Maintaining stable operating conditions and preventing the build-up of inhibitory substances is critical for consistent biogas production.

Moreover, biogas production is an energy-intensive process that requires electricity for running pumps, mixers, and other equipment, as well as heat the digesters. The energy input for maintaining optimal operating conditions and processing the feedstock may offset the energy output in the form of biogas. Achieving a positive energy balance and ensuring the overall environmental benefits of biogas production are critical considerations.

Biogas typically contains impurities, such as CO_2, moisture, H_2S, and siloxanes, which can hinder its direct use or injection into the natural gas grid. Upgrading biogas to remove these impurities and increase its methane content can be a technical and costly challenge. The choice of upgrading technology and its efficiency is crucial in producing high-quality biogas suitable for various applications.

Biogas production also requires efficient waste collection and logistics are essential for the continuous supply of feedstock to biogas plants. Proper waste segregation at the source is necessary to minimize contamination and enhance the quality of the feedstock. In areas with inadequate waste management infrastructure or limited waste segregation practices, securing a consistent supply of high-quality feedstock can be challenging.

Public perception of biogas and anaerobic digestion also plays a significant role in its acceptance and implementation of biogas production. Misconceptions or lack of awareness about the technology, its benefits, and its role in waste management and renewable energy can lead to resistance or opposition from local communities.

7 Opportunities and Considerations for Biogas Production from MSW

The production of biogas from MSW presents a significant opportunity for sustainable waste management, energy production, and reduction of GHG emissions. MSW typically contains a significant amount of organic waste, including food waste, yard trimmings, and other biodegradable materials. When these materials decompose, they generate methane, a potent GHG that contributes to global warming. However, this methane can be captured and used as a renewable energy source through anaerobic digestion (AD) or landfill gas (LFG) recovery.

AD is a biological process that breaks down organic waste in the absence of oxygen, producing biogas that is composed of approximately from 50–70% methane and 30–50% carbon dioxide, as well as trace amounts of other gases. The biogas can be used to generate electricity and heat or upgraded to biomethane, which can be injected into natural gas pipelines or used as a transportation fuel. AD also produces a nutrient-rich digestate that can be used as a fertilizer. LFG recovery involves the capture and utilization of the methane generated by the decomposition of MSW in landfills. As MSW decomposes, methane and other gases are released into the atmosphere. LFG recovery systems collect methane and other gases, which can be used to generate electricity or heat. LFG can also be upgraded to biomethane for injection into natural gas pipelines or used as a transportation fuel.

Due to these advancements, biogas production from MSW may create new market opportunities for the renewable energy industry. The use of biogas as a renewable energy source is gaining momentum globally, and there is increasing demand for renewable natural gas (RNG) as a low-carbon transportation fuel. Biogas production from MSW can provide a significant source of RNG, which can be sold to transportation companies or injected into natural gas pipelines. The development of markets for RNG can provide new revenue streams for the biogas industry, and can also support the transition to a low-carbon economy.

Biogas production from MSW also can contribute to the circular economy by closing the loop on waste management. The circular economy is an economic model that seeks to minimize waste and maximize resource efficiency. Biogas production from MSW is a key component of the circular economy, as it allows for the recovery of energy and nutrients from waste that would otherwise be discarded. The use of biogas as a renewable energy source also displaces fossil fuels, further contributing to the circular economy by reducing dependence on non-renewable resources.

7.1 *Economics Viability of Biogas Production*

The assessment presented in the paper focuses on comparing two waste management technologies, namely AD and LFG projects, in terms of electricity generation, economic feasibility, and environmental impact under different scenarios. Firstly, Huang and Fooladi (2021) analyze the amount of electricity generated by each technology in Tehran and Beijing. It is stated that the total electricity generated by AD technology is between 45.2% (4261.7 and 4571.7 GWh) and 41.9% (3149 and 3003.3 GWh) higher than that of the landfill gas project in Tehran and Beijing, respectively (Fig. 7). This suggests that AD technology is more efficient in generating electricity from the waste in both regions. Secondly, the economic feasibility of the two technologies is evaluated using two approaches: Levelized Cost of Energy (LCOE) and Payback Time (PBT) calculations (Fig. 8). LCOE refers to the average cost of generating electricity over the lifetime of a power plant, including investment, operation, and maintenance costs. PBT, on the other hand, represents the time required for the project to recover its initial investment. The assessment indicates that the average PBT value for 20 years in Tehran is 1.1 and 0.67 years higher than in Beijing under the landfill gas and anaerobic digestion technologies, respectively. LCOE for anaerobic digestion in Beijing and Tehran is \$0.0319–\$0.0424 and \$0.0335–\$0.0426 per kWh over 20 years, while landfill gas technology costs more at \$0.0603–\$0.0767 and \$0.0663–\$0.0819 per kWh in Beijing and Tehran, respectively. This implies that the anaerobic digestion technology has a shorter payback time and less costs, making it more economically feasible compared to the landfill gas project in both regions.

Another research aimed to evaluate the technical and economic feasibility of an AD plant in Panguipulli, a town in the south of Chile, which would use 12 tons of organic waste produced by households out of the 26 tons of solid waste produced daily (Pierre Doussoulin & Salazar Molina, 2022). The biogas plant would generate energy for the plant's consumption and for sale to the country's electricity grid, as well as producing biofertilizer from MSW. To evaluate the economic feasibility, the research analyzed the investment, costs, and income that make up the cash flow of the project over 15 years. The results showed a net present value (NPV) of 214.099.637 Chilean pesos and an internal rate of return (IRR) of 15% at a real discount rate of 10%, with a payback period of 6 years. The research concluded that it is feasible to design an AD plant in Panguipulli that would contribute to the mitigation of climate change and promote circular economy actions and the sustainable management of

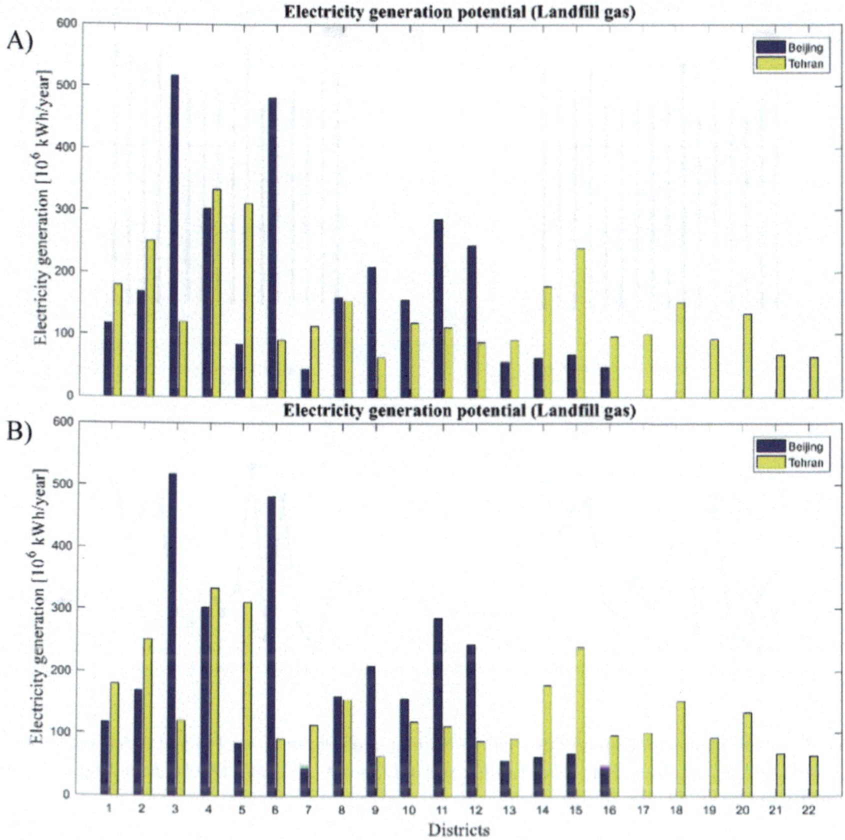

Fig. 7 Electricity generation potential for 20 years in Beijing and Tehran by **A** landfill gas technology and **B** anaerobic digestion Adopted from Huang and Fooladi (2021)

MSW in the south of Chile. By utilizing organic waste that currently goes to the landfill, the biogas plant would generate energy and produce biofertilizer, reducing greenhouse gas emissions and improving waste management practices in the region.

Dodo et al. (2022) investigated the feasibility of using MSW as a feedstock for biogas generators in a hybrid power system that also includes solar photovoltaic panels, battery banks, and a bi-directional converter. The researchers used the Hybrid Optimization Model for Electric Renewables (HOMER) to simulate and optimize the proposed standalone model for a remote village in Nigeria. The results of the study showed that the hybrid power system with MSW as feedstock for biogas generation and solar panels as the primary source of energy can meet the electricity demand of the isolated community (2,822.20 kWh per day). The optimal system comprises a 500-kW biogas generator, 800-kW photovoltaic panels, a 400-kW converter, and 5,000 strings of battery. The net present cost (NPC) of the optimal system is USD 8,510,723.00, with an LCOE of USD 0.2917 per kWh and an operating cost of USD

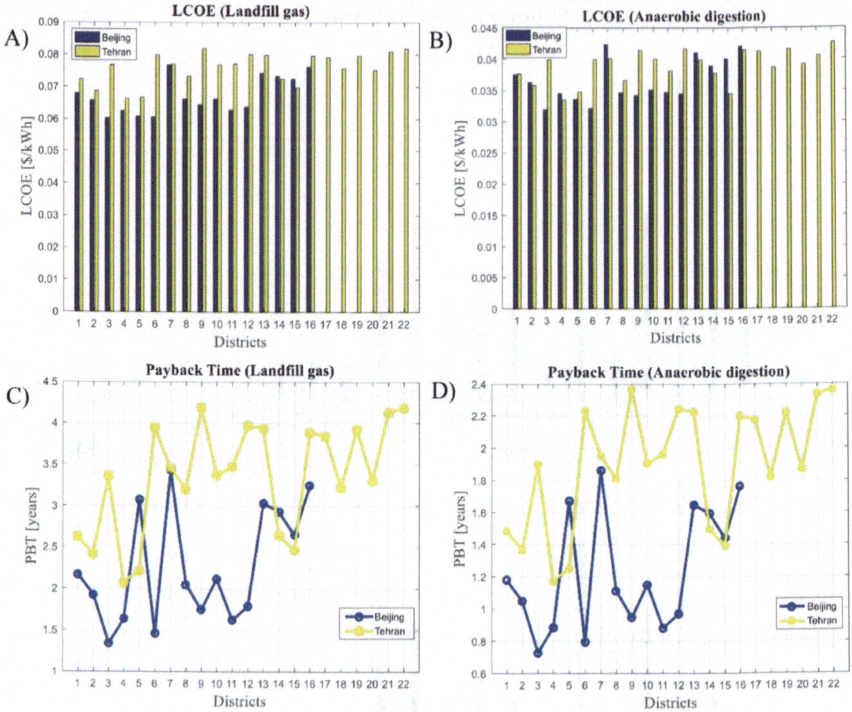

Fig. 8 LCOE values for different districts of Beijing and Tehran in **A** LFG and **B** AD technology. Payback time for different districts of Beijing and Tehran in **C** LFG and **D** AD technology Adopted from Huang and Fooladi (2021)

182,934.00 per year. According to the cost summary (Fig. 9), the battery bank is the most expensive part of the system (USD 5,440,640.04), followed by the biogas generator (USD 1,675,935.02) and the solar panels (USD 1,186,561.40). The system converter is the least expensive part of the system (USD 207,586.72).

Fig. 9 NPC summary of the components Adopted from Dodo et al. (2022)

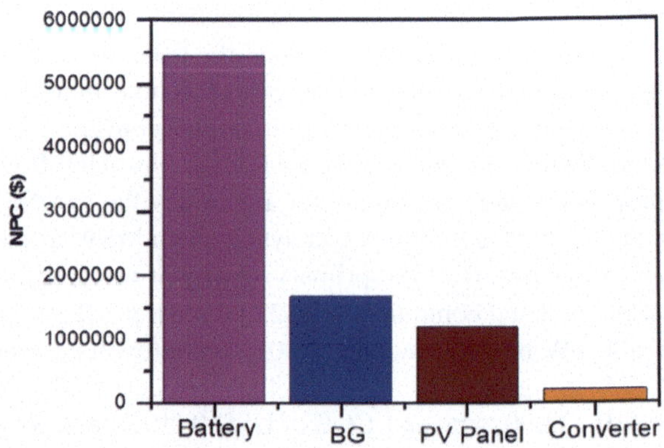

The study compared a base case scenario with five other scenarios that had different annual processing capacities. Biogas plant with an annual processing capacity of 55,000 m^3 MSW per year with two anaerobic digesters from landfill and wastewater treatment. The net energy consumption was between 7.09 to 7.74 GW h per year, which was from 64.8 to 80% higher compared to the base case due to the high energy consumption. Yet, it produced the highest net energy production of 38.21 GWh per year (21,041 m^3 per day biogas production), making it the most energy-efficient and economical scenario. The capital investments ranged from USD 38.4–49.2 million. The NPV was positive, indicating the profitability of biogas production from MSW (Fig. 10). The annual operating costs varied between USD 1.83 and 5.38 million, and the utility costs varied between 3 and 4% of the operational costs.

Ayodele et al. (2018) present a study on the biogas recovery from the organic fraction of municipal solid waste in Ibadan, Nigeria, to determine the economic and environmental benefits of the technology for the city. The study estimated the waste generation potential of the city between 477,001 and 966,897 tonnes over 20 years. The methane yield for AD and LFG technologies were determined to be about (104.66–212.15) × 106 m^3 per year and (22.65–127.65) × 106 m^3 per year, respectively. The study also estimated the electricity generation potential for both technologies, with AD having a range of 321.73 to 652.15 GWh and LFG having a range of 63.25 to 436.18 GWh. The economic analysis of the two technologies showed that they are economically viable, with both presenting positive net present values. The NPV for anaerobic digestion was estimated to be USD 834.12 million while that of landfill gas recovery was USD 489.26 million. The LCOE was estimated to be between USD 0.0681 and 0.0336 per kWh for anaerobic digestion technology and between USD 0.2411 and 0.0350 per kWh for landfill gas recovery technology. The payback period, internal return, and total life cycle cost were 5 years, 19.3%, USD 413.68 million for AD, while 7 years, 23.4%, and USD 288.05 million for LFG technology.

This study assessed the electricity generation potential of biogas from the organic fraction of municipal solid waste collected for disposal from 2004 to 2018 in 31 provinces of China using AD and LFG technologies (Cudjoe, 2020). Total electricity generated using anaerobic digestion technology ranged from 636.67 GWh to 33,610.22 GWh. Using landfill gas to energy technology, total electricity from 521.76 GWh to 23,452.42 GWh was generated. Economic feasibility assessment of the technologies was carried out using NPV, and LCOE methods. NPV of the AD project in the provinces over the project period ranged from USD 30.8 million to 1.6 billion, while that of LFG is from USD 28.5 million to 871.2 million. LCOE of LFG is from $0.1126 per kWh to $0.2663 per kWh, while that of AD is from $0.0712 per kWh to $0.0966 per kWh. Key findings showed that the electricity generation potential of AD technology was higher in all the provinces. Economically, the results showed that both projects are feasible in all 31 provinces. However, AD proved to be highly feasible, with more positive NPV, and LCOE.

Malaysia faces a persistent economic and environmental challenge when it comes to waste management, with over 80% of solid waste being disposed of in landfills

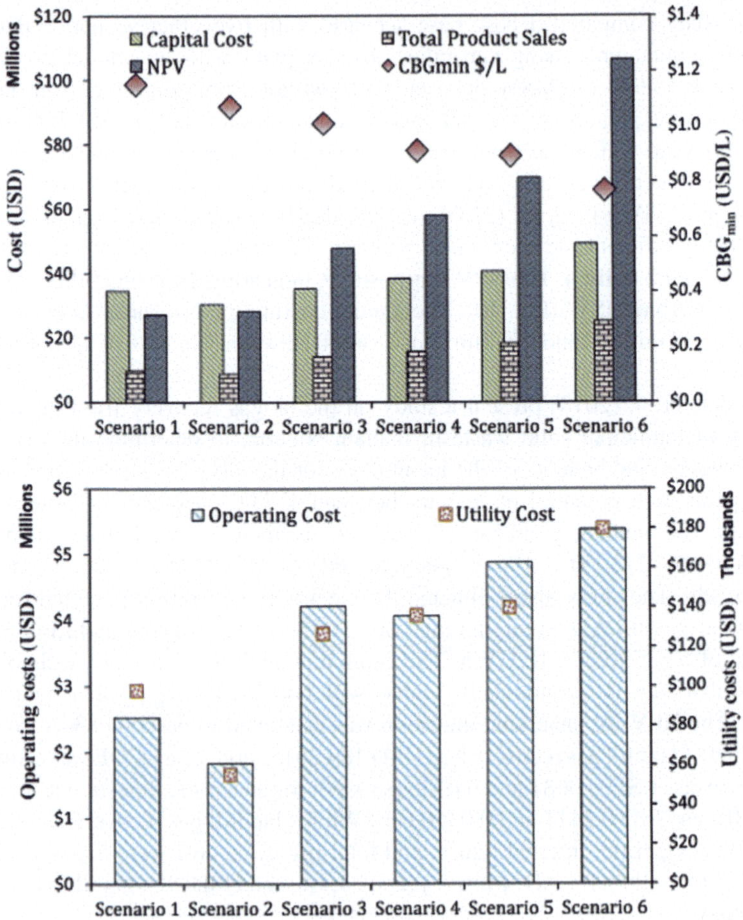

Fig. 10 Profitability of the biogas plant under different scenarios showing indicators such as annual product sales, capital investment, MMSP, operating and utility costs, and NPV Adopted from Rajendran et al. (2014)

and dumpsites. Jun et al. (2021) examine the feasibility of anaerobic digestion for recovering energy and biofertilizer from the organic fraction of MSW. It was discovered that organic waste contributes to approximately 45% of total MSW generated in Malaysia, and the anaerobic digestion of 50% of this waste could produce 3941 MWh per day of electrical energy and 2500 tonne per day of biofertilizer. In terms of revenue, renewable electricity generated from waste (biogas) is sold to the Malaysian utility company (Tenaga National Berhad, TNB) via the national electricity grid at a FiT rate of RM 0.3997 per kWh while the average market value of organic fertilizer is sold at RM 515 per tonne. The sales of electricity via Feed-in-Tariff and biofertilizer to local agricultural industries, combined with the savings generated from the reduction of MSW landfilling operations and leachate treatment at landfills, could generate a net revenue of RM 3300 million per year (RM 573.9 million per year for sales of biogas electricity, while RM 470 million per year sales for biofertilizer), the ROI in years was found to be 4 years.

7.2 Environmental Benefits

Lastly, the environmental evaluation is conducted based on the global warming potential (GWP) under three different scenarios (Fooladi & Huang, 2021). The results showed that scenario 1, which involves landfilling without energy recovery, has the highest greenhouse gas emissions in both Beijing and Tehran. The proposed project, which involves using biogas produced through the AD to generate electricity with agricultural fertilizer by-product, has the lowest greenhouse gas emissions in landfills without CH4 capturing system for electricity. GWP measures the potential pollutant emissions to contribute to global warming. Figure 11 states that, on average, the MSW management with AD in Tehran and Beijing reduced GWP between 79.16% and 92.65%, respectively. This indicates that the AD-integrated MWS management led to a positive environmental impact in both regions.

According to the life cycle analysis (LCA) found in the literature, it is estimated that power generated from MSW-derived biogas produces 11 gCO_2e per kWh, while power generated from diesel fuel produces 680 gCO_2e per kWh (Table 3) (Ghaem et al., 2015). This means that using a biogas engine instead of a diesel engine can save 670 gCO_2e per kWh, which is a significant environmental advantage. The design of a biogas hybrid system in Kenya proved to help avoid emitting 17 tonnes of CO_2 per year, compared to using a diesel engine as backup (48 tonnes of CO_2 per year). The integration of MSW resources also leads to the emission of 0.974 kg per year of NO, 140 kg per year of CO_2, and 1.56 kg per year of CO (Dodo et al., 2022). The optimized hybrid system generates 75,000 kWh of total power, with 25,000 kWh

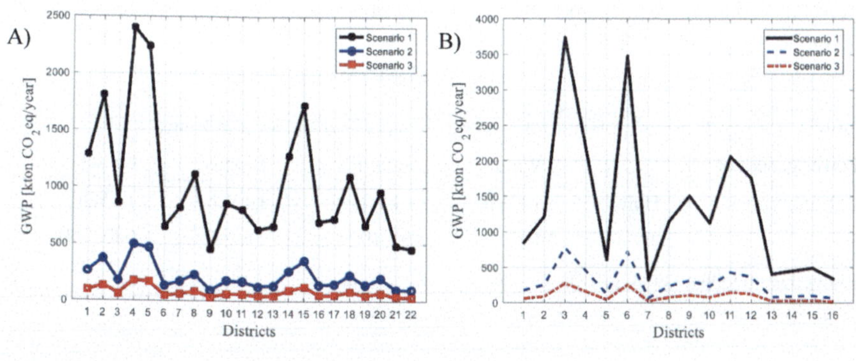

Fig. 11 Value of GWP under three scenarios for different districts of **A** Tehran and **B** Beijing Adopted from Huang and Fooladi (2021)

generated by the biogas engine. When compared to a diesel-based power generator, this system can save 48 tonnes of CO_2 during one year of operation.

In Ibadan, Nigeria, the GHG emissions for both AD and LFG technologies were assessed by evaluating the CO_2 equivalent of CH_4 generated from biogas (Ayodele et al., 2018). The study found that over the lifetime of the technology, AD would contribute 2770.8 kt of CO_2eq emissions, while LFG would contribute 10,441.5 kt of CO_2eq emissions. In addition, the study estimated the amount of diesel and CO_2eq that could be displaced or saved by utilizing biogas obtained from AD for electricity generation in an internal combustion engine over the lifespan of the system. The results indicated that for the city of Ibadan, 4685.176 thousand liters of diesel and 12,649.98 kt of CO_2eq could be displaced or saved as a result of using biogas from AD.

The environmental impact of waste management options based on global warming potential was assessed using AD and LFG technologies from 2004 to 2018 in 31 provinces of China (Cudjoe et al., 2020). It was realized that on average LFG could reduce global warming potential by 71.5%, while AD could reduce global warming potential by 92.7% (249.92–13,193.65 kt CO_2eq) in comparison to the total global warming potential without energy recovery (4402.50 to 197,884.39 kt CO_2eq).

Another study investigates the public medium-scale AD plant (BTA® technologies) of the Vallès Oriental Waste Treatment Centre situated in Granollers (Catalonia, Spain) and handles source-selected MSW from the neighboring areas (Abad et al., 2019). It processes 50,000 tonnes of source-selected MSW each year and yields 4,275,000 Nm^3 of biogas. This biogas is employed to generate heat and power through two 626 kW engines. The Plant produces 7500 MWh per year of electricity, which is sold to the electrical grid, generating an income of 18.64 € per tonne.

Resource recovery of organic fraction municipal solid waste in Malaysia via anaerobic digestion can significantly contribute towards environmental sustainability. Jun et al. (2021) examine the feasibility of anaerobic digestion for recovering energy and biofertilizer from the organic fraction of MSW. This approach could avoid the emission of 2735 tonnes per day of CO_2 (air pollution), 1128 m^2 per day of landfilling area (land contamination), and 481 m^3 per day of leachate pollution while generating clean electricity of 3941 MWh per day.

Table 3 Life cycle emission factor for power generators grid emission in Kenya

Power generator		Wind	Solar PV	Biogas	Diesel
CO_2 emissions	Unit	1 kWh	1 kWh	1 kWh	1 kWh
	g	10	30–100	11	680–780

Adopted from Ghaem et al. (2015)

7.3 Factors for the Profitable Biogas Industry

The profitability of biogas production from municipal solid waste (MSW) depends on various factors such as waste composition, management practices, scale of operation, acquisition, and transportation costs, as summarized in Table 4. The quantity and quality of the feedstock, especially the organic matter, play a significant role in the profitability of the operation. Organic waste such as food waste, yard waste, and agricultural residues has a higher potential for biogas production than non-organic waste such as plastics and metals.

Waste streams with consistent composition and stable organic matter content are preferred for biogas production to avoid fluctuations in biogas production that can reduce profitability. The cost of acquiring and transporting feedstock, such as the distance between the waste source and the biogas plant, the mode of transportation, and the volume of waste to be transported, can also impact profitability.

To minimize costs, biogas facilities may consider co-locating with waste sources or partnering with waste management companies. The size and scale of the biogas production operation also affect profitability. Small-scale systems are suitable for local waste management but may not be as profitable as large-scale systems that can benefit from economies of scale, lower operating costs per unit of biogas produced, and generate more significant revenue streams.

The market value of biogas, which can be influenced by regulatory policies and incentives and the supply and demand dynamics of the energy market, is another factor that can impact profitability. Policies that support renewable energy and incentives such as feed-in tariffs or tax credits can increase the market value of biogas by providing financial support for biogas production and reducing the cost of production.

8 Conclusion

This chapter study scrutinizes the significance of feedstock preparation, highlighting its pivotal role in optimizing biogas generation. Identifying suitable MSW categories targets the effective utilization of waste materials, redefining waste as a valuable resource. The article underscores the critical correlation between feedstock preparation techniques and biogas yield, emphasizing precise methodologies for achieving optimal gas production. Strategies such as co-digestion and innovative reactor designs are explored, amplifying biogas production from MSW and thus enhancing overall efficiency. Advancements in technology, encompassing physiochemical and adsorption techniques, are harnessed to maximize biogas output. The applications of biogas in energy and heat generation underscore its positive environmental impact. However, this chapter acknowledges challenges spanning financial constraints to technological complexities that must be acknowledged for the formulation of effective solutions. By recognizing these obstacles, the groundwork is laid for realizing the full potential

Table 4 Summary of factors for biogas industry improvement

Factors	Description	Impacts on profitability
Feedstock quantity and quality (Asam et al., 2018; Logan & Visvanathan, 2019)	• Organic matter in MSW is the primary source of biogas, and higher concentrations of organic matter led to greater potential for biogas production • Waste streams with high organic content are preferred for biogas production • Additionally, the quality of organic matter impacts biogas yield and efficiency • Moisture content and particle size of the waste are also important	• Positive impact on profitability if the feedstock is high quality and consistent • High moisture content can increase processing costs, while high organic matter content can increase biogas yield and revenue • Small particle sizes can improve digestion efficiency
Acquisition and transportation costs (Cong et al., 2020)	• The cost of acquiring and transporting feedstock can vary depending on availability, accessibility, and competition for waste sources • Transportation costs can also vary depending on distance, volume, and mode of transportation	High acquisition and transportation costs can negatively impact profitability
Size and scale of operation (Lantz et al., 2007)	• Biogas production plants can range in size and scale from small-scale to large-scale • Small-scale systems may be less profitable due to lower biogas production capacity and higher operating costs per unit of biogas produced, while large-scale systems can benefit from economies of scale but may face logistical challenges	Large-scale systems have the potential for greater profitability due to economies of scale, revenue streams from selling biogas to the grid, or large industrial customers
The market value of biogas (Ferella et al., 2019)	Regulatory policies and incentives, as well as supply and demand dynamics of the energy market, can influence the market value of biogas	The high market value of biogas can positively impact profitability

(continued)

Table 4 (continued)

Factors	Description	Impacts on profitability
Technology and equipment (Kalinichenko et al., 2016)	• Technology and equipment used in biogas production can impact the efficiency and yield of biogas production • For example, the type of anaerobic digestion system and pre-treatment methods used can impact biogas yield	• Efficient technology and equipment can improve profitability by optimizing biogas production and reducing costs • High energy input costs can reduce profitability, while renewable energy sources can reduce operating costs and improve sustainability
Market demand (Chinese & Patrizio, 2016)	The demand for renewable energy sources can impact profitability by influencing the market value of biogas	High demand can increase the value and profitability of biogas production as more customers are willing to pay a premium for renewable energy
Regulatory environment and incentive (Karlsson et al., 2017)	Government policies and regulations can impact biogas production and profitability by providing incentives (Feed-in tariffs, tax credits) or creating barriers to entry	• A favourable regulatory environment can positively impact profitability • Incentives such as feed-in tariffs and tax credits can reduce operating costs and increase revenue, thereby increasing profitability
Operating Conditions (Hagos et al., 2017)	Temperature, pH, and hydraulic retention time	Optimal operating conditions can improve digestion efficiency and increase biogas yield, while suboptimal conditions can reduce yield and increase processing time
Fossil Fuel Prices (Delzeit & Kellner, 2013)	Cost of fossil fuels	• High fossil fuel prices can increase the value of biogas relative to fossil fuels, making biogas more competitive and profitable • Low fossil fuel prices can decrease the value of biogas relative to fossil fuels, potentially reducing profitability

of biogas production. Balancing opportunities and concerns present a comprehensive outlook on the future trajectory of MSW-based biogas production. As research progresses, the integration of these concepts into sustainable waste management practices becomes increasingly viable. This work serves as a blueprint for optimized waste-to-energy strategies within the evolving landscape of technology, driving the path toward sustainable solutions.

References

Abad, V., et al. (2019). Promoting circular economy in the surroundings of an organic fraction of municipal solid waste anaerobic digestion treatment plant: Biogas production impact and economic factors. *Bioresource Technology, 283*, 10–17.

Abbasi, G., et al. (2022). Municipal solid waste management and energy production: A multi-objective optimization approach to incineration and biogas waste-to-energy supply chain. *Sustain Energy Technologies Assessments, 54*, 102809.

Abdel-Shafy, H. I., & Mansour, M. S. M. (2018). Solid waste issue: Sources, composition, disposal, recycling, and valorization. *Egyptian Journal of Petroleum, 27*, 1275–1290.

Angelidaki, I., et al. (2018). Biogas upgrading and utilization: Current status and perspectives. *Biotechnology Advances, 36*, 452–466.

Asam, Z., et al. (2018). How can we improve biomethane production per unit of feedstock in biogas plants? *Applied Energy, 88*, 2013–2018.

Augelletti, R., et al. (2017). Pressure swing adsorption for biogas upgrading. A new process configuration for the separation of biomethane and carbon dioxide. *Journal of Cleaner Production, 140*, 1390–1398.

Ayodele, T. R., et al. (2018). Economic and environmental assessment of electricity generation using biogas from organic fraction of municipal solid waste for the city of Ibadan Nigeria. *Journal of Cleaner Production., 203*, 718–735.

Baena-Moreno, F. M., et al. (2019). Biogas upgrading by cryogenic techniques. *Environmental Chemistry Letters, 17*, 1251–1261.

Benefits, B., et al. (2014). Biogas technology: Current trends, opportunities and challenges. *Water Intake Hydratation Physiology During Childhood, 30*, 166.

Beschkov, V. (2021). Biogas production: Evaluation and possible applications. In A. E.-F. Abomohra & E.-S. Salama (Eds), *Biogas—Basics, integrated approaches, and case studies compounds* (pp. 1–18). IntechOpen.

Bharathiraja, B., et al. (2018). Biogas Production—A review on composition, fuel properties, feed stock and principles of anaerobic digestion. *Renewable and Sustainable Energy Reviews, 90*, 570–582.

Carlsson, M., et al. (2015). Impact of physical pre-treatment of source-sorted organic fraction of municipal solid waste on greenhouse-gas emissions and the economy in a Swedish anaerobic digestion system. *Waste Management, 38*, 117–125.

Carrere, H., et al. (2016). Review of feedstock pretreatment strategies for improved anaerobic digestion: From lab-scale research to full-scale application. *Bioresource Technology, 199*, 386–397.

Chozhavendhan, S., et al. (2020). Enhancement of feedstock composition and fuel properties for biogas production. *Energy and Environmental Sustainability*. https://doi.org/10.1007/978-981-15-0410-5_9

Cong, R., et al. (2020). Is it beneficial to use biogas in the Danish transport sector? An environmental-economic analysis. *Journal of Cleaner Production, 165*, 1025–1035.

Cudjoe, D., et al. (2020). Electricity generation using biogas from organic fraction of municipal solid waste generated in provinces of China: Techno-economic and environmental impact analysis. *Fuel Processing Technology, 203*, 106381.

Dasgupta, A., & Chandel, M. K. (2020). Enhancement of biogas production from organic fraction of municipal solid waste using acid pretreatment. *SN Applied Science, 2*, 1–14.

Delzeit, R., & Kellner, U. (2013). The impact of plant size and location on profitability of biogas plants in Germany under consideration of processing digestates. *Biomass and Bioenergy, 52*, 43–53.

Dodo, U. A., et al. (2022). Techno-economic evaluation of municipal solid waste—Fueled biogas generator as a backup in a decentralized hybrid power system. *Process Integration and Optimization for Sustainability, 6*, 431–446.

EBA. (2019). Anaerobic digestion has the potential to reduce global GHG emissions with 10–13%. *European Biogas Association* [Online]. https://www.europeanbiogas.eu/new-study-shows-that-anaerobic-digestion-has-the-potential-to-reduce-global-ghg-emissions-with-10-13/. Accessed 10 May 2023.

Eze and Agbo. (2010). Maximizing the potentials of biogas through upgrading. *American Journal Scientific and Industrial Research, 1*, 604–609.

Ferella, F., et al. (2019). A techno-economic assessment of biogas upgrading in a developed market. *Journal of Cleaner Production, 210*, 945–957.

GASUM. (2021). Renewable biogas for efficient emission cuts. *GASUM* [Online]. https://www.gasum.com/en/our-operations/biogas-production/biogas-emissions/. Accessed 10 May 2023.

Ghaem, S., et al. (2015). Feasibility study of using a biogas engine as backup in a decentralized hybrid (PV/wind/battery) power generation system e Case study. *Energy, 90*, 1830–1841.

Giraudet, S., et al. (2014). Adsorption and electrothermal desorption of volatile organic compounds and siloxanes onto an activated carbon fiber cloth for biogas purification. *Energy and Fuels, 28*, 3924–3932.

Grande, C. A., & Blom, R. (2014). Cryogenic adsorption of methane and carbon dioxide on zeolites 4A and 13X. *Energy and Fuels, 28*, 6688–6693.

Hagos, K., et al. (2017). Anaerobic co-digestion process for biogas production: Progress, challenges and perspectives. *Renewable and Sustainable Energy Reviews, 76*, 1485–1496.

Huang, W., & Fooladi, H. (2021). Economic and environmental estimated assessment of power production from municipal solid waste using anaerobic digestion and landfill gas technologies. *Energy Reports, 7*, 4460–4469.

ICLEI—SEAS. (2021). Balikpapan city accelerates urban low emission development through methane gas capture pilot project. *Local governments for sustainability Southeast Asia* [Online]. https://icleiseas.org/index.php/2021/10/21/balikpapan-city-accelerates-urban-low-emission-development-through-methane-gas-capture-pilot-project/. Accessed 16 August 2023.

Jun, Z., et al. (2021). Biogas and biofertilizer production from organic fraction municipal solid waste for sustainable circular economy and environmental protection in Malaysia. *Science of the Total Environment, 776*, 145961.

Jürgensen, L., et al. (2015). Dynamic biogas upgrading based on the Sabatier process: Thermodynamic and dynamic process simulation. *Bioresource Technology, 178*, 323–329.

Kadam, R., & Panwar, N. L. (2017). Recent advancement in biogas enrichment and its applications. *Renewable and Sustainable Energy Reviews, 73*, 892–903.

Kalinichenko, A., et al. (2016). Evaluation of biogas production and usage potential. *Ecological Chemistry and Engineering S, 23*, 387–400.

Karlsson, N. P. E., et al. (2017). Success factors for agricultural biogas production in Sweden: A case study of business model innovation. *Journal of Cleaner Production, 142*, 2925–2934.

Kasinath, A., et al. (2021). Biomass in biogas production: Pretreatment and codigestion. *Renewable and Sustainable Energy Reviews, 150*, 111509.

Khan, M. U., et al. (2021). Current status of biogas upgrading for direct biomethane use: A review. *Renewable and Sustainable Energy Reviews, 149*, 111343.

Korbag, I., et al. (2020). Recent advances of biogas production and future perspective. In *Biogas—Recent advances and integrated approaches conversion* (pp. 1–40). IntechOpen.

Kougias, P. G., et al. (2017). Ex-situ biogas upgrading and enhancement in different reactor systems. *Bioresource Technology, 225*, 429–437.

Lantz, M., et al. (2007). The prospects for an expansion of biogas systems in Sweden—Incentives, barriers and potentials. *Energy Policy, 35*, 1830–1843.

Leitão, A., et al. (2022). Anaerobic co-digestion of food waste with sewage sludge from wastewater treatment plant of Sequele, Luanda-Angola. *Environmental Challenges, 9*, 100635.

Lisowyj, M., & Wright, M. M. (2020). A review of biogas and an assessment of its economic impact and future role as a renewable energy source. *Reviews in Chemical Engineering, 36*, 401–421.

Logan, M., & Visvanathan, C. (2019). Management strategies for anaerobic digestate of organic fraction of municipal solid waste: Current status and future prospects. *Waste Management & Research, 37*, 27–39.

Magazzino, C., et al. (2020). The relationship between municipal solid waste and greenhouse gas emissions: Evidence from Switzerland. *Waste Management, 113*, 508–520.

Meegoda, J. N., et al. (2018). *A review of the processes, parameters, and optimization of anaerobic digestion* (p. 15). Public Health.

Mohanty, M. K., & Das, D. M. (2019). Industrial applications of anaerobic digestion. In *Biogas Technology* (pp. 227–249).

Munoz, R., et al. (2015). A review on the state-of-the-art of physical/chemical and biological technologies for biogas upgrading. *Reviews in Environmental Science and Bio/technology, 14*, 727–759.

Pan, S. Y., et al. (2021). Anaerobic co-digestion of agricultural wastes toward circular bioeconomy. *iScience, 24*, 102704.

Paolini, V., et al. (2018). Environmental impact of biogas: A short review of current knowledge. *Journal of Environmental Science and Health—Part A Toxic/Hazardous Substance Environmental Engineering, 53*, 899–906.

Patrizio, P., & Chinese, D. (2016). The impact of regional factors and new bio-methane incentive schemes on the structure, profitability and CO_2 balance of biogas plants in Italy. *Renewable Energy, 99*, 573–583.

Pierre Doussoulin, J., & Salazar Molina, C. (2022). A case study for economic viability of biogas production from municipal solid waste in the south of Chile. *Biogas—Basics, Integrated. Approaches, Case Studies*. https://doi.org/10.5772/intechopen.104558

Rajendran, K., et al. (2014). Uncertainty over techno-economic potentials of biogas from municipal solid waste (MSW): A case study on an industrial process. *Applied Energy, 125*, 84–92.

Singh, D., et al. (2023). Biogas generation potential of discarded food waste residue from ultra-processing activities at food manufacturing and packaging industry. *Energy, 263*, 126138.

Singkhala, A., et al. (2021). Enhancement of thermophilic biogas production from palm oil mill effluent by pH adjustment and effluent recycling. *Processes, 9*, 1–17.

Song, C., et al. (2019). Cryogenic-based CO2 capture technologies: State-of-the-art developments and current challenges. *Renewable and Sustainable Energy Reviews, 101*, 265–278.

Tan, Y., et al. (2017). Cryogenic technology for biogas upgrading combined with carbon capture-a review of systems and property impacts. *Energy Procedia, 142*, 3741–3746.

Tippayawong, N., & Thanompongchart, P. (2010). Biogas quality upgrade by simultaneous removal of CO2 and H2S in a packed column reactor. *Energy, 35*, 4531–4535.

Tshemese, Z., et al. (2023). An overview of biogas production from anaerobic digestion and the possibility of using sugarcane wastewater and municipal solid waste in a South African context. *Applied System Innovation, 6*, 1–17.

UN. (2018). Renewable energy sources cut carbon emissions, efficiently increase electricity output worldwide. *United Nation* [Online]. https://press.un.org/en/2018/gaef3501.doc.htm. Accessed 10 May 2023.

USA EPA. (2022). How does anaerobic digestion work? *United States Environmental Protection Agency* [Online]. https://www.epa.gov/agstar/how-does-anaerobic-digestion-work. Accessed 10 May 2023.

Valli, L., Rossi, L., Fabbri, C., Sibilla, F., Gattoni, P., Dale, B. E., Kim, S., Ong, R. G., & Bozzetto, S. (2017). Greenhouse gas emissions of electricity and biomethane produced using the Biogasdoneright^{TM} system: Four case studies from Italy. *Biofuels, Bioproducts and Biorefining, 11*(5), 847–860

Walsh, J. L., et al. (1989). Utilization of biogas. *Biomass, 20*, 277–290.

Wang, S., et al. (2019). Influence of temperature on biogas production efficiency and microbial community in a two-phase anaerobic digestion system. *Water, 11*, 133.

Wilkie, C. A. (2019). Biogas uses. *Soil and water sciences department, university of florida* [Online]. https://biogas.ifas.ufl.edu/uses.asp. Accessed 16 August 2023.

WNA. (2021). Carbon dioxide emissions from electricity. *World nuclear association* [Online]. https://www.world-nuclear.org/information-library/energy-and-the-environment/carbon-dioxide-emissions-from-electricity.aspx. Accessed 10 May 2023.
Xia, A., et al. (2016). Innovation in biological production and upgrading of methane and hydrogen for use as gaseous transport biofuel. *Biotechnology Advances, 34*, 451–472.
Yousef, A. M., et al. (2018). New approach for biogas purification using cryogenic separation and distillation process for CO2 capture. *Energy, 156*, 328–351.
Zupančič, M., et al. (2022). Current status and review of waste-to-biogas conversion for selected European countries and worldwide. *Sustain, 14*, 1–25.

Advancing Circular Economy and Waste Management in Zanzibar Through Youth Entrepreneurship and Innovation

Intissar Doukali, Brijesh Mainali, Charafeddine Mokhtara, and Eleonora Gatti

Abstract This chapter seeks to examine the challenges and opportunities for waste management and circular economy innovation in the context of sustainable development in the Zanzibar islands, located in Africa. The primary goals of this study are to propose some strategies for overcoming significant obstacles and capitalising on favourable circumstances to enhance the waste management system. Additionally, it aims to demonstrate the significance of youth circular entrepreneurship in promoting and enhancing circularity in the islands. In order to achieve this objective, a PEST and SWOT analysis were conducted using data gathered from literature sources, as well as through online interviews with key stakeholders, such as experts, young entrepreneurs, policymakers, and relevant practitioners. The findings indicate that in order to shift from a linear "take-make-dispose" approach to a circular waste management system in Zanzibar, it is crucial to tackle gaps in collaboration, enhance funding prospects, reinforce policy frameworks, advance education, and investigate emerging areas of waste management research, such as e-waste management. In addition, it was discovered that the potential for beneficial transformation could only be achieved via the combined endeavours of different parties involved in order to establish sustainable and efficient waste management procedures in the area under investigation. While this study offers valuable insights, further research is required to gather more on-site data in order to formulate effective and practical strategies for advancing circularity in the waste management industry and attaining sustainability objectives in Zanzibar and Africa as a whole.

I. Doukali · B. Mainali (✉) · C. Mokhtara
Department of Built Environment and Energy Technology, Linnaeus University, Växjö, Sweden
e-mail: brijesh.mainali@lnu.se

I. Doukali · E. Gatti
Global Climate Innovation Portfolio, UNICEF, Office of Innovation Sweden, Stockholm, Sweden

A. Anouzla and S. Souabi (eds.), *Generation of Energy from Municipal Solid Waste*,
https://doi.org/10.1007/978-3-031-74334-4_15

1 Introduction

Zanzibar, like many other countries throughout the world, is struggling with climate change issues. This region of the United Republic of Tanzania (URT), which includes two major islands—Unguja and Pemba—as well as several smaller islets, is especially vulnerable to the effects of climate change, such as sea level rise, a higher intensity and frequency of windstorms and cyclones, and variations in rainfall patterns. These effects may have serious implications for the local community, such as struggles to food availability, facilities and public health (OMAR, 2019). Zanzibar had a population of 1.3 million in 2012, with a density of 530 persons per square kilometre. Most people in Zanzibar depend on weather-sensitive occupations such as agriculture, fishery, and tourism for their means of living. The current climate fluctuation has substantial effects on the population.

Moreover, the islands are currently seeing alterations in their climate, which, when coupled with socio-economic issues and transformations in lifestyle, are exacerbating detrimental consequences on the coastal areas, marine ecosystem, tourism, and agriculture.

The sectors mentioned are vital for maintaining both people's means of living and the overall economy in Zanzibar (Hendriksen, 2016). Wood and charcoal are the predominant fuel sources for cooking in almost all residences in Zanzibar, but electricity and petroleum are primarily supplied from mainland Tanzania (United Nations Industrial Development Organization [UNIDO], 2014). Although Zanzibar has relatively low per capita greenhouse gas (GHG) emissions, there are significant environmental concerns that need to be handled, particularly the inadequate waste management. Hence, it is imperative for the region to adopt a development trajectory that is both environmentally friendly and has reduced carbon emissions in order to address these environmental issues, while also yielding benefits for both the environment and the economy (Global Climate Adaptation Partnership, 2012). This change would also create possibilities for climate funding to aid this transition and allow for the integration of circular models into the islands' development plan.

1.1 An Overview of Waste Management in Zanzibar

Our analysis revealed that the predominant focus of the published articles on environmental research in Zanzibar is the inappropriate disposal of waste and the resulting poisoning of the ocean. An example of this is a study published in 2002 by the Institute of Marine Sciences in Zanzibar, which raised concerns about the deterioration of the environment in the islands. The untreated urban wastewater being released into coastal waterways, and in certain areas straight onto coastal ecosystems like mangroves, is causing significant harm to the natural environment and potentially affecting the economic livelihoods of many islands. In 2002, the waste management services provided by the Zanzibar Municipal Council were deemed insufficient and substandard. The analysis emphasises the lack of a thorough and coordinated strategy for managing pollution (Mohammed, 2002). The impact of tourism on the ecosystem has been often discussed in the studies. Commercial enterprises, including shops, hotels, and restaurants, contribute about 80% of the waste generated in Zanzibar. These establishments adopt procedures that are not environmentally suitable for waste disposal (Rotarou, 2013). Approximately 60% of the solid waste generated in Zanzibar is not adequately collected and sorted, resulting in the buildup of untreated and potentially hazardous materials in unauthorised landfills and dump sites (The Data Futures Platform—UNDP, 2022). This waste is either consumed by animals, incinerated, unlawfully discarded, or carried away by storm winds into the town, where it forms piles of accumulated debris (Bakari, 2016). Moreover, Environmentalists caution that Zanzibar has become one of the most imperilled islands globally due to the escalating devastation of water sources and the direct discharge of untreated liquid waste into the Indian Ocean (IRIN News, 2006).

In the last decade, there have been multiple projects that have arisen to tackle the problem of waste management. The Zanzibar Environmental Management Authority (ZEMA) launched a decentralised pilot project for trash management in 2017. The project's objective was to enhance solid waste management policies and practices on the self-governing island. By implementing a cost-effective and profit-generating segregation incentive approach, this initiative is helping to transform Zanzibar into an island with no landfills (Center for Science and Enviornment, 2018). In addition, a government-commissioned impact assessment in Romania has revealed that starting from 2018, cohorts of young and jobless individuals have been designated to undertake the responsibility of rubbish collection in the waste-to-energy initiative in Zanzibar. Wheelbarrows are used to convey the gathered rubbish to the collection sites. The waste collectors retrieve all recyclable materials of significance. From these collection/dump locations forward, it is the responsibility of the municipality to manage the final disposal of the garbage. In rural areas, families often dispose of their waste at specified locations inside their community. Afterwards, a municipal-owned tipper truck gathers and carries the rubbish to a designated landfill for disposal. Moreover, the vehicle is employed for fulfilling transportation requirements around the district. Currently, there are no operational youth organisations in the rural areas (Vasilica, 2018). Furthermore, an operational sanitary landfill named Kibele landfill was established in early 2021. However, the current state of waste treatment alternatives in the islands is somewhat restricted, mostly focussing on informal recycling and composting initiatives. In addition to the casual activity of salvaging and reusing scrap metals, cardboard, and other items that can be recycled, the majority of recycling initiatives are primarily centred around the reuse and recycling of plastic bottles. White or green plastic bottles are collected by private firms and people in the community, who then carry them to recycling facilities. Two indigenous private enterprises are engaged in the recycling of plastic bottles, with each company recycling around 40–45% of the bottles collected on the island of Unguja. The plastic bottles undergo a series of steps including stripping, cleaning, shredding, and packing before being transported by truck for sale in Tanzania mainland or Dar Es Salaam. According to reports, the companies operate with little profits and do not have the financial resources to grow or improve their recycling operations. However, the current solid waste and septic sludge disposal facilities are inadequate, and there is no designated area for waste management. The solid waste and sewage sludge are irresponsibly disposed of at informal locations around the municipality, including ecologically and culturally sensitive regions such as wetlands, residential areas, heritage sites, and mangrove forests. The management of waste collection and disposal is inadequate, resulting in community complaints and significant environmental deterioration and health risks. Although the Kibele site is now being utilised as a temporary managed solid waste disposal facility, it is insufficient and not a viable long-term option for waste management on the island. Currently, septic waste is being informally disposed of in a mangrove region that is in close proximity to a protected heritage site (Zanzibar, 2016).

1.2 Aim, Objectives, and Research Questions

In developing nations such as Zanzibar, the concept of Circular Economy (CE) has not been included into the overall macro policies, rules, and plans, unlike in China and the European Union (EU) where it has been effectively applied. Nevertheless, these countries can serve as exemplars to acquire knowledge from and implement Circular Economy practices in emerging nations, taking into account their distinct circumstances, socio-economic conditions, and cultural influences. Hence, the objective of this book chapter is to pinpoint the existing obstacles and crucial prospects for improving waste management and adopting the circularity of waste practices in Zanzibar. By conducting a thorough and concentrated investigation, we have successfully addressed key research questions:

- What obstacles hinder the integration of circular economy principles in the waste management industry in Zanzibar?
- What potential exist for improving the waste management system in Zanzibar and how could young entrepreneurship contribute to this?
- What effective measures may be applied, specifically in utilising youth entrepreneurship, to enhance circularity in Zanzibar's waste management sector?

2 Materials and Methods

In order to address the research problems listed earlier, we employed a PEST (Policy, Environment, Social, and Technical) analysis and a SWOT (Strengths, Weaknesses, Opportunities, and Threats) analysis. Initially, a thorough examination of scientific publications and studies were conducted to assess information and data regarding circular economy (CE), entrepreneurship, and waste management in Zanzibar. Next, the PEST framework is utilised to get insight into the context and pinpoint the obstacles impeding the waste management system and the associated ecosystem on the island. Next, a SWOT analysis is conducted to identify the strengths, weaknesses, opportunities, and threats associated with the adoption of CE in waste management in Zanzibar. Ultimately, interviews are conducted with young entrepreneurs involved in waste management, as well as representatives from UNDP and UNICEF who are actively engaged in waste management initiatives in Zanzibar. These interviews involve community members and experts and aim to provide support in developing strategies to promote circularity by utilising youth entrepreneurship. All of the interviewees are situated in Zanzibar. All of them have participated in the innovation ecosystem established by UNICEF and UNDP in Zanzibar. The selection of young innovators in Zanzibar is done through an innovation challenge organised by Waste X lab, which is a joint initiative of the UNICEF office of innovation and UNDP. The policy makers are stakeholders who play a role in the implementation or establishment of the Waste X lab.

3 Results and Discussions

3.1 PEST Analysis

3.1.1 Policy Context in Zanzibar

The policy framework in Zanzibar serves as a foundation for organising waste management. In 2020, the Organisation for Economic Cooperation and Development released a paper focussing on the concept of circular economy as it applies to regions and cities (Organisation for Economic Cooperation and Development, 2020) where policies and regulations are the core of the governance of circular economy. The circular economy is enabled by the use of systems thinking and the construction of a system-oriented approach. It is accomplished by identifying the regulatory instruments that require adjustment or implementation and by mobilising financial resources and allocating them. To promote a circular economy, governments at various levels must establish clear objectives and implement strategic plans. Zanzibar is implementing diverse sustainable business models that integrate circular economy principles, including sustainable tourism, waste management, renewable energy, and sustainable farming (Zanzibar Patent No. No. 6, 2009).

The research demonstrates that the environmental policy lacks a thorough and all-encompassing waste policy. At present, the government is formulating a nationwide strategy plan for waste management, but there is no focus on systematically encouraging the practice of source separation. Furthermore, despite the existence of policies and regulations, individuals nevertheless have scepticism regarding their execution and enforcement (Knape, 2019). Zanzibar, an island located in the Indian Ocean, with a coastline spanning approximately 1,424 km. The life of the inhabitants of Zanzibar mostly depends on coastal and marine services due to the dominance of tourism (including beach activities), small-scale fishing, and maritime trade in the economy (Semboja, 2021). The primary goals of the blue economy are to implement sustainable strategies for utilising ocean resources, including the adoption of safe methods and technologies, the promotion of renewable energy, and the establishment of circular material flows to minimise waste and encourage material recycling. Additionally, the blue economy aims to regulate and prevent over-exploitation of these resources. Additionally, the UNDP and UNICEF have conducted a comprehensive analysis to gain a thorough understanding of the solid waste ecosystems in Zanzibar. This initiative aims to promote a circular economy and align with the blue economy concept (The Data Futures Platform—UNDP, 2022). As the tourism business grows, more garbage is generated from tourism activities and is improperly disposed of, leading to its accumulation in the ocean. In research conducted by Carol Maione on the pollution affecting the seas of Zanzibar, it was found that plastic constituted nearly half (48.5%) of the waste entering the environment, with single-use packaging being the most prevalent kind of litter (Maione, 2021).

Furthermore, as of June 1, 2019, the United Republic of Tanzania has become a member of the Southern African Development Community (SADC) and has implemented a comprehensive ban on plastic bags. This restriction restricts the import, export, manufacture, selling, storage, providing, and use of plastic bags. One of the key aspects contributing to the success of the single-use plastic bag policy was the inclusion of manufacturers as essential stakeholders in the approved total ban (Bezerra, 2021). Zanzibar has enforced a comprehensive prohibition on single-use plastic bags, driven by its waste management strategy. This ban, which took effect on June 1, 2019, encompasses the import, export, manufacturing, sale, storage, supply, and use of plastic bags. The 2019 urban municipal council laws for solid waste management in Zanzibar have the objective of fostering sustainable and ecologically responsible practices for handling solid waste. The primary objective of the policy is to diminish the quantity of trash produced, encourage the practice of reusing, recycling, and recovering materials, and guarantee the appropriate disposal of garbage. Furthermore, the policy establishes a structure for the transportation, disposal, and diversion of trash. Additionally, there are explicit regulations and penalties in place for instances of mishandling the trash. The policy addresses the unlawful employment of waste scavengers and establishes a legal framework for their employment (Juma, 2019).

Since 2015, Zanzibar has strengthened its institutional framework for waste management through the implementation of the Zanzibar Environmental Management Act. This act led to the establishment of the Zanzibar Environmental Management Authority (ZEMA), which is responsible for coordinating, regulating, monitoring, and supervising environmental management issues (Zanzibar Environmental Management Authority [ZEMA], 2023).

The effectiveness of national pollution prevention initiatives relies on the active engagement of all stakeholders, particularly the public sector, which includes Non-governmental Organisations (NGOs) and Community-Based Organisations (CBOs) (Mohammed, 2002). Additionally, one of the systems that show potential for poor countries is the Locally Managed Decentralised Circular Economy (LMDCE). This technique aims to enable local communities in locations with inadequate infrastructure to assume control of their plastic waste management supply chain, while ensuring that the social, economic, and environmental advantages are retained within the local region (Chandni Joshi, 2019).

This notion bears resemblance to grassroot inventions, which exemplify a model that originates from the bottom-up.

Zanzibar boasts a robust presence of non-governmental agencies and associations that actively engage in garbage management on the island. The majority of the efforts are community-driven, such as Zanzibar Scraps and Environment Association, which are actively involved in recycling plastic garbage to prevent incineration and landfill disposal (Juma, 2023).

- The Zanzibar Recycling Project, a community-based initiative started in 2011 (Zanzibar Scraps and Environment Association (ZASEA), 2023).
- ZANREC private company, ZANREC operates waste collection services for approximately 50–60 hotels located along the East coast of Unguja in Zanzibar (ZANREC, 2023).
- Permaculture Institute Zanzibar, FTSC, the Fumba Town Service Center, serves as a hub for collecting waste from residential houses and restaurants primarily in the West B region using trucks (permaculture Institute of Zanzibar, 2023).
- TakaNiAjira Foundation is a non-governmental organization (Foundation) aiming to strengthen financial and technical support to individuals and organizations involved in environmental management in the islands with the support of Zaidi Recyclers Company (Foundation T., 2023).

The effectiveness of national pollution prevention initiatives relies on the active engagement of all stakeholders, particularly the public sector, which includes Non-governmental Organisations (NGOs) and Community-Based Organisations (CBOs) (Mohammed, 2002). In addition, the 2012 Strategic Action Plan for Climate Change integrated many strategies and initiatives to address the severe consequences of climate change and enhance the islands' ability to withstand its effects. The ZCCS, also known as the Zan-zibar Climate Change Strategy, outlines the formulation of a comprehensive strategy to tackle the issue of climate change in Zanzibar. In light of the extensive array of risks and opportunities, as well as the constraints in terms of capacity and funds, it is imperative to ascertain strategic priorities. These measures aim to enhance capacity, offer options that have minimal or no negative consequences, address the effects of current climate variability, and increase resilience for the future. They also integrate climate change into development efforts, plan for future impacts over a longer time frame, and identify initial actions to mitigate those impacts. The Strategy has identified a specific group of priority sectors for Zanzibar. These sectors include: enhancing climate information and capacity, improving disaster risk management, developing resilient coastal and marine areas, promoting climate-smart agriculture and better natural resource management (including land and water use), ensuring sustainability in forests and energy/electricity, and fostering sustainable and low carbon tourism (Global Climate Adaptation Partnership, 2012).

In 2020, the Zanzibar Planning Commission initiated a comprehensive development plan for the next three decades, specifically targeting the socio-economic components of the island and placing significant emphasis on the environment. The environment and climate change are encompassed within the strategic pillars of this growth strategy. Furthermore, goal 4.2.4 of the latter specifies: "The incorporation of sustainable management actions for the environment and natural resources, along with climate resilience values, into development objectives to achieve the highest possible contribution to long-term socio-economic development in accordance with the circular economy, while also considering the best interests of children." (Zanzibar Planning Commission, 2020). This underscores the island's emphasis on adopting circular economy business models.

Summary of the policies related to waste management and circularity in Zanzibar are given in Table 1.

Zanzibar has made substantial progress in improving its legislative and regulatory framework for waste and environmental management on a worldwide level over the last two decades. The Zanzibari government acknowledges the vital significance of the environment in fostering economic development through tourism and other industries and focusses significant emphasis on its conservation. Indeed, enhancing political dedication to the circular economy by means of governmental policies and regulations pertaining to waste reduction and Extended Producer Responsibility (EPR) will additionally offer greater assurance for the advancement of financial institutions and bolster the confidence of private investors (Schroeder, 2021).

3.1.2 Social Situation in Zanzibar

Zanzibar's population is expected to exceed 1.6 million in 2019, with nearly half of the population being under 18 years old. According to the Zanzibar Environment Management Authority (ZEMA), Zanzibar produces around 663 tonnes per day of municipal solid trash. However, only about 45% of this rubbish is collected and properly disposed of in a designated dumping site (The Wold Bank, Urban, Resilience and Land Global Practice, 2021). As a component of the urban service initiative, the Zanzibar municipal council established a routine system for collecting solid trash and constructed the first sanitary landfill on the islands, which now receives approximately 320 tonnes of waste daily. On the other hand, the pressures associated with expanding populations are expected to escalate fast. As per the most recent household budget survey conducted by UNICEF before COVID-19, the current poverty rate stands at 25.7% of the population, with children being more significantly affected at a rate of 30.1% (UNICEF, 2020). According to Wei's analysis, poverty hindered the positive interactions and amplified the negative trade-offs among the Sustainable Development Goals (SDGs) (Wei, 2023). Hence, poverty has a significant influence on the countries' advancement towards attaining the Sustainable Development Goals (SDGs). Moreover, a study conducted by World Vision in New Zealand has revealed a strong correlation between environmental issues and the poverty rates of countries.

Table 1 Recap of the policy framework

Title	Description	Area	Implemented in year
Zanzibar Development Vision 2050	A 30-year plan, to ensure sustainable economic growth. The development plan is centred on human with the aim to achieve a higher overall standard of living as well as the attainment of near-zero extreme poverty in Zanzibar	Economy—Sustainability—social development—institutional strengthening	2020
Blue Economy policy in Zanzibar	Sustainable practices and sector specific interventions to ensure circular material flows, reduce waste and promote recycling of materials, as well as controlling over-exploitation	Ocean pollution—strengthening institutional framework	2020
The Urban Municipal Council Solid Waste Management Regulations	The policy aims to promote sustainable and environmentally sound management of solid waste in Zanzibar. The main goal of the policy is to reduce the amount of waste generated, promote reuse, recycling and recovery, and ensure the proper disposal of waste	Solid waste management—strengthening regulation framework	2019
The Zanzibar Environmental Management Authority (ZEMA)	Is responsible for coordinating, regulating, monitoring and supervising the environmental management concerns	Regulatory—Environmental management	2015

(continued)

Table 1 (continued)

Title	Description	Area	Implemented in year
Community Based organisations	Several initiatives and associations active in recycling plastic litter and creating value from waste	Waste management—circular business	2013
Strategic Action Plan for Climate Change	To identify a set of priority sectors based around: Building climate information and capacity and disaster risk management; Resilient coastal and marine areas; Climate smart agriculture and improved natural resource management (including land and water use management); Sustainable forests and energy/electricity; Sustainable and low carbon tourism	Environment preservation	2012

The report states that environmental concerns can hinder individuals from attaining a satisfactory level of living (Ruck, 2016).

Collecting, transporting, and disposing of municipal solid waste is a significant cost for cities in underdeveloped countries. Waste management often represents 30–50% of the operational expenditures of municipalities (Medina, 2010). Hence, the accumulation of waste in urban areas of emerging nations may be attributed to the exorbitant expenses associated with waste management.

There is now widespread recognition that climate change is significantly impeding efforts to reduce poverty (The General Assembly—United Nations, 2015). Societies heavily reliant on natural resources are particularly affected by the effect of global climate change on their livelihoods. On a global scale, the growing occurrence and intensity of natural hazards and extreme weather events, along with the increasing number of impoverished individuals impacted by these disasters, substantiate this hypothesis (Kabir, 2019).

Historically, the economy of Zanzibar has heavily depended on natural resources, specifically agriculture and fishing. Agriculture, fishing, and forestry account for 23% of the island's Gross Domestic Product (GDP), while tourism, which heavily depends on the conservation of landscapes and beaches, contributes approximately 27% to its GDP (World Bank, 2022).

In regard to population and capacity development in Zanzibar, the youth of Zanzibar make up a significant portion of the population, with 35.3% of the population falling between the ages of 15 and 35 years (International Labour Organization,

2022). This asset plays a crucial role in fostering sustainable development. However, there is a noticeable lack of public understanding regarding the attainment of sustainable municipal solid waste management (Alfredsson, 2011). Ally contends that the general public possesses limited understanding of waste management concerns and lacks awareness of the environmental and health risks linked to inadequate waste management methods. He further states: "There is a lack of initiatives to engage the public and raise awareness regarding the significance of segregating household waste at its origin and implementing proper waste management practices." The survey findings from specific residential regions suggested a low level of segregation (8%) at the home level, with the common practice of using food leftovers for animal feeding (Ally, 2014).

A field research conducted by Victoria Rice and Johan Knape supports Ally's assertion that the people in Zanzibar have a restricted understanding of the environmental conditions in the islands. In addition, they emphasise that public knowledge is crucial for the success of grassroots initiatives in trash management in Zanzibar (Knape, 2019).

A study conducted by Kalin and Skoog supports this claim, as it reveals that the farmers they encountered during their field study in Zanzibar hold the belief that all waste may be utilised as fertiliser. This belief suggests a lack of access to information and training (Skoog, 2012).

During our research, we identified numerous groups and community-based organisations in Zanzibar that are actively promoting awareness of waste management among the general public, youth, government officials, and private sectors. The German Federal Ministry for Economic Cooperation and Development has initiated a program that provides guidance to companies regarding their sustainable involvement in developing countries and emerging economies across 40 countries, including Tanzania. This program will run from 2021 to 2024 and aims to prioritise investment and growth in three key areas: climate, sustainable supply chain management, and digitalisation (Gross, 2023). In addition, a waste management awareness campaign targeting young individuals was conducted in 2013, funded by the European Union. As part of this initiative, booklets were distributed around the population (Kloiber, 2013).

Although these programs play a crucial role in reducing the negative impacts of cities on climate change, environmental degradation, and urban poverty, their long-term viability remains a considerable challenge (Charles Goodluck, 2019).

An investigation conducted by Nkech examined households' attitudes towards solid waste services and practices, specifically focussing on their awareness of waste disposal and segregation methods, waste generation rates, willingness to pay for waste management, and waste disposal methods. The findings revealed that 71.1% of respondents were aware of waste disposal methods at the source, while 67% of respondents paid for solid waste management. Furthermore, it was observed that the trash generation rate in Zanzibar is moderate, accounting for 39.5% of the total. Additionally, a significant proportion of families (47.6%) resort to burning rubbish as an alternative means of disposal (Nkech, 2017).

The author's conclusion is that although there is a general awareness of sustainable waste disposal methods and the importance of a sustainable environment, the respondents in the survey showed a serious lack of commitment to voluntarily putting this knowledge into practice for the benefit of society.

Waste pickers in Zanzibar are individuals that clean the streets, gather waste from residences, or collect recyclable waste from landfills. These individuals often work autonomously, without affiliation to any particular organisation. In a select few nations with available data, the informal sector's proportion to total urban employment is less than 1% (International Labour Organization, 2013). Prevailing theories suggest that there are four main factors that contribute to engagement in informal employment. These factors include the exclusion of the urban poor from formal employment due to structural barriers, the exploitation of formal firms that use informal employment arrangements to avoid costs, the voluntary decision of enterprise operators to leave formal employment in order to bypass state regulations, and the strategic choices made by individuals (Chen, 2012). WIEGO (Women in Informal Employment Globalising and Organising) has a global mission to enhance the conditions of informal workers by facilitating policy discussions, conducting research, and generating data and statistics that illuminate the situation of workers in the informal sector (Charles Goodluck, 2019). Several associations and institutions in Zanzibar have made significant efforts to build a regulatory framework in order to formalise and safeguard the rights of informal workers in the region.

3.1.3 Economic Situation in Zanzibar

The informal sector is where a significant amount of circular economy activity in African countries is taking place. This has impeded the allocation of funds to a demographic that lacks financial services, credit options, banking facilities, and collateral, hence constraining the expansion of value-chains (Schroeder, 2021).

In 2016, the United Nations Development Programme (UNDP) evaluated the accessibility and availability of financing instruments and processes for supporting climate projects in Tanzania and Zanzibar. The report states that in 2016, none of UNDP's development partners, including governments, the United Nations system, international financial institutions, the private sector, foundations, civil society organisations, goodwill ambassadors, and advocates, have made any specific financial commitments to support circular economy strategies. Nevertheless, Zanzibar has already allocated funds from its own budget and received assistance from development partners to engage in adaptation and mitigation measures. According to the research, approximately 6–8% of the total Government budget was designated for initiatives related to climate in 2016 (Hendriksen, 2016).

Regarding the availability of funds, there is a shortage of varied options for obtaining long-term finance. The main sources of climate finance in Tanzania and Zanzibar are public sector funding, which includes Official Development Assistance and other forms of development assistance (The United Republic of Tanzania—Vice President's Office, 2021).

Although the private sector is making investments in climate adaptation and mitigation, they may not always be cognisant of the fact that their activities are helping to these endeavours. Furthermore, there is presently an absence of a monitoring framework to oversee the climate change investments made by the private sector.

The United Nations' department of economic and social affairs published a study on the climate financing situation in Tanzania and Zanzibar. The paper indicates that at present, budgets related to climate change are primarily directed towards technical help rather than actual on-the-ground actions. The government of Zanzibar depends on development partners who serve as Accredited Entities, Implementing Entities, or Partner Agencies to access multilateral money. This is due to the fact that certain funds necessitate collaboration between the government and its partners, and government organisations may have limited capacity and resources to fulfil the conditions set by these funds. Creating applications for UNFCCC money can be a demanding and resource-intensive endeavour. As an illustration, it took four years for a GCF project in Tanzania to receive approval for its request (Department of Economic & Social Affairs, 2021).

The African Development Bank is specifically dedicated to investing in circular economy projects throughout Africa as part of the Development of a Green Growth Investment Program in Africa, with a particular emphasis on waste management and the circular economy. This is achieved through an improved policy and regulatory environment, initiatives focused on strengthening capability, and efforts to mobilise resources (Contify Banking News, 2020).

In recent years, Zanzibar has garnered the attention of numerous international investors, namely those interested in the tourism industry. However, investors have lately taken part in climate action by founding Sustainable Zanzibar, a hotel platform that exclusively features hotels and resorts who have made the commitment to embrace eco-friendly tourism. A significant aspect of fulfilling this promise involves establishing a sustainable waste management methodology. This entails the involvement of Zanrec, MTI investments, a Swedish venture firm that holds a significant stake in Zanrec Plastics Company. They have also invested in the sustainable Zanzibar initiative (Innovation, 2023).

In 2016, the Zanzibar president's office submitted a paper to the United Nations Development Programme (UNDP) outlining a Zanzibar climate change funding structure. The administration emphasised the urgent need to establish a Zanzibar Climate Fund as a top priority. The studies suggest that the Government should evaluate the possibilities for accreditation as an Implementing Entity for the Green Climate Fund (Hendriksen, 2016).

According to our research, Tanzania is recognised as a National Designated Authority (NDA) and the CRDB bank of Tanzania has been recognised as a direct access entity since 2019. One could contend that this certification was a direct consequence of the report released by the President of Zanzibar's office in 2016.

In June 2022, CRDB and the Tanzanian government signed a Funded Activity Agreement (FAA) to launch the Tanzania Agriculture Climate Adaptation Technology Development Programme. This program is a component of a substantial funding commitment of $200 million (equal to about Sh459.9 billion) by CRDB to provide support for climate-resilient and adaptation projects throughout the country (Mwananchi Communication Limmited, 2021).

The Zanzibar Tourism Act (Zanzibar Patent No. No. 6, 2009) explicitly assigns the responsibility of promoting and maintaining eco-tourism to the Commission for Tourism. In addition, the Vision 2020 places persistent emphasis on the development of sustainable tourism, recognising that the tourism industry is now the largest and most important sector in Zanzibar. According to the Zanzibar Investment Promotion Authority, the tourist industry is the mainstay of the economy, contributing 15% to the Gross Domestic Product (GDP) (ZIPA, 2023). Nevertheless, a field research conducted by Himany has demonstrated that Zanzibar prioritises economic sustainability over ecological sustainability. Therefore, it is advisable for Zanzibar to acknowledge the significance of not just rationalising the utilisation of land in sustainable tourist development, but more importantly, ensuring that land and resources are managed ecologically for both current and future generations (Hikmany, 2015).

In order to fully harness the opportunities presented by the circular economy, it is crucial to allocate resources towards developing waste management infrastructure that enables the efficient reuse and extended lifespan of products and materials. Additionally, it is essential to promote and support innovative practices and adopt circular business models in both formal and informal sectors. These approaches may encompass repair, reuse, and sharing activities (Foundation E. M., 2021).

3.1.4 Technical and Technological Situation in Zanzibar

The utilisation of innovation and technology can have a substantial impact on the advancement of circular economy and the effective management of trash. Technology can enhance the efficiency and efficacy of recycling operations, simplifying and reducing the cost of trash recycling. Waste-to-energy technologies offer innovative solutions to decrease waste by converting it into a valuable energy resource. Technology can also enable the advancement of circular business models, which involve the intentional design of products and materials for the purpose of reuse and recycling. Advanced technology can be utilised to oversee and control products throughout their complete lifecycle, spanning from manufacturing to disposal.

Utilising digitalisation and technology in the circular economy can transform the manufacturing mindset from focussing on product/process innovation to system innovation that includes various stakeholders, different portions of the value chain, and a holistic perspective on the product lifespan. Connected devices and online platforms offer immediate data on the condition of components, suggesting when to do maintenance, replace, or repair them. This leads to improved resource utilisation and extended product lifespans (Annata, 2022).

The Waste X lab has established a digital platform that demonstrates how technology plays a crucial role in improving waste management methods and promoting innovation. This novel method enhances the exchange of information on inventions and provides assistance to young folks who are interested in circular innovation, hence expanding the scope of circular education.

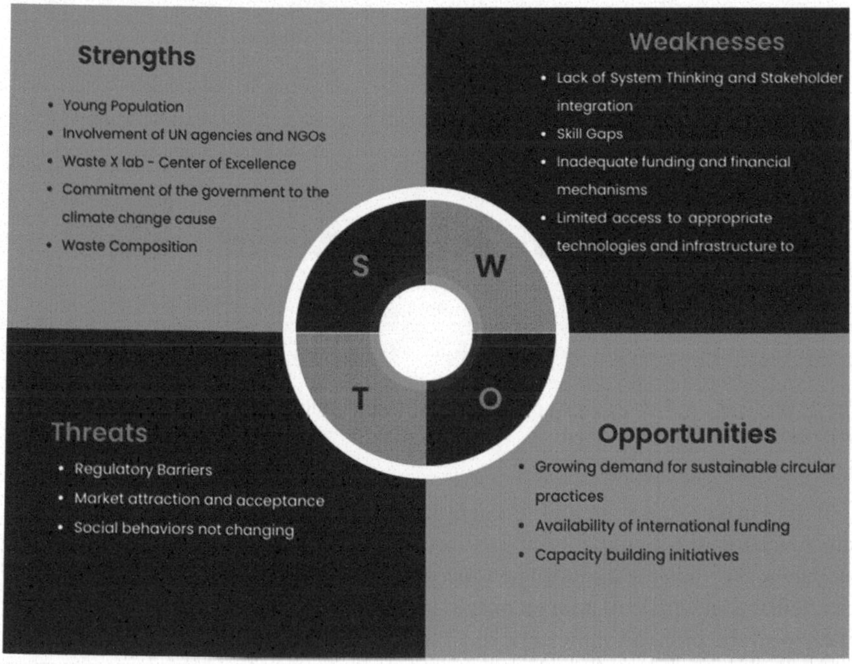

Fig. 1 Summarized SWOT matrix

3.2 Key Results from SWOT Analysis

The key strengths, weaknesses, opportunities and challenges that we highlighted within waste management in the Zanzibar context are summarised in Fig. 1.

3.3 Key Insights from Interviews

3.3.1 Factors Impacting the Efficiency of Circular Innovation in Zanzibar

Our research indicates that multiple factors influence the circular innovative ecosystem in Zanzibar. The majority of interviewees provided comparable responses about the elements that impact the progress of circular innovation and enterprises in Zanzibar (Fig. 2). Unlike other interviewers, community people assert that social behaviour and awareness are the sole factors that can influence circular innovation in Zanzibar. The World Business Council for Sustainable Development highlights that policy is one of the crucial factors that might expedite the expansion of the circular economy (World Business Council for Sustainable Development, 2019).

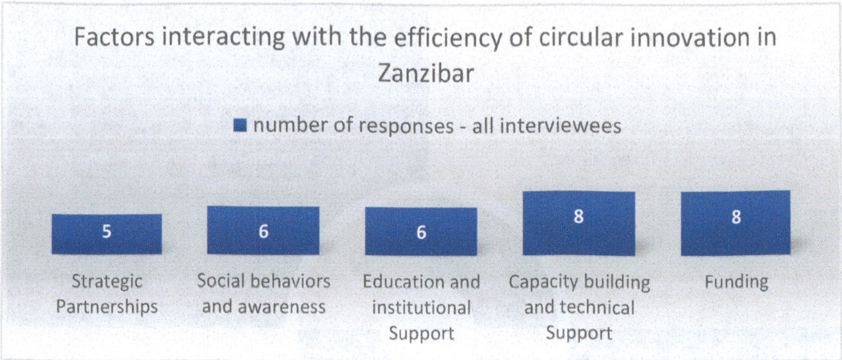

Fig. 2 Chart of the number of responses for the factors influencing the efficiency of circular innovation in Zanzibar

Policy makers have the ability to ensure that the circular economy is integrated into mainstream programs, establish further partnerships and collaborations, and encourage the implementation of new incentive mechanisms. Hence, it is intriguing to ascertain the viewpoint of policymakers regarding the aspects that influence the effectiveness of the circular innovation ecosystem (as depicted in Fig. 3). However, the young innovators themselves play a crucial role in the circular innovation ecosystem as they are the driving force behind the circular firms in Zanzibar. Hence, it is intriguing to ascertain their viewpoint on measures that can augment the effectiveness of circularity in Zanzibar (Fig. 4).

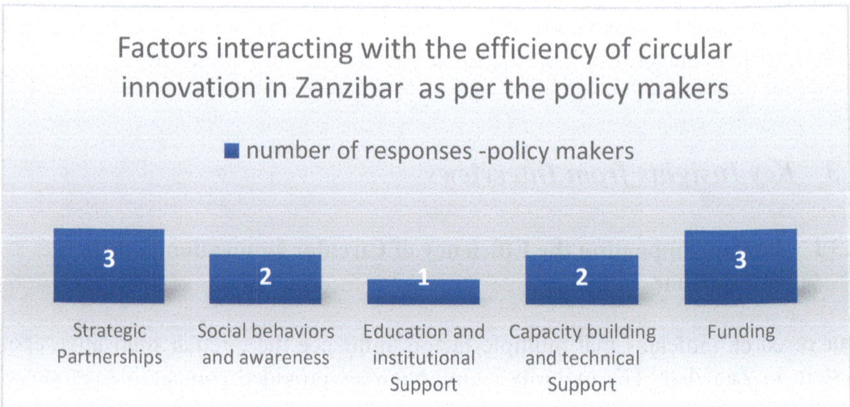

Fig. 3 Policy makers opinion on the factors impacting the efficiency of circular innovation in Zanzibar

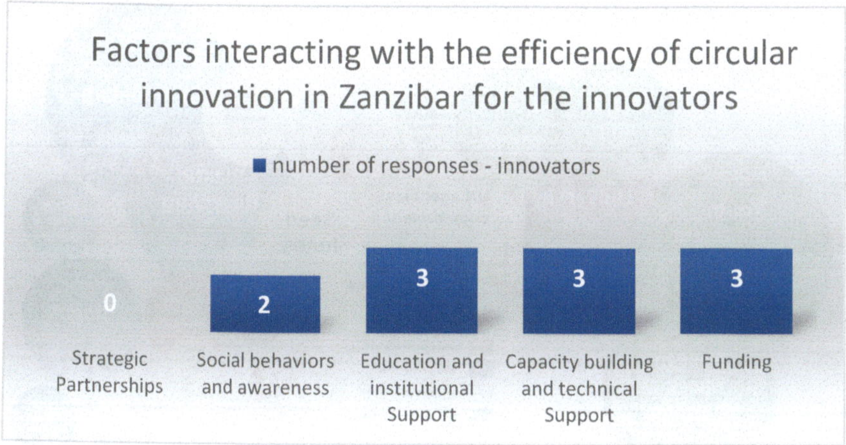

Fig. 4 Innovators opinion on the factors impacting the efficiency of circular innovation in Zanzibar

The World Economic Forum Annual meeting in 2022 concluded that better collaboration can result from cross-sector solutions, research and development partnerships and strengthening commercial viability (Poh, 2022). The role of different contributing organizations is highlighted in Fig. 5.

Fig. 5 Partnerships and collaborations for circular economy eco-system in Zanzibar

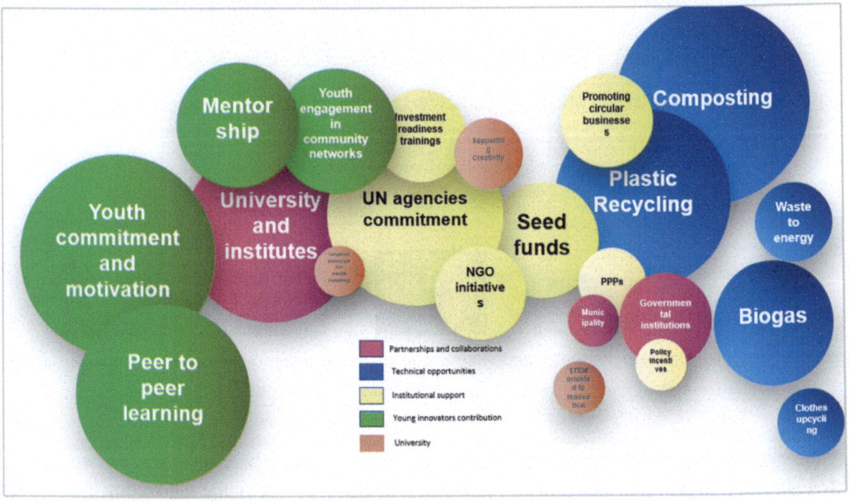

Fig. 6 Interactive components of the innovation eco-system on circular economy in Zanzibar

3.4 *Opportunities for Innovation in Circular Economy and Waste Management in Zanzibar*

3.4.1 Innovation Eco-System in Zanzibar

The investigation uncovers numerous interconnected and interdependent components of the entire value chain. This encompasses the current prospects available within the innovative eco-system of Zanzibar. The size of the bubbles in Fig. 6 corresponds to the dominance of the components. The size of the bubble directly correlates with the level of existence and success of the component inside the innovation eco-system. The interconnections among the bubbles symbolise the linkage and synergies among the components.

3.4.2 Preferred Technology and Circular Economy Solutions in Zanzibar

All the interviewers in Zanzibar were almost in complete agreement that composting is the most preferable approach for utilising waste as a resource. Furthermore, the waste composition in Zanzibar consists predominantly of organic matter. The interviewees have also shown that plastic recycling on the island has exhibited encouraging outcomes (Fig. 7).

Private companies remunerate garbage collectors for their services in collecting plastic bottles. The plastic is either recycled by a plastic manufacturer in mainland Tanzania or reused within the community for water, milk, or other items. Our

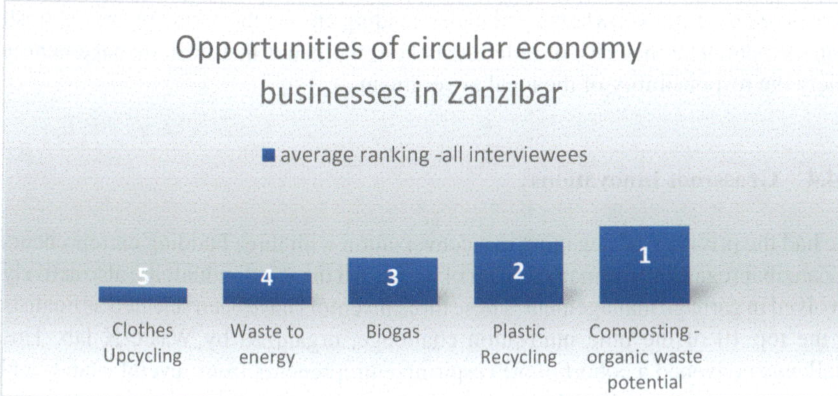

Fig. 7 Ranking of circular business solutions corresponding to their financial viability

interviews with the policy makers have uncovered that there was an endeavour to investigate waste to energy projects in Zanzibar, but it was considered impractical given the circumstances of the islands. According to the policy makers we interviewed, this project necessitates enhancing the energy distribution infrastructure on the islands. Additionally, the high capital expenditure of waste to energy solutions was identified as another reason.

3.4.3 Kibele Landfill and Waste Handling by Society

The process of rehabilitating the Kibele landfill began in early 2018, and garbage management activities commenced at the site in 2020. The initiative was implemented with the assistance of the World Bank. The projected lifespan of the Kibele dump is around 10 years. At first, three sites for trash landfilling were taken into account. Presently, the landfill is reaching its capacity in the third designated garbage disposal zone. The primary activities include the treatment of wastewater, leachate, and landfilling. In the future, the landfill will have drainage systems to collect rainwater and facilitate the storage of sorted products.

Our investigation found that there is minimal cooperation between the Kibele Landfill and the waste management initiatives undertaken by Waste X lab. The waste gathered by the teams involved in the solid waste management component of the Waste X laboratory solution is being collected from households with the purpose of either recycling it or disposing of it at designated collection sites. However, the interviewees have emphasised that the frequency of rubbish collection by the municipality is insufficient, resulting in the accumulation of waste on the streets for several days. Furthermore, the respondents residing in Zanzibar have emphasised that the communal waste receptacles for residential areas are occasionally situated at a considerable distance from residences. Consequently, individuals are occasionally compelled to discard their rubbish in the streets. In addition, several interviewers have noted the little knowledge and understanding among the public regarding trash management. The majority of individuals hold the belief that trash management is solely the responsibility of the local government.

3.4.4 Grassroot Innovations

We had the privilege of engaging in a conversation with three budding entrepreneurs in Zanzibar regarding the management of waste. All three individuals are also actively involved in garbage management. These three inventors have been selected as finalists in the top 10 of the blue innovation challenge, organized by Waste X lab. This challenge convened a cohort of 500 aspiring entrepreneurs from several islands and successfully identified 40 individuals with innovative potential for intensive training in enhancing their skills, refining their ideas, and validating their concepts.

The first young entrepreneur is **Arafa B. creator of Asili Fertilizers.** After graduating from gender and development studies, Arafa has participated in innovation incubator processes, including Waste X lab that has enabled her to scale up her business. Currently Arafa's business is successful and financially viable. In fact, She recruited recruit three other people that help her with waste collection and transformation. She's acting as a leader in her community supporting other women in joining her business. Arafa is conducting meetings in a regular basis, with other young people that come to her to learn how to produce organic products from waste.

The second young innovator in this study is **Justin B, owner of recyclable energy solutions.** He is also a student in electromechanics in the university of Dar Essalam in Tanzania. His startup company aims to reduce plastic pollution by transforming waste plastic bottles to fuel. The fuel produced can be used as source of energy for transportation and various industrial combustion systems as a furnace fuel. Justin has been able to develop a functioning prototype for his solution. Plastic pyrolysis process is a well know technology in other countries to process plastic waste however it is still not yet introduced to Tanzania. Justin is adapting this technology to Zanzibar context and has been able to involve other young people in collecting plastic bottles. Justin, like other young innovators in the island, has difficulties accessing funding and loans to improve his business.

The third and last young entrepreneur in this study is **Abdulhamid**. Along his studies in Business information technology, he also started a business, **E.F Labels & packages,** specializing in ecologic packages. He is active in campaigns to plant mangroves in the islands in Zanzibar as part of his climate activist commitments. In fact, it was his participation in the Waste X lab innovation challenge that led him to come up with the ecofriendly packages' idea. He later developed it to a successful startup. Abdulhamid makes sustainable handcraft packaging from banana leaves and softwood. With the presence of tourists in Zanzibar, many people want to buy local made packages for their gifts. Interviewees have praised the efforts of the young entrepreneurs and have emphasized on the necessity of more financial and capacity building support for these startups. Youth entrepreneurship is seen as an important pillar for the transformation of waste handling system in Zanzibar, Implementing paradigm shift projects and economic growth.

4 Discussions

4.1 *Challenges Related to Waste Management and Circular Economy Innovation in Zanzibar*

4.1.1 Lack of an Integrated Innovation Eco-System in Waste Management and Circular Economy

Our research has uncovered a deficiency in the coordination among the many parties involved in trash management in Zanzibar. The ministry of health is responsible for trash management in Zanzibar. The Ministry of Health has been involved in the implementation process of the Waste X lab. However, its ability to combine the efforts of Waste X lab innovators with the companies responsible for collecting waste from collection locations has been restricted. There is consistently a disparity between the residential areas and roads and the final destination at the landfill. Furthermore, there is a lack of communication between the entrepreneurs who utilise garbage as a valuable resource and the Kibele dump. In order to transition from basic waste management practices to adopting a circular economy approach, it is crucial to incorporate a system thinking mindset. The systems-based approach can develop the necessary capacities to recognise and comprehend enduring linear patterns, hence facilitating proactive and strategic investment in promoting sustainable transformations (Lacovidou, 2021). The holistic approach also promotes collaboration among the various stakeholders.

4.1.2 Limited Capacity Building Activities for the Young Innovators

The innovators in our study have generally had a favourable experience with the capacity building training provided by the Waste X lab's centre of excellence. The training they got empowered them to conceptualise their business idea and examine its environmental and financial viability.

The innovators are proactive participants in their communities; yet, they have always emphasised the necessity for technical assistance to improve their ideas. These young innovators and students could get valuable insights by exchanging and adopting successful strategies used in other nations. Their ideas are grassroots innovations that are facilitated by the local context. However, in order to expand the reach and effectiveness of these innovations, technical support is required. For instance, Justine has successfully developed a prototype for his pyrolysis technology. However, further technical assistance is needed to expand its scale and perhaps implement it in an industrial setting. Abdulhameed's packaging business seeks assistance in determining the optimal material to employ and implementing an efficient processing method for the products.

The little technical assistance in waste management and circular economy in Zanzibar can be attributed to the lack of prioritisation of this field in the education system. All of our participants said that they did not receive any formal education on circular economy practices. Nevertheless, they were exposed to waste management practices such as beach cleaning and environmental cleaning initiatives from an early age in order to enhance their consciousness. The university's participation in the Waste X lab operations and establishment is crucial for ensuring its success and long-term viability. The State University of Zanzibar (SUZA) university successfully provided the necessary physical infrastructure to build the Waste X lab and ensured the presence of qualified people in the lab. Additionally, the institution serves as a centre for young individuals with exceptional abilities and creative thinkers, facilitating their seamless absorption into the laboratory.

4.1.3 Limitations of Funding and Bankability

Poverty reduction is a fundamental aspect of Zanzibar Development vision 2050. According to the conducted study, the interviews have revealed that young individuals face challenges in accessing financial resources. According to Dr. Abubakar, the head of the Waste X lab at SUZA University, obtaining loans is challenging, even for individuals with a stable income. The inventors have attempted to obtain cash that has been allocated for young entrepreneurs as part of the economic stimulus measures implemented by the government during the Covid-19 pandemic. Nevertheless, Arafa has indicated that the procedure has been protracted, and the criteria were intricate for her to meet. In addition, the young entrepreneurs lack the requisite abilities to create proposals that can effectively justify the allocation of funds for their ideas. A policymaker who is part of the team responsible for formulating Zanzibar's blue economy plan has confirmed our research findings, which suggest that the CRDB Bank is interested in providing financial support for green economy projects in Zanzibar. Nevertheless, there appears to be a communication deficit between banks and young entrepreneurs, which has prevented them from accessing potential financial prospects. The dearth of financial opportunities directly affects the level of sophistication exhibited in the company ideas of young entrepreneurs. On the other hand, Arafa, one of the three inventors who received financial support to develop their company concepts, has seen her firm prospering. She is earning revenue and creating work chances.

4.1.4 People's Behaviour Hindering Waste Management:

Individuals' behaviour and attitudes towards waste play a pivotal role in the advancement of sustainable waste management systems. In order to effectively tackle this issue, it is crucial to employ social behaviour change tactics and incorporate the notion of a paradigm shift when devising systematic solutions. The changing consumption habits in Zanzibar have led to a shift in waste composition, characterised by an increased presence of plastic, metal, glass, and other types of garbage together with organic waste. During an interview, an expert emphasised the significant influence that altering individuals' mindsets may have on improving waste management techniques. Harnessing the potential of young entrepreneurs who are interested in circular innovation is a significant move towards achieving a fundamental change in waste management. Strengthening the regulations and ensuring their strict implementation can effectively promote a shift in social behaviour among individuals. Moreover, adopting a participatory waste management paradigm that recognises and appreciates individuals' efforts in environmental conservation might effectively influence their actions.

4.1.5 Policy Implications

All the interviewees concur that the authorities have established regulations, however there are challenges in effectively enforcing these restrictions in practice. Enforcing regulations in a strict manner will enhance individuals' responsibility in relation to pollution.

Intellectual property (IP) serves as a catalyst for economic expansion and competitiveness, while also fostering innovation (WIPO, 2022).

IP laws incentivise creativity by providing individuals or organisations with exclusive rights to their creations or innovations, so rewarding their inventive endeavours. IP frameworks facilitate the establishment of trust among parties. However, one expert expressed the view that Zanzibar does not have a developed intellectual property policy. This has already resulted in confrontation with one of his pupils who has devised a novel technical procedure for managing the sand from the beaches, which is utilised for construction purposes. The student found out that her idea had been accepted by an organisation and contracted to another company without her involvement. Hence, it is imperative to establish rules that promote a transparent and reliable innovation ecosystem in the islands. Furthermore, the innovators have collectively conveyed the challenges they have in initiating their own startup ventures, which encompass securing financial resources and cultivating robust customer relationships. An inhibiting factor is the requirement to have a physical office in order to initiate a business. By circumventing such bureaucratic processes, young entrepreneurs will be able to officially form a startup or corporation and expand their client base significantly.

4.2 Opportunities and Strategies for Promoting Circularity and Enhancing Waste Management in Zanzibar.

4.2.1 Tourism Industry

In 2022, a study conducted by the World Bank found that expanding the tourism industry can serve as a potential means to expedite poverty alleviation (World Bank, 2022). Our literature review, interviews, and research have all emphasised the significance of the tourism industry in the economy of Zanzibar. As the tourism sector grows, more garbage is being generated from tourism activities and is being improperly managed once it is used, resulting in it ending up in the ocean (Maione, 2021). Uniting the tourism business and waste management sector in the islands could create a mutually beneficial opportunity to promote innovation. Zanrec, a company actively involved in utilising community-based solutions on the island, has been intensifying its collaboration with local hotels during our ongoing study. Zanrec engages in collaborations with approximately 106 hotels. Nevertheless, Zanzibar boasts a staggering number of over 700 hotels, each offering ample possibilities for waste valorisation. Moreover, the recent introduction of Sustainable Zanzibar, a reservation platform that advocates for ecotourism by showcasing eco-friendly accommodations, is anticipated to encourage a greater number of hotels to implement measures aimed at minimising their environmental footprint (Madho, 2023). A research conducted by the Stewarding Association International found that approximately 70 to 75% of the garbage generated by hotels is biodegradable. However, when this waste is mixed with other types of waste and disposed of at collection sites, it represents a missed opportunity for waste conversion. We strongly believe that including garbage from hotels, either through Zanrec or by directly involving hotels, in the operations of young entrepreneurs might significantly enhance their productivity and consequently, their overall impact.

4.2.2 University Involvement as a Key Stakeholder

The SUZA played a pivotal role in facilitating the implementation of the Waste X lab solution by serving as a central point for young individuals to engage in discussions about their innovative ideas and seek support from within the university. Furthermore, the university has the ability to initiate a fundamental change in the way circularity is approached in general. This may be achieved by leading the way in research, education, and innovation, influencing a new way of thinking and offering solutions to facilitate the transition towards a circular economy. In the case of Zanzibar, we propose incorporating the concept of circular economy into their educational programs. This would help develop a new cohort of professionals who possess the necessary skills and mindset to promote sustainable and regenerative practices in various industries and sectors. Ultimately, this would contribute to a more circular and sustainable future.

4.2.3 Policies in Movement

Our interviewees have also addressed the issue of e-waste in the islands. In the past decade, the second-hand market for electronic goods in Zanzibar has experienced significant growth. However, in order to reduce the environmental impact caused by these used gadgets, the government has implemented higher taxes on their imports. The Zanzibar investment promotion authority is now investigating the possibility of implementing a system where manufacturers are required to take back their products. Currently, Asian firms have expressed interest in reclaiming the electronic garbage in the islands.

The primary measures for improving circularity in the waste management sector in Zanzibar are outlined in Table 2.

5 Conclusion

This study employs a PEST-SWOT approach to develop some key strategies for fostering circularity in the waste management sector in Zanzibar, Africa. Adopting circular economy innovation is essential for developing a sustainable waste management system in Zanzibar. Zanzibar may contribute to a more sustainable future by shifting from a linear "take-make-dispose" strategy to a regenerative approach. In order to achieve this goal, it is crucial to tackle any existing gaps in collaboration, enhance funding options, reinforce policy frameworks, raise awareness, and investigate new areas of study in waste management, such as e-waste management. Integrating youth participation in the supply chain has been recognised as a viable approach to progress in this regard. The realisation of good change in waste management methods on the islands is contingent upon the collaborative efforts of diverse stakeholders to ensure sustainability and effectiveness. Promoting young entrepreneurship and innovation is considered essential in order to achieve the objective. The research findings have been verified using expert interviews. While this study offers valuable insights with broad sectoral strategies, further research is required, and additional on-site data must be gathered to formulate effective and practical strategies for promoting youth entrepreneurship in the waste management sector and achieving sustainability goals in Zanzibar and Africa.

Table 2 Key strategies to enhance circularity in Zanzibar

Key strategy	Impact description
Infrastructure development	Ensuring the availability of collection and sorting systems will facilitate access both to households that want to dispose of the waste and the entrepreneurs that are interested in using the waste as a resource. This will also improve the environmental sustainability
Stakeholder collaboration and system innovation	The holistic approach of engaging all the different entities and organizations in the waste management efforts can speed up the process of closing the loop and enhance the value chain
Funding opportunities	Seed funding provided by Waste X lab has provided tangible impacts on the growth of the innovations. Sharing information on the climate finance opportunities facilitated by the government will enhance the young entrepreneurs' businesses
Encouraging public–private partnerships	In Zanzibar, waste management is primarily carried out by the Ministry of Health in collaboration with municipalities. However, establishing public–private partnerships (PPPs) could strengthen waste management efforts and bridge the existing gap. Such partnerships would enable the pooling of resources, expertise, and knowledge from the private sector, enhancing the overall effectiveness of waste management initiatives
Monitoring and evaluation	The collaboration between the Waste X lab and the innovators has created an opportunity for ongoing support and follow-up, fostering accountability and independence among the entrepreneurs involved
Promoting innovation, circular businesses, and research	The innovations within the Waste X lab have been able to create new market niches such as the one for the biodegradable packaging. It increases competitiveness; Arafa has been able to connect with her own clients for her organic fertilizer by enhancing the quality of her product and providing competitive pricing. In general, the youth in Zanzibar with the support of the university have the capability to implement activities that will drive innovation in waste management and circularity
Increasing awareness and education	Raising awareness about waste management in the islands has a direct impact on reducing the presence of waste in the streets. Additionally, education plays a pivotal role in fostering circular mindsets and driving innovation in the region
Organic waste management and Resource recovery	Organic waste management: Businesses engaged in organic waste transformation are generating revenue, and the organic waste itself holds potential as a future source for waste-to-energy in Zanzibar. However, recent studies have indicated that implementing waste-to-energy solutions requires a significant initial capital expenditure (CAPEX), which currently poses financial viability challenges Resource recovery: Zanzibar has seen a thriving plastic recycling industry, resulting in reduced landfill waste and the creation of employment opportunities

Acknowledgements This book chapter is part MSc thesis work of the first author and further contribution by supervisors to enhance as the chapter. Authors are grateful to UNICEF's Office of Innovation for the opportunity to conduct MSc thesis research on circular economy and innovation. The experience and insights gained during this thesis work has been instrumental in shaping this book chapter. The authors would also like to thank all the respondents for providing some key insights for this chapter. Besides, thanks to the UNICEF Office of Innovation and Linnaeus University for their assistance with this research work.

References

Act, Z. T. (2009). *Zanzibar Patent No. No. 6.*

Alfredsson, F. (2011). *Sustainable management of land and environment in Zanzibar.*

Ally, B. (2014). Municipal solid waste management of Zanzibar: Current practice, the challenges and the future. *International Journal of Current Research and Academic Review*, 5–19.

Annata. (2022). *Leveraging technology towards a circular economy.*

Bakari, A. A. (2016). Challenges and problems of solid waste management in three main markets in Zanzibar. *Advances in Recycling & Waste management*.

Bezerra, J. C. (2021). Single-use plastic bag policies in the Southern African development community. *Environmental Challenges, 3.*

Center for Science and Enviornment. (2018, June 14). *Decentralized solid waste management in Zanzibar, Tanzania.* Center for Science and Enviornment.

Chandni Joshi, J. S. (2019). Building momentum for sustainable behaviors in developing regions using Locally Managed Decentralized Circular Economy principles. *Chinese Journal of Chemical Engineering, 27*, 1566–1571.

Charles Goodluck, J. G. (2019). *Sustainable waste collection and recycling for inclusive cities.* Dar Essalam.

Chen, M. A. (2012). *The informal economy: Definitions, theories and policies.* Women in Informal Employment Globalizing and Organizing.

Contify Banking News. (2020, December 23). African Development Bank takes steps to accelerate the circular economy in Africa. *Gale Business: Insights.*

Department of Economic and Social Affairs. (2021). *Developing the strategic guidelines for climate financing, considering possible linkages between climate financing and post COVID-19 agricultural development in Tanzania.*

Foundation, E. M. (2021). *Circular economy in Africa: examples and opportunities.*

Foundation, T. (2023, April 23). *TakaNiAjira Foundation.* https://takaniajira.org/

Global Climate Adaptation Partnership. (2012). *Assessing the benefits of a lower carbon development pathway for Zanzibar.*

Global Climate Adaptation Partnership. (2012). *The economics of climate change in Zanzibar—Final report.*

Gross, R. (2023, March 22). *Advising companies on sustainable commitment in developing countries and emerging economies.* Retrieved from GIZ: https://www.giz.de/en/worldwide/93161.html

Hendriksen, G. (2016). *Zanzibar climate change financing mechanism and resource mobilisation plan.* United Nations Development Programme.

Hikmany, A.-N. H. (2015). A legal analysis of the land planning authorities and sustainable tourism in Zanzibar. *Journal of Asian and African Social Science and Humanities*, 58–74.

Innovation, E. (2023, February 10). *MTI participates in Zanzibar Tourism Investment and Travel Exhibition 2023 (ZTITE).* MTI Investments.

International Labour Organization. (2013). *Measuring informality: A statistical manual on the informal sector and informal employment.*

International Labour Organization. (2022). *ToR_Youth Zanzibar.*

IRIN News. (2006, December 18). New drive for the environment in Zanzibar. *The New Humanitarian.*
Juma, M. (2023, March 16). *Zanzibar scraps and environment association.* Retrieved from 2011 SEED Award Winner Waste Management Zanzibar, Tanzania. https://seed.uno/enterprise-profiles/zanzibar-scraps-and-environment-association
Juma, M. K. (2019). *Solid waste management regulations of urban municipal council.* Department of Social and Development Services.
Kabir, M. E. (2019). Climate change effects on people's livelihood. *Climate Action*, 169–179.
Kloiber, U. (2013). *Waste in Zanzibar education and awareness guide.*
Knape, V. R. (2019). *Towards a societal transformation—A qualitative study of grassroots innovations within waste management in Zanzibar*. University of Gothenburg.
Lacovidou, E. (2021). A systems thinking approach to understanding the challenges of achieving the circular economy. *Environmental Science and Pollution Research*, 24785–24806.
Madho, J. (2023, April 14). General manager for Zanrec. (M. I. SE, Interviewer).
Maione, C. (2021). Quantifying plastics waste accumulations on coastal tourism sites in Zanzibar, Tanzania. *Marine Pollution Bulletin*, 168.
Medina, M. (2010). *Solid wastes, poverty and the environment in developing country cities.* world Institute for Development Economics Research.
Mohammed, S. (2002). Pollution management in Zanzibar: The need for a new approach. *Ocean & Coastal Management*, 301–311.
Mwananchi Communication Limmited. (2021, November 3). CRDB unveils Sh460 billion facility for financing tech on climate. *The Citizen.*
Nkech, B. M. (2017). Attitudes of households toward solid waste management services and practices in Zanzibar. *International Journal of Development Research* , 16154–16161.
Omar, M. K. (2019). *Climate change adaptation in Zanzibar and the implications for evaluation.* Zanzibar.
Organisation for Economic Cooperation and Development . (2020). *The circular economy in cities and regions: Synthesis report.*
Permaculture Institute of Zanzibar. (2023, April 23). *Permaculture institute of Zanzibar*. https://permaculture-zanzibar.co.tz/
Poh, J. (2022). *3 Ways we can collaborate better for a stronger circular economy.* World Economic Forum Annual Meeting.
Rotarou, E. S. (2013). Environmental impacts of toursime on Zanzibar. *Revista Interamericana de Ambiente y Turismo*, 2–17.
Ruck, B. (2016). *Poverty & the enviornment.*
Schroeder, P. (2021). *Circular economy in Africa: Examples and opportunities.* Ellen Mac Arthur Foundation.
Semboja, J. (2021). *Realizing the blue economy in Zanzibar: Potentials, opportunities and challenges.* The Institute of African Leadership for Sustainable Development (UONGOZI Institute).
Skoog, K. S. (2012). *Asessment of the waste management system on Zanzibar.*
The Data Futures Platform—UNDP. (2022). *Crowdmapping informal urban infrastructure to improve waste management.*
The General Assembly—United Nations. (2015). *Transforming our world: The 2030 agenda for sustainable.*
The United Republic of Tanzania—Vice President's Office. (2021). *Mationally Determined Contribution.*
The Wold Bank, Urban, Resilience And Land Global Practice. (2021). *Implementation completion and results report.*
UNICEF. (2020). *National Budget Brief 2019/2020.*
United Nations Industrial Development Organization (UNIDO). (2014). *Assessment for potential of ethanol for household.*
Vasilica, I. A. (2018). *Waste to energy project—Zanzibar island.*

Wei, Y. (2023). Exploring the impact of poverty on the sustainable development goals: Inhibiting synergies and magnifying trade-offs. *Sustainable Cities and Society, 89*, 104367.
WIPO. (2022, July 21). *Discussion on the role of IP in supporting innovation ecosystems.*
World Bank. (2022). *Zanzibar poverty assessment: More diversification in tourism will help accelerate poverty reduction.* Dar Es Salam.
World Business Council for Sustainable Development. (2019). *Policy enablers to accelerate the circular economy scaling up actions across regions and stakeholders.* WBCSD.
ZANREC. (2023, April 23). *About us.* Retrieved from ZANREC: https://www.zanrec.com/home#about_us
Zanzibar Environmental Management Authority (ZEMA). (2023, March 18). *Introduction.* Retrieved from The Zanzibar enviornmental management authority: https://zema.go.tz/introduction.php
Zanzibar Planning Commission . (2020). *Zanzibar Development Vision 2050.* Zanzibar.
Zanzibar Scraps and Environment Association (ZASEA). (2023, April 23). *2011 SEED award winner waste management Zanzibar, Tanzania.* Retrieved from SEED: https://seed.uno/enterprise-profiles/zanzibar-scraps-and-environment-association
Zanzibar, R. G. (2016). *Environmental and Social Management Framework (ESMF).*
ZIPA. (2023, February 18). *Tourism.* Retrieved from Zanzibar Investment Promotion Authority (ZIPA): https://www.zipa.go.tz/sectors/tourism/

Emerging Environmental Contaminants from Landfills

Md.Mostafizur Rahman, Sadia Sikder, and Mohammad Toha

Abstract Emerging contaminants (ECs) refer to compounds that have recently been discovered to exist in the environment and have been identified as potential environmental hazards. Emerging contaminants (ECs) refer to compounds recently found to be widespread and potentially detrimental to the environment or public health. However, due to limited evidence, a comprehensive assessment of their risk needs to be improved. Therefore, it is vital to comprehend the makeup of emerging concern contaminants (CECs) in landfills. The landfill gas comprises a variety of harmful gases and compounds that contribute to global warming. On the other hand, leachate contains ECs, which pose a severe threat to the environment and public health. This chapter aims to put more emphasis on the emerging environmental contaminants from landfills to enlighten the issues and risks associated with them. This chapter discussed the origin, disposition, and movement of emerging pollutants in landfills, particularly in landfill leachate and surrounding habitats. Also, the threats that these compounds pose to the environment or human health were discussed in this chapter.

1 Introduction

Landfills are often the last stop for the wide variety of solid and liquid municipal garbage generated from households, businesses, and factories that have been abandoned. Landfill use is expected to increase as the global population and the economies of individual countries continue to develop (Masoner et al., 2014).

Landfills are locations created to house trash. They are made to reduce the impact of waste on the environment and human health (National Geographic Society, 2022).

M. Rahman (✉) · S. Sikder · M. Toha
Department of Environmental Science, Bangladesh University of Professionals, Mirpur, Cantonment, Dhaka 1216, Bangladesh
e-mail: rahmanmm@juniv.edu

M. Rahman
Laboratory of Environmental Health and Ecotoxicology, Department of Environmental Sciences, Jahangirnagar University, Savar, Dhaka 1342, Bangladesh

A. Anouzla and S. Souabi (eds.), *Generation of Energy from Municipal Solid Waste*,
https://doi.org/10.1007/978-3-031-74334-4_16

These days, landfills are carefully monitored and designed to dispose of solid waste safely. Federal regulations require careful placement, planning, operation, and monitoring of landfills. In addition, they are manufactured to prevent harmful chemicals from entering the waste stream. Landfills cannot be built in ecologically fragile areas, and on-site environmental monitoring systems determine their sites. Groundwater contamination and landfill gas are two potential dangers that can be detected by these systems, which also serve as an added layer of protection (US EPA, 2016).

A landfill's trash remains there for a very long time. A dump contains extremely little oxygen and moisture. In these circumstances, garbage decomposes slowly. Landfills are just intended to bury waste; they are not intended to decompose it. Whenever a landfill reaches capacity, it is often covered. Different capping technologies are used for general and hazardous waste (Rinkesh, 2020).

Landfills make a great source of energy. Carbon dioxide and methane are created as rubbish builds up and starts to decompose. These gases can be extracted, purified, and used to produce energy. Landfills are the third source of methane generation in the United States. The nation's landfills generate about 95.6 million of this Carbon (IV) oxide each year. Landfills cost little.

Additionally, eco-friendly are modern landfills. Any nation's efficient waste management system is a multimillion-dollar endeavour that requires labour at every stage. More regional landfills equate to more jobs and improved living conditions for the local population. Although landfills have a lot of potential for managing trash, they have also given rise to several worrying problems, such as developing pollutants (Rinkesh, 2020).

The emergence of a novel category of contaminants has elicited significant concern due to its extensive environmental distribution and heightened public awareness. The compounds referred to as "emerging contaminants" (ECs) are artificial pollutants that lack regulation and pose significant ecological concerns, potentially requiring immediate legislative action. Persistent environmental pollutants are substances that can survive in the environment for long periods and can cause disruptions to the physiological functioning of specific receptors. These pollutants are found in minimal quantities in humans' and animals' air, soil, water, food, and biological tissues. Environmental pollutants of significant emerging concern are present. According to the study conducted by Ramakrishnan et al. (2014), Certain environmental chemicals have been classified as endocrine-disrupting substances (EDCs).

The potential production of leachate containing diverse organic chemicals, including new pollutants such as medicines, plasticizers, personal care products, and steroid hormones, poses a risk to landfills that accept complex and heterogeneous garbage. The existing landfill monitoring criteria outlined in Part 258 of the Code of Federal Regulations still need to encompass the examination of various Contaminants of Emerging Concern (CECs). The increasing body of evidence showcasing the presence of CECs in aquatic and terrestrial ecosystems originating from different urban, industrial, agricultural, and other human-related sources contributes to the growing recognition of their significance. While the worldwide occurrence of CECs is widely recognized, there is still a significant amount of knowledge to be gained regarding their destiny and effects. However, increasing research indicates that CECs could adversely affect ecosystem health (Masoner et al., 2014).

As defined by the EPA, an emerging contaminant is a chemical or material that poses an actual, potential, or perceived risk to human health or the environment or lacks recognized health criteria. A pollutant can be "emerging" due to the identification of a fresh source or a fresh route to reach people (Connecticut, 2023).

Emerging contaminants, also known as contaminants of emerging concern, are artificial or naturally occurring chemicals or microorganisms that are not regularly monitored in the environment but have the potential to harm the environment and human health. Since landfills are frequently used as the final disposal method for municipal solid wastes, which may contain significant amounts of emerging contaminants, they have been recognized as one of the primary environmental sources of emerging contaminants. It is also a considerable concern that the growing pollutants in municipal solid wastes could slowly seep into landfill leachates and be released unintentionally or intentionally into the environment (Sui & Yu, 2023).

Most of these compounds have not yet been researched for their environmental and human toxicology, and many of them are not or cannot be tested for in municipal water systems, making them a threat. Additionally, these contaminants cause the creation of by-products whose chemical composition is yet unknown as they transit through drinking water treatment systems (Rosenfeld & Feng, 2011).

Researchers are concentrating on these new landfill contaminants so that appropriate management measures can be implemented. These contaminants can be controlled more effectively if data and findings are available. This paper aims to investigate the emerging contaminants coming from landfills. This study also explored how these emerging contaminants affect the environment and human health.

2 Different Environmental Contaminants of Landfill

2.1 Conventional Contaminants

Landfill leachate is considered dangerous due to its high amounts of pollutants. These pollutants include organic waste, heavy metals, nitrogen compounds (mainly ammonia nitrogen), a range of ions (e.g., chlorides, carbonates, and sulfates), and persistent organic pollutants (POPs). The hydrogeology of the landfill area, the weather, the age and rate of landfill degradation, the type and composition of municipal solid waste deposited, and the stage of waste degradation all influence the chemical composition of leachate (Mandal et al., 2017; Naveen et al., 2017). There may be substantial changes in the levels of some contaminants as the landfill ages and is influenced by various factors (Torretta et al., 2017).

2.2 Emerging Contaminants

Leachates contain organic and inorganic pollutants, some toxic to the environment. The composition of landfill leachate depends on the rubbish discharged. Underdeveloped nations lack waste segregation, management, and treatment systems. Landfills are also poorly equipped to stop leachate discharge. Mixed garbage at landfill dump sites may occur from improper waste segregation. Food waste, residential rubbish, plastics, paints, mercury, batteries, dangerous chemicals and heavy metals are all part of municipal solid waste (MSW). Generally, dissolved organic matter contains acids, alcohols, aldehydes, sugars, and inorganic compounds, including calcium, magnesium, sulphate, chloride, and ammonia. These elements are much more abundant than in ordinary aquifers. Leachates contain inorganic pollutants such as ammonium, phosphorus, sulfate, and heavy metals. Young leachates are predominantly volatile fatty acids, which decrease with landfill age. Humic and fulvic acids are common in old landfills and leachate plumes. Most heavy metals are Fe, Pb, Ni, Cd, As, Cr, Cu, and Hg. Landfill leachates contain aromatic hydrocarbons like benzene, toluene, ethylbenzene, xylene, phenols, insecticides, polyethene, plasticizers, and halogenated organic compounds like PCBs and dioxins. Pathogenic microorganisms, including coliform bacteria and a few viruses, live in landfills. However, pH and temperature may cause these bacteria to become inactive (Salam & Nilza, 2021).

Landfill leachates contain dissolved and suspended debris that pollutes surface and groundwater. They may harm the environment by transporting toxic substances. Many of their components are refractory and dangerous. Landfill leachates contain dissolved organic molecules, inorganic macro compounds, heavy metals like copper, lead, cadmium, chromium, nickel, and xenobiotic substances like polychlorinated biphenyls. In many landfill leachates, perfluorooctanoic acid and benzothiazole are complex organic compounds released by industrial effluents. Many municipal solid waste landfill leachates contain 100–1000 m microplastics, another dangerous component. These have been made for a long time, and polyethene and polypropylene have grown in popularity due to their utility. Xenobiotics such as halogenated organic chemicals and heavy metals like lead, arsenic, cadmium, and mercury have long been harmful (Salam & Nilza, 2021).

Landfill leachate contains microplastics, pharmaceuticals, PFAS, antibiotic resistance genes, and per- and polyfluoroalkyl chemicals. Table 1 lists emerging landfill pollutants. MSW leachate comprises dissolved organic matter. TOC, COD, and BOD are vital metrics in DOM. The geochemical and biological effects of DOM affect landfills significantly. Inorganic macro components contain calcium, magnesium, sodium, potassium, ammonium, iron, chloride, sulfate, nitrate, and hydrogen carbonate. One of the most significant leachate contaminants is heavy metal. Heavy metals can damage soil, groundwater, surface water, and human health due to their mobility, solubility, and transfer in water and plants. Cadmium, chromium, mercury, copper, zinc, lead, and arsenic dominate waste disposal leachate. Xenobiotic organic compounds contain monoaromatic hydrocarbons like benzene, toluene, ethylbenzene, and xylenes and halogenated hydrocarbons like tetrachloroethylene and trichloroethylene. Food additives, antioxidants, pigments, and packaging materials also contribute xenobiotic organic compounds to landfills (Iravanian & Ravari, 2020).

Alternative reality games (ARGs), mobile genetic elements (MGEs), and antibiotic-promoting chemicals are detected in 350 million tons of municipal solid waste (MSW). These materials may spread to groundwater, rivers, soil, and airborne particle matter. MSW landfills dispose of the worst garbage, including used diapers with antibacterial chemicals and antibiotic residues. Additionally, weak regulatory measures may lead to incorrect disposal of expired medications and medical waste. Most MSW dumps in underdeveloped nations had more antibiotics than those in developed countries. Antibiotics were also lower in older landfills. Landfills are ARG hotspots for organic waste, anaerobic respiration, and fermentation. The distribution of these hotspots was defined by landfill stage (leachates vs. garbage) and age (Wu et al., 2017). The most common ARG subtypes were the sulfonamides, macrolides, aminoglycosides, and multidrugs. The incidence of antimicrobial resistance genes (ARGs) in landfill leachate increased with age. However, leachates from new landfills have higher antibiotic resistance gene (ARG) concentrations (Wu et al., 2023).

3 Classification of Mostly Found Emerging Contaminants in Landfill

3.1 *Pharmaceuticals and Personal Care Products*

The human health industry, as well as animal husbandry, make extensive use of pharmaceutical items. These chemicals have been created to have biological activity and produce extremely particular effects. According to an initial evaluation, sulfonamide antibiotics were detected in the groundwater downstream of a municipal solid waste (MSW) landfill site in Denmark. The amounts of sulfonamide in landfill leachate exceeded 18 mg/L, a significantly higher level than the average concentrations in untreated urban and agricultural wastewater supplies (Zhou et al., 2013).

3.2 *Surfactants*

Surfactants are organic substances that lower surface tension in water and other liquids. They are also known as surface active agents or wetting agents. Surfactants are commonly found in shampoos, dishwashing solutions, laundry detergents, and soaps. Lubricants, emulsion polymerization, textile processing, mining flocculates, petroleum recovery, and a wide range of other goods and procedures are a few examples of major industrial applications. Perfluoro- (PF) surfactants, including

Table 1 Mostly found Emerging Contaminants (ECs) in landfills

Emerging Contaminant	Frequently found ECs in landfills
Pharmaceuticals	Ibuprofen
	Propyphenazone
	Phenazone
	Clofibric acid
	Paracetamol
	Naproxen
	Carbamazepine
	Gemfibrozil
Pesticides	Ampa
	Atrazine
	Bentazon
	Chloridazon
	Chlorpropham
	Dichlobenil
	Endosulfan
	Endrin
	Fenpropimorf
	Glyphosate
	Hydroxysimazin
	Hydroxyatrazin
	Isoproturon
	Malathion
	Mecoprop
	Methyl parathion
	Simazine
	Propoxur
	Tridimefon
Surfactants	Perfluorooctane sulfonate
	Perfluorooctanoic acid
Plasticizers	Phthalate esters
	Di-(2-ethylhexyl) phthalate
	Di-butyl phthalate
	Bis(2-ethylhexyl)-phthalate
	Benzyl butyl phthalate
Flame retardants	Bisphenol
	Diphenyl ethers
	Cyclohydrocarbons
	Phenols and phthalic acid derivatives

(continued)

Table 1 (continued)

Emerging Contaminant	Frequently found ECs in landfills
Organic compounds	Cholesterol
	Tri(2-chloroethyl) phosphate
	Triclosan
	Cotinine
	2,6-di-tert-butyl-1,4-benzoquinone
	Anthracene
	Bis(2-ethylhexyl) phthalate
	Bisphenol a
	Lincomycin
	Naphthalene
	4-Nonylphenol
	4-Octylphenol monoethoxylate
	Stigmastanol

Source Ramakrishnan et al. (2014)

perfluoro octane sulfonate (PFOS, C8F17SO3H) and perfluorooctanoic acid (PFOA, C7F15COOH), represent a novel category of emerging surfactants. The unique chemistry, durability, bioaccumulative nature, toxicological effects, possible water and oil repellency, and ability to reduce surface tension have garnered significant attention.

Additionally, these anionic perfluorinated (PF) surfactants have been identified as the most noteworthy ones present in the environment. Ramakrishnan et al. (2014) reported that Eggen et al. (2010) conducted a quantitative analysis of several perfluorinated (PF) chemicals, which were classified as emerging contaminants (ECs). The study revealed that most PFs were found in the aqueous phase.

3.3 *Plasticizers*

Plastic compositions can be processed and altered effectively with plasticizers and organic molecules with low molecular weight. Flexible polymers can be used in various applications, including automotive, medical, and consumer goods industries. Plasticizers are well known to be dangerous in varied degrees, with some varieties showing a risk for endocrine disruption. The notable class of plasticizers known as phthalate esters (PAE) and several environmentally prevalent plasticizers have been acknowledged for their potential to change endocrine function. The intermediates produced by the breakdown of plasticizers, including the particularly harmful substance known as 2-ethylhexanol, have been the subject of extensive research. Phthalate-containing plasticizers linger in the environment and build up in living things via biological processes. The research done by Schwarzbauer et al. (2002)

serves as the primary justification for the inclusion of these criteria on priority pollution lists. According to Ramakrishnan et al. (2014), phosphates and sulfonamides have been recognized as plasticizers that fall under the category of environmental contaminants. Flame retardants:

Chemicals known as flame retardants are capable of delaying or preventing combustion. The most typical flame retardant is water. More than 175 distinct kinds of substances are employed as flame retardants. Brominated flame retardants (BFRs) are the most popular fire retardants. The largest market for BFRs is the electronic sector. BFRs can be divided into five groups: phthalic acid derivatives, cyclohydrocarbons, phenols, and diphenyl ethers. The first three of these BFRs have the most industrial production, and as a result, they have a significant environmental impact. Only three substances are given precedence regarding their dispersal across the environment and leaching from landfills. There have been few studies on BFR leaching from landfill processes (Ramakrishnan et al., 2014).

3.4 Pesticides

The most frequently discovered chemicals in MSW landfill leachate were herbicides and insecticides (Slack et al., 2005). Eggen et al. (2010) provided a list of the substances most frequently discovered in landfill leachate.

3.5 Nanomaterials

Zhang and Surampalli (2009) define nanomaterials as particles with dimensions ranging from 1 to 100 nm. Particles exhibit a diverse range of morphologies, encompassing spherical structures known as nanocrystals, rod-like geometries, carbon nanotubes, and various other geometric configurations. The transportation of these substances can be achieved through their dissolution in a liquid medium or their utilization in a dehydrated form, hence facilitating their utilization in various domains. These entities can form covalent bonds with surfaces, exhibit binding by van der Waals forces, or be incorporated into an adhesive matrix, such as a plastic polymer. Nanoparticles exhibit distinct movement patterns in soils, bodies of water, and water treatment techniques, contingent upon their particle type and size compared to their corresponding bulk form. The inclusion of these emerging contaminants (ECs) in leachate from landfills can potentially complicate the treatment processes due to the altered physical, chemical, and biological characteristics of the pollutants of interest resulting from the incorporation of nanomaterials into everyday objects (Van Tassel & Goldman, 2011).

Consumer products typically contain NMs that have been surface-modified, functionalized, or coated to obtain desired surface qualities; it is highly unusual to come across free NMs. The nature of the product matrix and the features of the NM surface will probably influence what happens to NMs after disposal. The fate of NMs during waste processing and disposal, such as shredding, biological treatment, burning, and landfilling, needs to be better understood. It is essential to comprehend how landfilled NMs behave to evaluate probable particle releases from landfills. NMs contained in landfilled commercially available products will likely be released. The harsh environmental conditions (such as low pH and strongly reducing conditions) present in landfills are believed to facilitate the release of NMs bound to plastics and other polymers. The NMs most easily transported to the leachate stream are those contained in liquid or gel items (such as cosmetics, sunscreens, or hair products), wastewater biosolids, or waste streams from NM manufacturing operations. When predicting the fate of NMs as they age, their surface changes and their reactivity is altered, it is essential to take into account changes in situ landfill conditions (such as leachate characteristics, moisture content, and temperature) as these factors will probably have a significant impact on NM behaviour. When artificial barriers end their useful life in older landfills, NMs that typically collect and precipitate in new landfills may be released by mobilizing humic and fulvic acids (Handy et al., 2008). Additionally, the surface chemistry of different NMs will significantly impact how they behave. The properties of NMs must be established before they are published to guarantee that the pertinent state is toxicology evaluated (Reinhart et al., 2010).

3.6 *Organic Compounds*

A study conducted by experts from the United States Geological Survey (USGS) in 2004 examined the migration of a leachate plume from a landfill, which was found to contain emerging contaminants (ECs) (Barnes et al., 2004). In this study, a total of 76 organic pollutants originating from wastewater-effluent leachate and the adjacent groundwater system were investigated. A total of 22 pollutants, as identified by Barnes et al., were detected among the 76 substances analyzed. These contaminants include stigmastanol, triclosan, 2,6-di-tert-butyl-1,4-benzoquinone, anthracene, bis(2-ethylhexyl) phthalate, lincomycin, naphthalene, 4-nonylphenol, 4-octylphenol monoethoxylate, and cholesterol. Chemicals such as steroids derived from plants, animals, and faeces, metabolites of nicotine, antioxidants, polycyclic aromatic hydrocarbons (PAHs), plasticizers, antibiotics, disinfectants, insect repellents, metabolites of non-ionic detergents, antimicrobials, fire retardants are among the various classes of chemicals identified by Barnes et al. (2004). Similar chemicals were identified in a second experiment conducted by Buszka et al. (2009), which examined the occurrence of ECs in groundwater in Indiana that had been impacted by landfill leachate. The study authors conducted by Ramakrishnan et al. (2014) have inferred that the prevalence of these compounds in landfills and landfill leachate can be attributed to their extensive usage in various routine activities.

3.7 Microplastics

Microplastics are a newly discovered contaminant that may be a carrier for infectious microorganisms and ARGs. They account for around 20% of municipal solid waste. MPs have been found in various locations, including landfills (Anand et al., 2021).

Research into the occurrence of MPs in landfill leachates has been carried out in China, Iran, Finland, Iceland, Norway, and Indonesia. Different MP size distributions were used in those tests. Independent of landfill age, Su et al. (2019) discovered MPs with big particle sizes ranging from 70 mm to 3.67 mm, while He et al. (2019) found MPs with a range of 100 to 1000 mm in examined samples. Landfill leachate contained MPs of varying concentrations and types of polymer and MPs of varying sizes. The age of the landfill can provide insight into the connection between MP particle quantity and size (Maja Petrovi et al., 2023)].

Micro- and mesoplastics may attract and carry persistent organic pollutants (POPs), heavy metals, and diseases, which can have devastating effects on human health and aquatic ecosystems (Nurhasanah et al., 2021).

4 Sources and Pathways of Emerging Contaminant Release

Organic chemicals, inorganic metals, and metalloids are all potential contaminants in landfill leachate. Leachate from landfills can contain a wide variety of contaminants, the presence (concentration) of which is mainly determined by the physical and chemical properties of the waste. These contaminants permeate the landscape, poisoning the soil, groundwater, and surface water. Anand et al. (2021) state that the characteristics of the pollutants, the presence of organic matter, and the ion exchange behaviour and mineralogy of the sediment all play a role in the mobility and final fate of inorganic contaminants in the natural environment.

4.1 Wastewater

Municipal wastewater, along with nonpoint and point sources, industry, stormwater runoff, residential sewage, and water treatment facilities, is one of the principal sources of CEC. People consume CEC, which then accumulate in sewer pipes, which are not designed to remove them, and eventually make their way to a MWWTP. The bulk of effluent samples collected by Wu et al. from four Shanghai (PRC) WWTPs contained ten different antibiotics. Antibiotics were concentrated near WWTPs and then released into receiving water bodies (Lofrano et al., 2020).

Most drugs are found in hospital garbage. Ciprofloxacin concentrations in hospital effluent ranged from 0.70 to 124.5 mg/L. Ampicillin was found in concentrations of 20 to 80 mg/L in the wastewater of a large hospital in Germany (Lofrano et al., 2020).

Polybrominated diphenyl ethers (PBDEs) in wastewater treatment plants can be attributed to the introduction of wool, carpets, and grey water. Polybrominated diphenyl ethers (PBDEs) can also be detected in the leachate originating from discarded goods and the flowing water discharged from industrial materials within wastewater treatment plants (WWTPs). Engineered nanoparticles, also known as engineered NPs, are introduced into the environment by releasing wastewater from many sources, such as cosmetics, sunscreen formulations, medications, paints, and other consumer goods (Lofrano et al., 2020).

4.2 Sewage Sludge

The treatment processes at wastewater treatment plants produce sewage sludge. According to Martín-Pozo et al. (2019), the sample may contain liquid-phase bacteria and various organic and inorganic chemicals, including CEC. Due to physical, chemical, and biological processes, wastewater treatment plants (WWTPs) generate sludge that contains pathogenic microorganisms like viruses and bacteria, heavy metals, and trace organic compounds that resist degradation. Due to the discovery of various emerging contaminants (ECs) in digested sewage sludge, extensive research has been done on its use. The ongoing working document by Lofrano et al. (2020) includes threshold values for concentrations of seven organic contaminants, including dibenzofurans (PCDD/F), polychlorinated dibenzodioxins (PCDD/F), and chlorinated dioxins.

4.3 Municipal Solid Waste and Landfill Leachate

Wastewater is the primary Contaminant of Emerging Concern (CEC) source for the environment. In contrast, municipal solid waste (MSW) landfill leachate has yet to receive more investigation. "Municipal solid waste" (MSW) includes organic garbage and empty plastic containers. Economic incentives drive modern municipal solid waste (MSW) landfill disposal. Batteries, paints, oils, electronics, and medications are often dumped in landfills. Human pharmaceuticals enter the aquatic environment through household channels enabled by municipal sewage treatment facilities. The infiltration of precipitation through waste materials, metabolic activities in waste cells, and intrinsic moisture content of wastes produce landfill leachate. Surface area, geography, local climate (such as rainfall), and cover type affect landfill leachate output. The chemical makeup and concentration of chemicals in a landfill are affected by the rubbish dumped and the leachate produced. Vast amounts of CEC (chemical oxygen demand) could be made from municipal solid waste disposed

Table 2 The studies showing the detection of Emerging Contaminants in landfill leachate and nearby landfill area

Landfill-country	Detected ECs	References
City Xi'an, China	Oxytetracycline. Tetracycline and Sulfamethoxazole	Song et al. (2016)
USA MSW landfills	N,N-diethyl-m-toluamide	Clarke et al. (2015)
Singapore	Acetaminophen, Bisphenol A, N,N-diethyl-m-toluamide, Salicylic acid, Gemfibrozil, Sulfamethazine	Yi et al. (2017)
Japan	Bisphenol A	Urase and Miyashita (2003)
Spain	Azithromycin, Salicylic acid	López-Serna et al. (2013)
Western Sweden	Di-(2-ethylhexyl) phthalate	Paxéus (2000)
Norway	Pesticides, Brominated compound, Mercury, Polychlorinated biphenyls	Haarstad and Borch (2008)
North Italy	Ciprofloxacin, Sulfamethoxazole, Erythromycin, Sulfadiazine, Enrofloxacin	He et al. (2021)
EU	Nano-TiO2, Ag, ZnO	Sun et al. (2016)
Turkey	TiO2 and Ag nanoparticles	Yazici Guvenc et al. (2017)
Iran	Iron nanoparticle	Kashitarash et al. (2012)

of in landfills. Therefore, landfill leachate is both a sink and a potential environmental source of CECs. Research into the toxicity of landfill leachates using a wide range of model organisms highlights the importance of cleaning them up so they may be safely discharged into the water supply. Older landfills without protective liners have a higher risk of leachate seepage into groundwater (GW), yet leachate is rarely collected. Traditional (biological, physical/chemical) or membrane methods are widely used for on-site leachate treatment from recently constructed and lined landfills.

Additionally, it can go through treatment at an STP (sewage treatment plant). Concentrated biosolids or wastewater effluent may contain CECs that are not effectively removed by biological processes, as Lofrano et al. (2020) stated. Table 2 shows the list of detected ECs in landfill leachate in different studies.

5 Fate and Transport of Emerging Contaminants in Landfill

The presence of environmental contaminants (ECs) within landfill leachate has the potential to result in long-term groundwater contamination spanning from decades to centuries. Various biogeochemical and microbial processes and the hydrodynamic framework influence the transport of pollutants in the subsurface and leachate-affected aquifers around landfills. These processes include groundwater recharge events, seasonal water variations, and the availability of aqueous electron acceptors, such as constituent concentrations.

If CEC is not degraded or destroyed through sewage treatment, soil, or other environmental compartments, it may permeate surface, ground, or drinking water. Contaminants of emergent concern (CEC) may vary in surface waters (SWs) received from wastewater treatment plants (WWTPs). These pollutants can penetrate groundwater (GW) or remain attached to soil and sediment depending on their physical and chemical properties and interaction with the environmental matrix. Chemical and emerging contaminants (CEC) for surface water (SW) and groundwater (GW) come from untreated sewage and improper disposal of expired pharmaceuticals and cosmetics in landfills or toilets. Environmental contaminants (ECs) or their metabolites or transformation products can enter an environmental compartment and undergo adsorption, absorption, dilution, hydrolysis, photolysis, biodegradation, volatilization, oxidation, or complexation. The process determines whether the molecule is destroyed, altered, or left in the environment. Kummerer et al. found that several antibiotics biodegraded partially in aquatic tests. A large majority persevered. The genotoxicity of quinolones and metronidazole cannot be eliminated. Quinolones like those indicated adsorb onto soils, sediments, and sewage sludge without degrading. Loffler et al. classified ten drugs and pharmaceutical metabolites into low-, moderate-, and high-persistence groups. The persistence of these chemicals in water and soil tests determined this classification. Ibuprofen, 2-hydroxyibuprofen, paracetamol, and CBZ-diol had limited persistence. Clofibric acid, diazepam, and carbamazepine are highly persistent, while oxazepam, iopromide, and ivermectin are intermediate. The rapid adsorption of polybrominated diphenyl ethers (PBDEs) onto organic and particulate matter increases their likelihood of detection and distribution in wastewater treatment plant sewage sludge. Biosolids in terrestrial environments emit PBDEs. Pollutant-contaminated soil runoff can transfer polybrominated diphenyl ethers (PBDEs) to waterways. The complex composition of matrices like sewage sludge and biosolids might cause geographical variations in organic component concentrations like PBDEs. Nanoparticles (NPs) may be removed from water during treatment, but aggregation or sorption may increase their concentration in the sludge matrix. Sludge application on agricultural land causes leaching, runoff, and erosion, which can harm aquatic life. Sludge disposal to agrarian soil, incineration, and landfilling release nanoparticles (NPs) (Lofrano et al., 2020).

In landfills, PPCPs have the potential to decompose, absorb, or release leachate. Pharmaceuticals and personal care products (PPCPs) are present in landfill leachates, albeit at much lower concentrations than ammonia and metals. Even in trace amounts, pharmaceuticals and personal care products (PPCPs) in landfill leachates can have a negative impact on human health and ecosystems. There may be unintended consequences for other types of wildlife. Leachates from landfills typically contain a high concentration of pharmaceuticals and personal care products. More pharmaceuticals and personal care products (PPCPs) were found in landfills than in any other setting, according to research by Yu et al. (2020).

One of the newest waste pollutants identified is microplastics. Due to weathering, photolysis, mechanical abrasion, and microbiological degradation, MPs are formed in landfills from municipal garbage such as plastic bags, bottles, tableware, packing products, textiles, and construction materials. MPs can spread both laterally and vertically (through wind and leaching, respectively). MPs spread throughout the ecosystem through leaching from waste soil to groundwater, transfer of aerosols, and landfill mining, especially in filthy landfills and dumpsites. According to published reports, the majority of MPs in examined landfill sections are composed of polymer fibres and fragments, particularly those made of polyethene (PE), polyethene terephthalate (PET), polyvinyl chloride (PVC), polypropylene (PP), and polystyrene (PS). Heavy metals, antibiotics, and tiny pollutants may be absorbed by MPs from landfills (Maja Petrovi et al., 2023).

Contaminants from both abiotic (such as antibiotics and MPs; MPs are particles larger than 5 mm) and biotic (such as ARGs, viruses, and pathogens) sources significantly impact landfills. One of these, MP pollution, has been declared a global crisis. The presence of both antibiotics and MPs in the environment poses a threat to wildlife and ecosystems. MPs are considered antibiotic transporters that modify the toxicity of individual MPs or antibiotics, as stated by Song et al. (2023). Figure 1 shows the origin, fate, transport, and pathways of ECs in landfills.

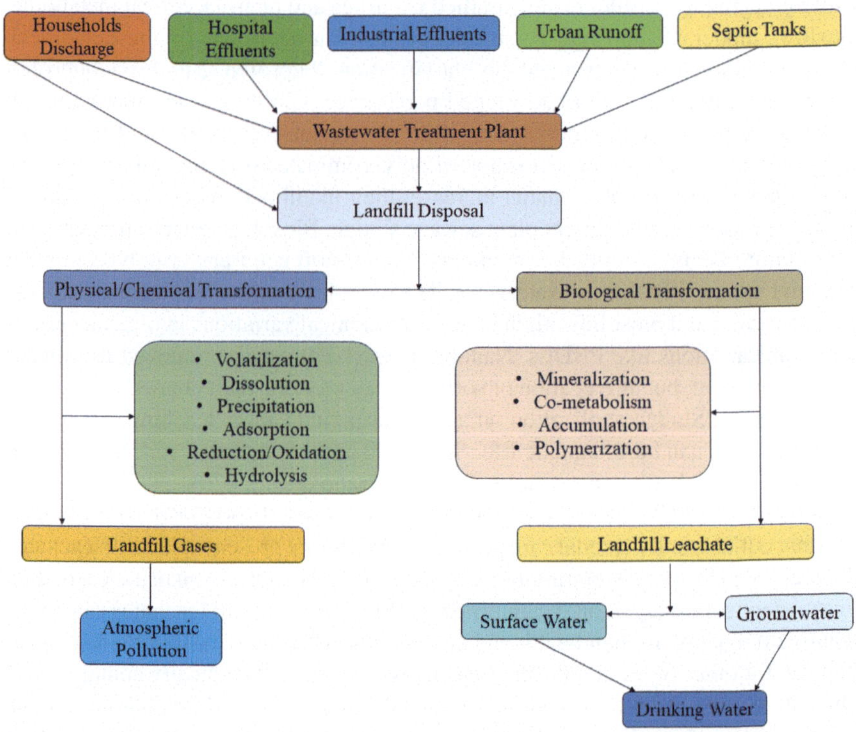

Fig. 1 Origin, fate, transport and pathways of emerging contaminants in landfills

6 Health and Environmental Impacts

6.1 *Potential Risks to Human Health*

Bacillus cereus, Staphylococcus lugdunensis, and *Pseudomonas aeruginosa* are three examples of airborne microorganisms linked to landfills. The presence of these bacteria causes microbiological air pollution because of their connection to aerosol/bioaerosol emissions. Bacteria, viruses, fungi, pollen, poisons, allergies, insects, mammalian hair, organic wastes, and plant fragments are all bioaerosols. Toxic bioaerosols and endotoxins can be produced in landfills due to the decomposition of waste items, promoting the growth of microorganisms like bacteria and fungi. Acute harmful allergies, respiratory illnesses, infectious infections, and malignancies are the critical health concerns or unfavorable effects of bioaerosols. Due to their frequent interaction with the waste site, landfill workers are at high risk of contracting infectious diseases due to bacterial aerosols. Endotoxins are a common component of bioaerosols, which increase the likelihood of contracting respiratory infections. It is essential to evaluate the airborne bacteria from MSW dumps because many airborne bacteria are antibiotic-resistant (AR). It can release AR genes (ARB and ARG) when interacting with other environmental bacteria. Accurate identification of the microorganisms involved with bioaerosols is essential for a thorough response to the problem. In addition, it is critical to suggest efficient management measures for lowering the likelihood of adverse outcomes (Anand et al., 2021).

The presence of rats and mice at landfills has been linked to the spread of Hantavirus pulmonary syndrome and asthma in nearby communities. Organic pollutants such as PCDDs and PCDFs (polychlorinated dibenzo-para-dioxins and polychlorinated dibenzofurans) are frequently found in garbage dumps. Human exposure to these chemicals, which have carcinogenic qualities, has been linked to the growth of cancers in the lungs and skin. In addition, being around them raises the risk of having a heart attack. It has been discovered that breathing in polychlorinated biphenyls (PCBs) and the hydrogen sulfide (H2S) that is released from landfills can cause respiratory problems, including asthma. Non-respiratory diseases, such as diabetes, have also been linked to these emissions. Toxic and hazardous substances such as lead, zinc, nickel, barium, and chromium can be found in electronic waste (e-waste), typically dumped in landfills or recycled. Adverse effects on human blood, kidneys, and the central nervous system have been linked to the inappropriate recycling of electronic waste in low- and middle-income nations, as Anand et al. (2021) reported.

Martí et al. (2014) examined soil gas concentrations and emissions of volatile organic compounds to determine the risk to human health posed by the landfill. Due to the potential consumption of dangerous compounds by aquatic plants and animals, which may then be transferred down the food chain, dumping trash straight into bodies of water without first treating it could pose serious health risks. Exposure to hazardous waste can harm human health, which has been discussed extensively in the existing literature. This is especially true for kids, who are more likely to be affected negatively by exposure to harmful chemicals. Chemical poisoning has been documented due to the release of chemical waste into the environment, and exposure to these chemicals has been connected to the onset of multiple diseases (Martí et al., 2014).

Rezapour et al. (2018) discovered that uncontrolled landfill leachate leaks significantly raised the content of different PTEs in the soil, interacting with the crops planted there. Except for Cd, which was above limits and provided a moderate intensity noncarcinogenic danger to those ingesting the wheat, they reported finding a variety of metals in average concentrations. However, according to this study, there is little chance that a local person will develop cancer. This study demonstrates the degree of contamination caused by landfilling. The PTEs might negatively impact the human population if they interact with the soil system and enter the food chain (Rezapour et al., 2018).

Other studies carried out by several researchers revealed that women who lived close to hazardous dump sites in Washington State were at an elevated risk of having kids with malformations, and this risk was higher for those living in urban areas as opposed to those in rural areas (Siddiqui et al., 2022). Figure 2.depicts the impacts of ECs on human health.

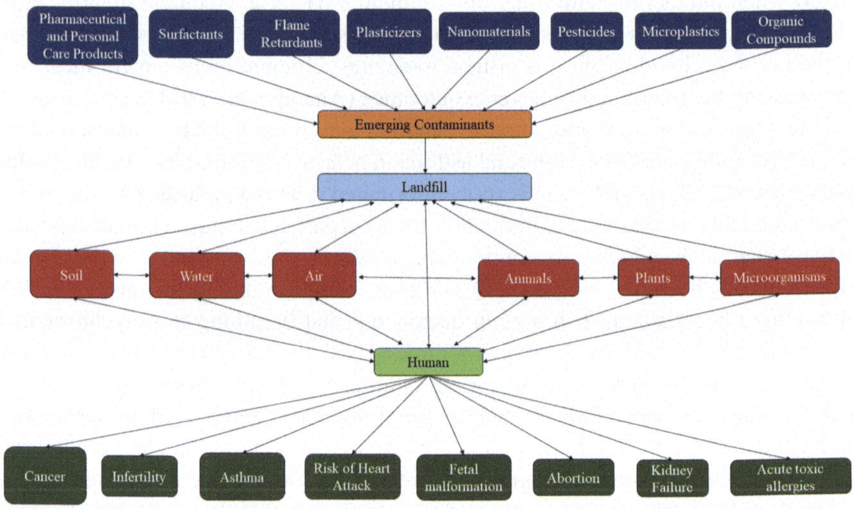

Fig. 2 Impacts of Emerging Contaminants in human health

6.2 Impact on Environment

Groundwater is one of the most crucial resources for human survival. The leakage of toxic compounds from municipal solid waste disposal sites has made them a significant source of water pollution. After a rainstorm, leachate may seep into the ground or pool on the surface, polluting the water and making it unsafe for human consumption (Alghamdi et al., 2021). Phosphate, heavy metals, ammonium, chloride, and sodium are some contaminants that seep into surface and groundwater from landfills (Han et al., 2016). The rising concentration of antibiotics in landfill leachate, such as ciprofloxacin, -lactams, vancomycin, cefepime, erythromycin, sulphonamide, and tetracycline, poses a threat to groundwater resources (Anand et al., 2021).

Municipal solid waste (MSW) landfills have been identified as a well-established contributor to groundwater contamination. Polluted groundwater might affect various aspects, such as drinking and irrigation wells, underground dwellings by vapor intrusion, surface water bodies that receive groundwater discharge, and potentially terrestrial phreatophytes. Wastewater treatment facilities (WWTPs) often receive leachate or leachate-affected groundwater through on-site designed collecting systems, which may have elevated levels of certain compounds compared to typical WWTP influent. The treated effluent discharged into surface waterways may pose a risk due to leachate chemicals inadequately eliminated by wastewater treatment plants (Propp et al., 2021).

Open dumping of MSW involves various complex components, including heavy metals, humic compounds, and organic contaminants that can influence the stability and strength of the soil due to leachate seeping into the ground. It should be considered that leachate-contaminated groundwater is frequently used to irrigate crops in nearby agricultural areas, which could raise the danger of leachate toxicity and allow it to enter the food chain through nearby plants (Turki & Bouzid, 2017). Plants' evaporation, infiltration, and uptake of nutrients, metals, and organics can all be impacted by producing toxic fumes and polluted soil. Therefore, it is predicted that any alteration in the soil's physicochemical and biological markers, such as its microbial biomass and enzymatic activities (nutrient transformation and microbial decomposition), may eventually result in a decline in the composition and health of the soil. Because of the importance of leachates to soil physicochemical and biological processes (nutrient transformation), toxicological studies focusing on these interactions are essential. Soil contamination is caused by heavy metals being taken up by plants or leached into groundwater after being mobilized in soil solutions via a variety of heavy metal ion or compound solubility mechanisms. Soil that is moderately to significantly contaminated by heavy metals from MSW disposal sites has been shown to harm environmental and neighboring vegetation health. Therefore, optimizing the soil-crop-human ecosystem requires understanding the characteristics of heavy metal pollutants in soil habitats affected by landfill leachate (Anand et al., 2021).

Landfills have long been a significant source of GHG emissions. This is because landfills emit large amounts of CO2 and CO4 oxide. These gases harm the environment during degradation (Papargyropoulou et al., 2015). Leachate from landfills pollutes subsurface water supplies. This is especially true for poor landfill liners. Residents near landfills are severely affected by air pollution from odors. Dust, litter, and rodents are among landfill contaminants, according to Ilankoon et al. (2018). According to Hossain et al. (2014), there are several traditional landfill pollution types. The most common types are emissions, effluent, and dumps and disposals. A more detailed description would distinguish surface and groundwater, troposphere and stratosphere, and inland and coastal waterways. Satellites and other debris should be included in outer space.

Most pollution discussions and laws focus on these categories. Still, cross-medial effects are gaining attention, such as air pollution acidifying lakes and streams or the disposal of sludge and other waste products from air and water pollution control measures on soil or in the ocean. Landfills pollute the air due to complex microbiological and chemical processes that produce gaseous pollutants. Landfills release SO_2 and NO_2 into the air. Any of these gases can cause asthma by irritating the throat and nostrils. Some waste gases can irritate the respiratory system, endangering neighboring residents (Cucchiella et al., 2017). When it rains on landfills, inorganic and organic trash disintegrates. Thus, toxic substances enter subsurface water systems. Due to its high metal content, this water is dangerous to drink. These pollutants may harm aquatic life in lakes and rivers (Zhang et al., 2016). Waste disposal sites cause air pollution worldwide. About two-thirds of landfills are expected to contain organic, biodegradable materials. These chemicals degrade into methane (Babayemi et al., 2016). As a GHG, methane gas helps retain atmospheric heat. Dumping trash affects biodiversity (Siddiqui et al., 2022).

Landfill leachate degrades soil and water. Leachate immediately contaminates soil and groundwater without proper containment, making it worse when it rains. Rainfall runoff pollutes surface rivers, putting aquatic life and vegetation at risk. Heavy metals and other toxicants can contaminate surface rivers, threatening human health in potable and non-potable water (Bandala et al., 2021).

7 Removal Techniques of Emerging Contaminants from Landfill

Despite the increased focus on ECs in landfills, specialized or established technology has yet to be used to remove ECs from landfill leachate. Alternative technologies, such as built wetlands, biotechnology, and physical–chemical treatment technology, eliminate ECs associated with landfill leachate (Teng & Chen, 2023).

With its comparatively low mechanical complexity, maintenance costs, and energy requirements, constructed wetlands have been hailed as cutting-edge technology for their ability to intercept and degrade pollutants through plants and microbes in aquatic environments. Constructed wetlands offer an immense potential to remove ECs from landfill leachate because aquatic vascular plants can immediately collect and stabilize ECs and then degrade them through the metabolism of plant root biofilm. Recent studies have shown that artificial wetlands can remove 97% of nanoscale plastics and 88%e98% of total microplastics (Chen et al., 2020).

Leachate from landfills has a complicated quality that varies widely, making remediation difficult. Mainly, MPs in landfill leachate can be a carrier for infections, resistance genes, heavy metals, organic pollutants, and organic pollutants, which can help these contaminants spread (Shi et al., 2021). Additionally, removing toxins from MPs presents more of a challenge than removing them from landfill leachate. As a result, it is crucial to create efficient treatment systems to take ECs and common contaminants out of landfill leachate. Generally, the effluent can be ensured to satisfy the discharge standard using biological techniques (i.e., membrane bioreactor) and membrane filtration (such as nanofiltration and reverse osmosis).

In contrast, nano-plastics/MPs can worsen membrane fouling in the presence of ECs by obstructing the pores on the membrane surface, reducing membrane filtration efficiency. As a result, particular efforts have been made to reduce the occurrence of membrane fouling. For instance, membrane fouling has been dramatically reduced by adding biochar filters between the membrane bioreactor and the membrane filtering system, which remove and immobilize more than 95% of MPs with a particle size of 10 mm. Due to their powerful oxidation capabilities, photocatalysis, advanced oxidation, and electrochemical catalysis have all been identified as viable technologies for removing contaminants from landfill leachate (Teng & Chen, 2023).

The term "biological treatment technology" refers to using native or inoculated microorganisms to metabolize ECs in landfill leachate and transform them into low/nontoxic products, which can be utilized independently of or in addition to the currently used conventional procedures to improve treatment efficiency. From wastewater, activated sludge may remove 3.6% to 42.9% of the MPs. The treatment of ECs also triggered anaerobic processes. Anaerobic sludge digestion can eliminate ARGs (Teng & Chen, 2023).

Chemically active and electrically charged surface functional groups interact to remove molecular and ionic toxicants from landfill leachate by adsorption (Bandala et al., 2021). Hamid et al. (2020) emphasize one of the aspects of zeolites that deserves more investigation: their ability to extract ammoniacal nitrogen (NH3-N) from landfill leachate. The popularity of clay minerals, including many different adsorbents, has increased in recent years because of their low cost and environmental friendliness (Kjeldsen et al., 2002). The many beneficial qualities of clay minerals include a high specific surface area, high chemical and mechanical stability, a laminar structure, and a remarkable ionic exchange capacity. As Luo et al. (2019) pointed out, these features help explain why they are considered very efficient sorbents. Kjeldsen et al. (2002) found that these compounds can adsorb a wide range of contaminants. Ion exchange, adsorption, and combining these two processes allow clays to remove cations and anions from the environment (Luo et al., 2019).

Amor et al. (2015) and Ghanbari et al. (2020) showed that combining coagulation-flocculation (C/F) with other chemical processes, like advanced oxidation, is an effective method for treating mature landfill leachate.

8 Case Studies Regarding the Presence of Emerging Contaminants in Landfills

8.1 Emerging Contaminants in the Fresh Landfill Leachate in the Conterminous United States

A landfill disposes of residential, industrial, and commercial garbage. Thus, landfills can produce leachate containing complex CECs from consumer items. Our investigation found 129 of 202 Contaminants of Emerging Concern (CECs) in new leachate samples from 19 landfills, supporting the idea. Recently collected landfill leachate contained a variety of Contaminants of Emerging Concern (CECs), including industrial and domestic pollutants, medications with and without prescription, plant and animal sterols, and steroid hormones. Unlike landfills with mostly homogenous waste mixtures, those with heterogeneous municipal and industrial rubbish had higher home chemical detection and total measured levels. Of 202 samples, 129 chemical and environmental contaminants (CECs) were found. These CECs included 62 prescription medications, 23 industrial chemicals, 18 OTC drugs, 16 home chemicals, six steroid hormones, and four plant/animal sterols. The overall count of emerging concerns (CECs) in each leachate sample ranged from 6 to 82 (median 31). The three CECs found in 95% leachate samples were BPA, cotinine, and N, N-diethyltoluamide. Lidocaine was discovered in 89% of samples and camphor in 84%. Masoner et al. (2014) found CECs like benzophenone, naphthalene, and amphetamine in 79% of leachate samples.

8.2 Emerging Contaminants of Historic Landfill Sites in Canada

Per- and poly-fluoroalkyl substances (PFAS) and other pollutants of increasing concern (PECs) have been detected in leachate originating from active municipal landfills. Nevertheless, it is worth noting that many historic landfills, despite being closed for thirty years, frequently lack engineered liners or mechanisms for collecting leachate. Furthermore, there is a shortage of information regarding the potential existence of contaminants of emerging concern (CECs) in the leachate, specifically at levels that may threaten neighboring wells and surface water habitats. Propp et al. (2021) conducted a study in Ontario, Canada, wherein they gathered 48 samples of groundwater contaminated with leachate from 20 decommissioned landfills. Several examples of chemical environmental contaminants (CECs) that have been extensively tested include artificial sweeteners (AS), perfluorinated substances (PFAS), organophosphate esters (OPE), pharmaceuticals, bisphenols, sulfamic acid, perchlorate, and substituted phenols. The prevalence of AS saccharin, a marker for aged landfill leachate, and the generally insignificant concentrations of AS acesulfame, a marker for contemporary wastewater, indicate that the majority of the samples were predominantly impacted by leachate and not subject to cross-contamination from wastewater (which may contain these same contaminants of emerging concern). Many chemical constituents (CECs) assessed in this study were frequently detected at higher levels (ranging from 0.1 to 10 mg/L) in the groundwater contaminated by leachate in the 20 selected historical landfill sites. These CECs include AS saccharin, PFAS compounds such as PFOA, and PFOS, OPE compounds as well as BPA and cotinine. Propp et al. (2021) observed that there was no discernible decrease in the concentrations of these CECs across landfills that had been closed for a period ranging from 30 to 60 years. This assertion holds, particularly from the initial introduction or widespread adoption of specific compounds. This observation indicates they have endured for an extended duration within historical dump sites.

8.3 Emerging Contaminants in Landfills of China

Chinese researchers have been studying 180 CEC chemicals in landfill leachate since 1996. APEOs/BPA, cVMS, FRs, OCPs, PFCs, PPCPs, PAEs, Sterols, and sPAHs are the nine Contaminants of Emerging Concern (CECs). We created China's first complete and systematic repository of Contaminants of Emerging Concern (CECs) in landfill leachate after reviewing 43 research articles. The southeastern coastal regions of China have done many studies on landfill leachate Contaminants of Emerging Concern (CECs) throughout the past decade. Phthalate esters (PAEs), pharmaceuticals and personal care products (PPCPs), alkylphenol ethoxylates/bisphenol A (APEOs/BPA), and several PAHs are examined as emerging pollutants. Di-n-butyl phthalate, di (2-ethylhexyl) phthalate, bisphenol A, sulfadiazine, sulfamethoxazole, sulfamethazine, dimethyl, diethyl, erythromycin, and trimethoprim are the 10 most-studied CECs. In Contaminants of Emerging Concern (CECs), 15 studies have examined phthalate esters (PAEs) and 13 on pharmaceuticals and personal care products (PPCPs). The Contaminants of Emerging Concern (CECs) included nine categories: organochlorine pesticides at 0.03 mg/L and alkylphenol polyethoxylates/bisphenol analogue at over 4500 mg/L. Doxycycline was below the detection threshold, while bisphenol reached 4500 mg/L in CECs. Qi et al. (2018) found significant differences in the concentrations of phthalate esters (PAEs) like diethyl, di-n-butyl, and di (2-ethylhexyl) phthalate and pharmaceuticals and personal care products (PPCPs) like diclofenac and gemfibrozil across sampling sites.

8.4 *Emerging Contaminants of the Galuga Landfill Area of Indonesia*

Micro- and mesoplastics are emerging contaminants frequently found in aquatic ecosystems. Research on landfill micro- and mesoplastic pollution, however, is scarce. In a study, micro- and mesoplastics were found in all surface water samples taken from Galuga landfill leachate influent and effluent. There were predicted to be 618,240 1905.45 mesoplastic particles and 80,640 604.80 microplastic particles released into the aquatic environment daily. Following input from the leachate drain, the amount of microplastic increased three-fold, and the mesoplastics increased nine-fold. Polyethene, polypropylene, polystyrene, polyethylene terephthalate, polyester, and cellophane were the principal chemical components of micro- and mesoplastics. According to this study's findings (Nurhasanah et al., 2021), leachate may contaminate the aquatic environment with micro- and mesoplastic materials.

8.5 *Emerging Contaminants at a Closed and Operating Landfill in Oklahoma*

Landfills have the potential to produce leachate containing many organic compounds found in consumer products such as pharmaceuticals, plasticizers, disinfectants, cleaning agents, fire retardants, flavorings, and preservatives, which are known as emerging contaminants (ECs). Landfills are the final depositories for various solid waste from residential and commercial sources. Three different age ranges of landfill cells from two landfills in Central Oklahoma were analyzed for landfill leachate. To examine the chemical variability and persistence of specific ECs in landfill leachate from sources of varying ages, samples were taken from an old cell with solid waste older than 25 years, an intermediate-age cell with solid waste older than 16 years, and an operating cell with solid waste younger than 5 years. 28 of the 69 ECs examined were found in one or more samples from the three leachate sources. They detected ECs comprised of 4 faecal and plant sterols, 13 industrial and domestic, 7 hydrocarbon, and 4 pesticide compounds. Their concentrations varied from 0.11 to 114 g/L. The oldest leachate sample included four ECs, the intermediate leachate sample contained two ECs, and the youngest had no ECs. The presence of eleven ECs, which were consistently found in all three leachate samples, provides insight into the nature of the solid waste that has been deposited over many years and the relative resistance of some ECs to the natural attenuation processes that occur in and around landfills (Andrews et al., 2011).

9 Conclusion

The matter of landfill pollution highlights the significance of implementing proactive and all-encompassing waste management strategies. The escalating population increase and rapid urbanization have resulted in a significant strain on landfills, potentially leading to the release of harmful substances that pose risks to the environment and human health. The emergence of pollutants that pose a hazard to conventional disposal technologies and societal systems can be attributed to the intricate interactions between waste components and the dynamic structure of waste streams. Microplastics, medicines, heavy metals, and other hazardous substances necessitate further emphasis on training, research, and preventive measures.

The effective management of recently identified landfill poisons necessitates implementing a multidisciplinary approach that involves the collaboration of waste management experts, environmental scientists, policymakers, and business professionals. A practical solution requires utilizing state-of-the-art waste treatment apparatus, streamlined recycling processes, and environmentally conscious waste management practices. To foster responsibility in garbage generation and disposal, it is imperative to cultivate public participation and enhance educational efforts. Promote the adoption of sustainable living practices among individuals as a means to mitigate landfill waste and improve the environment for subsequent generations.

To mitigate the release of landfill gas emissions, communities must prioritize preserving environmental well-being over immediate and transient benefits. Gaining insight into the interconnections between waste management practices and ecological well-being can facilitate the development of a more pristine, secure, and robust global environment. Collaborative efforts, enhanced research methodologies, and informed decision-making have the potential to mitigate the repercussions associated with the proliferation of toxins, thereby fostering the development of a more sustainable global environment.

References

Alghamdi, A. G., Aly, A. A., & Ibrahim, H. M. (2021). Assessing the environmental impacts of municipal solid waste landfill leachate on groundwater and soil contamination in western Saudi Arabia. *Arabian Journal of Geosciences, 14*(5). https://doi.org/10.1007/s12517-021-06583-9

Amor, C., De Torres-Socias, E., Peres, J. A., Maldonado, M. I., Oller, I., Malato, S., & Lucas, M. S. (2015). Mature landfill leachate treatment by coagulation/flocculation combined with Fenton and solar photo-Fenton processes. *Journal of Hazardous Materials, 286*, 261–268.

Anand, U., Reddy, B., Singh, V. K., Singh, A. K., Kesari, K. K., Tripathi, P., Kumar, P., Tripathi, V., & Simal-Gandara, J. (2021). Potential environmental and human health risks caused by Antibiotic-Resistant Bacteria (ARB), Antibiotic Resistance Genes (ARGs) and Emerging Contaminants (ECs) from Municipal Solid Waste (MSW) landfill. *Antibiotics, 10*(4), 374. https://doi.org/10.3390/antibiotics10040374

Andrews, W. J., Masoner, J. R., & Cozzarelli, I. M. (2011). Emerging contaminants at a closed and an operating landfill in Oklahoma. *Ground Water Monitoring & Remediation, 32*(1), 120–130. https://doi.org/10.1111/j.1745-6592.2011.01373.x

Babayemi, J. O., Ogundiran, M. B., & Osibanjo, O. (2016). Overview of environmental hazards and health effects of pollution in developing countries: A case study of Nigeria. *Environmental Quality Management, 26*(1), 51–71. https://doi.org/10.1002/tqem.21480

Bandala, E. R., Liu, A., Wijesiri, B., Zeidman, A. B., & Goonetilleke, A. (2021). Emerging materials and technologies for landfill leachate treatment: A critical review. *Environmental Pollution, 291*, 118133. https://doi.org/10.1016/j.envpol.2021.118133

Barnes, K. K., Christenson, S. C., Kolpin, D. W., Focazio, M. J., Furlong, E. T., Zaugg, S. D., Meyer, M. T., & Barber, L. B. (2004). Pharmaceuticals and other organic waste water contaminants within a leachate plume downgradient of a municipal landfill. *Groundwater Monitoring & Remediation, 24*(2), 119–126. https://doi.org/10.1111/j.1745-6592.2004.tb00720.x

Buszka, P. M., Yeskis, D. J., Kolpin, D. W., Furlong, E. T., Zaugg, S. D., & Meyer, M. T. (2009). Waste-indicator and pharmaceutical compounds in landfill-leachate-affected ground water near Elkhart, Indiana, 2000–2002. *Bulletin of Environmental Contamination and Toxicology, 82*(6), 653–659. https://doi.org/10.1007/s00128-009-9702-z

Chen, J., Tong, T., Jiang, X., & Xie, S. (2020). Biodegradation of sulfonamides in both oxic and anoxic zones of vertical flow constructed wetland and the potential degraders. *Environmental Pollution, 265*, 115040.

Clarke, B. O., Anumol, T., Barlaz, M., & Snyder, S. A. (2015). Investigating landfill leachate as a source of trace organic pollutants. *Chemosphere, 127*, 269–275. https://doi.org/10.1016/j.chemosphere.2015.02.030

Connecticut. (2023). *Contaminants of emerging concern*. CT.gov—Connecticut's Official State Website. https://portal.ct.gov/DEEP/Remediation--Site-Clean-Up/Contaminants-of-Emerging-Concern/Contaminants-of-Emerging-Concern#:~:text=According%20to%20EPA%2C%20an%20emerging

Cucchiella, F., D'Adamo, I., & Gastaldi, M. (2017). Sustainable waste management: Waste to energy plant as an alternative to landfill. *Energy Conversion and Management, 131*, 18–31. https://doi.org/10.1016/j.enconman.2016.11.012

Eggen, T., Moeder, M., & Arukwe, A. (2010). Municipal landfill leachates: A significant source for new and emerging pollutants. *Science of the Total Environment, 408*(21), 5147–5157. https://doi.org/10.1016/j.scitotenv.2010.07.049

Ghanbari, F., Wu, J., Khatebasreh, M., Ding, D., & Lin, K. Y. A. (2020). Efficient treatment for landfill leachate through sequential electrocoagulation, electrooxidation and PMS/ UV/CuFe2O4 process. *Separation and Purification Technology, 242*, 116828. https://doi.org/10.1016/j.seppur.2020.116828

Haarstad, K., & Borch, H. (2008). Halogenated compounds, PCB and pesticides in landfill leachate, downstream lake sediments and fish. *Journal of Environmental Science and Health, Part A, 43*(12), 1346–1352. https://doi.org/10.1080/10934520802231966

Hamid, M. A. A., Aziz, H. A., Yusoff, M. S., & Rezan, S. A. (2020). Optimization and analysis of zeolite augmented electrocoagulation process in the reduction of high-strength ammonia in saline landfill leachate. *Water, 12*, 247. https://doi.org/10.3390/w12010247

Han, Z., Ma, H., Shi, G., He, L., Wei, L., & Shi, Q. (2016). A review of groundwater contamination near municipal solid waste landfill sites in China. *Science of the Total Environment, 569–570*, 1255–1264. https://doi.org/10.1016/j.scitotenv.2016.06.201

Handy, R. D., von der Kammer, F., Lead, J. R., Hassellöv, M., Owen, R., & Crane, M. (2008). The ecotoxicology and chemistry or manufactured nanoparticles. *Ecotoxicology, 17*, 287–314.

He, P., Chen, L., Shao, L., Zhang, H., & Lü, F. (2019). Municipal solid waste (MSW) landfill: A source of microplastics? -Evidence of microplastics in landfill leachate. *Water Research, 159*, 38–45. https://doi.org/10.1016/j.watres.2019.04.060

He, P., Huang, J., Yu, Z., Xu, X., Raga, R., & Lü, F. (2021). Antibiotic resistance contamination in four Italian municipal solid waste landfills sites spanning 34 years. *Chemosphere, 266*, 129182. https://doi.org/10.1016/j.chemosphere.2020.129182

Hossain, M. K., Hossain, Md. L., & Das, S. R. (2014). Impact of landfill leachate on surface and ground water quality. *Journal of Environmental Science and Technology, 7*(6), 337–346. https://doi.org/10.3923/jest.2014.337.346

Ilankoon, I. M. S. K., Ghorbani, Y., Chong, M. N., Herath, G., Moyo, T., & Petersen, J. (2018). E-waste in the international context—A review of trade flows, regulations, hazards, waste management strategies and technologies for value recovery. *Waste Management, 82*, 258–275. https://doi.org/10.1016/j.wasman.2018.10.018

Iravanian, A., & Ravari, S. O. (2020). Types of contamination in landfills and effects on the environment: A review study. *IOP Conference Series: Earth and Environmental Science, 614*, 012083. https://doi.org/10.1088/1755-1315/614/1/012083

Kashitarash, Z. E., Taghi, S. M., Kazem, N., Abbass, A., & Alireza, R. (2012). Application of iron nanoparticles in landfill leachate treatment—Case study: Hamadan landfill leachate. *Iranian Journal of Environmental Health Science & Engineering, 9*(1). https://doi.org/10.1186/1735-2746-9-36

Kjeldsen, P., Barlaz, M. A., Rooker, A. P., Baun, A., & Ledin, A. (2002). Present and long-term composition of MSW landfill leachate: A review. C. T.H. *Critical Reviews in Environment Science and Technology, 32*, 297–336.

Lofrano, G., Libralato, G., Meric, S., Vaiano, V., Sacco, O., Venditto, V., Guida, M., & Carotenuto, M. (2020). Occurrence and potential risks of emerging contaminants in water. *Visible Light Active Structured Photocatalysts for the Removal of Emerging Contaminants*, 1–25.https://doi.org/10.1016/b978-0-12-818334-2.00001-8

Luo, H., Cheng, Y., He, D., & Yeng, E. H. (2019). Review of leaching behavior of municipal solid waste incineration (MSWI) ash. *Science of the Total Environment, 668*, 90–103.

López-Serna, R., Jurado, A., Vázquez-Suñé, E., Carrera, J., Petrović, M., & Barceló, D. (2013). Occurrence of 95 pharmaceuticals and transformation products in urban groundwaters underlying the metropolis of Barcelona, Spain. *Environmental Pollution, 174*, 305–315. https://doi.org/10.1016/j.envpol.2012.11.022

Petrović, M., Mihajlović, I., Tubić, A., & Novaković, M. (2023). Microplastics in municipal solid waste landfills. *Current Opinion in Environmental Science & Health, 31*, 100428–100428. https://doi.org/10.1016/j.coesh.2022.100428

Mandal, P., Duber, B. K., & Gupta, A. K. (2017). Review on landfill leachate treatment by electrochemical oxidation: Drawbacks, challenges and future scope. *Waste Management, 69*, 250–273.

Martí, V., Jubany, I., Pérez, C., Rubio, X., De Pablo, J., & Giménez, J. (2014). Human Health Risk Assessment of a landfill based on volatile organic compounds emission, immission and soil gas concentration measurements. *Applied Geochemistry, 49*, 218–224. https://doi.org/10.1016/j.apgeochem.2014.06.018

Martín-Pozo, L., de Alarcón-Gómez, B., Rodríguez-Gómez, R., García-Córcoles, M. T., Çipa, M., & Zafra-Gómez, A. (2019). Analytical methods for the determination of emerging contaminants in sewage sludge samples. A review. *Talanta, 192*, 508–533. https://doi.org/10.1016/j.talanta.2018.09.056

Masoner, J. R., Kolpin, D. W., Furlong, E. T., Cozzarelli, I. M., Gray, J. L., & Schwab, E. A. (2014). Contaminants of emerging concern in fresh leachate from landfills in the conterminous United States. *Environmental Science: Processes Impacts, 16*(10), 2335–2354. https://doi.org/10.1039/c4em00124a

National Geographic Society. (2022). *Landfills | National Geographic Society*. Education.nationalgeographic.org. https://education.nationalgeographic.org/resource/landfills/

Naveen, B., Mahapatra, D. M., Sitharam, T., Sivapullaiah, P., & Ramachandra, T. (2017). Physicochemical and biological characterization of urban municipal landfill leachate. *Environmental Pollution, 220*, 1–12.

Nurhasanah, Cordova, M. R., & Riani, E. (2021). Micro- and mesoplastics release from the Indonesian municipal solid waste landfill leachate to the aquatic environment: Case study in Galuga

Landfill Area, Indonesia. *Marine Pollution Bulletin, 163*, 111986. https://doi.org/10.1016/j.marpolbul.2021.111986

Papargyropoulou, E., Colenbrander, S., Sudmant, A. H., Gouldson, A., & Tin, L. C. (2015). The economic case for low carbon waste management in rapidly growing cities in the developing world: The case of Palembang, Indonesia. *Journal of Environmental Management, 163*, 11–19. https://doi.org/10.1016/j.jenvman.2015.08.001

Paxéus, N. (2000). Organic compounds in municipal landfill leachates. *Water Science and Technology, 42*(7–8), 323–333. https://doi.org/10.2166/wst.2000.0585

Propp, V. R., De Silva, A. O., Spencer, C., Brown, S. J., Catingan, S. D., Smith, J. E., & Roy, J. W. (2021). Organic contaminants of emerging concern in leachate of historic municipal landfills. *Environmental Pollution, 276*, 116474. https://doi.org/10.1016/j.envpol.2021.116474

Qi, C., Huang, J., Wang, B., Deng, S., Wang, Y., & Yu, G. (2018). Contaminants of emerging concern in landfill leachate in China: A review. *Emerging Contaminants, 4*(1), 1–10. https://doi.org/10.1016/j.emcon.2018.06.001

Ramakrishnan, A., Blaney, L., Kao, J., Tyagi, R. D., Zhang, T. C., & Surampalli, R. Y. (2014). Emerging contaminants in landfill leachate and their sustainable management. *Environmental Earth Sciences, 73*(3), 1357–1368. https://doi.org/10.1007/s12665-014-3489-x

Reinhart, D. R., Berge, N. D., Santra, S., & Bolyard, S. C. (2010). Emerging contaminants: Nanomaterial fate in landfills. *Waste Management, 30*(11), 2020–2021. https://doi.org/10.1016/j.wasman.2010.08.004

Rezapour, S., Samadi, A., Kalavrouziotis, I. K., & Ghaemian, N. (2018). Impact of the uncontrolled leakage of leachate from a municipal solid waste landfill on soil in a cultivated-calcareous environment. *Waste Management, 82*, 51–61. https://doi.org/10.1016/j.wasman.2018.10.013

Rinkesh. (2020, July 23). *Advantages and disadvantages of landfills*. Conserve Energy Future. https://www.conserve-energy-future.com/advantages-disadvantages-landfills.php

Rosenfeld, P. E., & Feng, L. G. H. (2011). Emerging contaminants. *Risks of Hazardous Wastes*, 215–222.https://doi.org/10.1016/b978-1-4377-7842-7.00016-7

Salam, M., & Nilza, N. (2021). *Hazardous components of landfill leachates and its bioremediation*. www.intechopen.com. IntechOpen. https://www.intechopen.com/chapters/74164

Schwarzbauer, J., Heim, S., Brinker, S., & Littke, R. (2002). Occurrence and alteration of organic contaminants in seepage and leakage water from a waste deposit landfill. *Water Research, 36*(9), 2275–2287. https://doi.org/10.1016/s0043-1354(01)00452-3

Shi, J., Wu, D., Su, Y., & Xie, B. (2021). Selective enrichment of antibiotic resistance genes and pathogens on polystyrene microplastics in landfill leachate. *Science of the Total Environment, 765*, 142775. https://doi.org/10.1016/j.scitotenv.2020.142775

Siddiqua, A., Hahladakis, J. N., & Al-Attiya, W. A. K. A. (2022). An overview of the environmental pollution and health effects associated with waste landfilling and open dumping. *Environmental Science and Pollution Research, 29*.https://doi.org/10.1007/s11356-022-21578-z

Slack, R. J., Gronow, J. R., & Voulvoulis, N. (2005). Household hazardous waste in municipal landfills: Contaminants in leachate. *Science of the Total Environment, 337*(1–3), 119–137. https://doi.org/10.1016/j.scitotenv.2004.07.002

Song, L., Li, L., Yang, S., Lan, J., He, H., McElmurry, S. P., & Zhao, Y. (2016). Sulfamethoxazole, tetracycline and oxytetracycline and related antibiotic resistance genes in a large-scale landfill, China. *Science of the Total Environment, 551–552*, 9–15. https://doi.org/10.1016/j.scitotenv.2016.02.007

Song, L., Yang, S., Gong, Z., Wang, J., Shi, X., Wang, Y., Zhang, R., Wu, Y., & Wager, Y. Z. (2023). Antibiotics and antibiotic-resistant genes in municipal solid waste landfills: Current situation and perspective. *Current Opinion in Environmental Science & Health, 31*, 100421. https://doi.org/10.1016/j.coesh.2022.100421

Su, Y., Zhang, Z., Wu, D., Zhan, L., Shi, H., & Xie, B. (2019). Occurrence of microplastics in landfill systems and their fate with landfill age. *Water Research, 164*, 114968. https://doi.org/10.1016/j.watres.2019.114968

Sui, Q., & Yu, X. (2023). *Emerging contaminants in landfills.* Environmental Toxicology. https://www.sciencedirect.com/journal/current-opinion-in-environmental-science-and-health/special-issue/10MP384NNPW

Sun, T. Y., Bornhöft, N. A., Hungerbühler, K., & Nowack, B. (2016). Dynamic probabilistic modeling of environmental emissions of engineered nanomaterials. *Environmental Science & Technology, 50*(9), 4701–4711. https://doi.org/10.1021/acs.est.5b05828

Teng, C., & Chen, W. (2023). Technologies for the treatment of emerging contaminants in landfill leachate. *Current Opinion in Environmental Science & Health, 31*, 100409. https://doi.org/10.1016/j.coesh.2022.100409

Torretta, V., Ferronato, N., Katsoyiannis, I. A., Tolkou, A. K., & Airoldi, M. (2017). Novel and conventional technologies for landfill leachates treatment: A review. *Sustainability, 9*, 9.

Turki, N., & Bouzid, J. (2017). Effects of landfill leachate application on crops growth and properties of a mediterranean sandy soil. *Journal of Pollution Effects & Control, 5*, 186. https://doi.org/10.4176/2375-4397.1000186

Urase, T., & Miyashita, K. (2003). Factors affecting the concentration of bisphenol A in leachates from solid waste disposal sites and its fate in treatment processes. *Journal of Material Cycles and Waste Management, 5*(1), 77–82. https://doi.org/10.1007/s101630300012

US EPA. (2016, March 24). *Basic information about landfills.* US EPA. https://www.epa.gov/landfills/basic-information-about-landfills#:~:text=Modern%20landfills%20are%20well%2Dengineered

Van Tassel, K. A., & Goldman, R. (2011, March 15). *The growing consumer exposure to nanotechnology in everyday products: Regulating innovative technologies in light of lessons from the past.* Social Science Research Network. https://papers.ssrn.com/sol3/papers.cfm?abstract_id=1556683

Wu, D., Fang, R., & Xie, B. (2023). Viewing landfill-borne antibiotic resistance genes through the kaleidoscope of "One Health": Essential legacies and research needs from the environmental dimension. *Current Opinion in Environmental Science & Health, 31*, 100424.

Wu, D., Huang, X.-H., Sun, J.-Z., Graham, D. W., & Xie, B. (2017). Antibiotic resistance genes and associated microbial community conditions in aging landfill systems. *Environmental Science & Technology, 51*(21), 12859–12867. https://doi.org/10.1021/acs.est.7b03797

Yazici Guvenc, S., Alan, B., Adar, E., & Bilgili, M. S. (2017). The impact of nanoparticles on aerobic degradation of municipal solid waste. *Waste Management & Research: The Journal of the International Solid Wastes and Public Cleansing Association, ISWA, 35*(4), 426–436. https://doi.org/10.1177/0734242X17695884

Yi, X., Tran, N. H., Yin, T., He, Y., & Gin, K.Y.-H. (2017). Removal of selected PPCPs, EDCs, and antibiotic resistance genes in landfill leachate by a full-scale constructed wetlands system. *Water Research, 121*, 46–60. https://doi.org/10.1016/j.watres.2017.05.008

Yu, X., Sui, Q., Lyu, S., Zhao, W., Liu, J., Cai, Z., Yu, G., & Barcelo, D. (2020). Municipal solid waste landfills: An underestimated source of pharmaceutical and personal care products in the water environment. *Environmental Science & Technology, 54*(16), 9757–9768. https://doi.org/10.1021/acs.est.0c00565

Zhang, B., Li, G., Cheng, P., Yeh, T.-C. J., & Hong, M. (2016). Landfill Risk Assessment on Groundwater Based on Vulnerability and Pollution Index. *Water Resources Management: An International Journal*, Published for the *European Water Resources Association (EWRA), 30*(4), 1465–1480. https://ideas.repec.org/a/spr/waterr/v30y2016i4d10.1007_s11269-016-1233-x.html

Zhang, T., & Surampalli, R. Y. (2009). Nanotechnologies for water environnement applications. *ASCE Reston.* https://doi.org/10.1061/9780784410301

Zhou, L.-J., Ying, G.-G., Liu, S., Zhao, J.-L., Yang, B., Chen, Z.-F., & Lai, H.-J. (2013). Occurrence and fate of eleven classes of antibiotics in two typical wastewater treatment plants in South China. *Science of the Total Environment, 452–453*, 365–376. https://doi.org/10.1016/j.scitotenv.2013.03.010